Cognitive Sciences and Education in Non-WEIRD Populations

Marcus Vinicius Alves • Roberta Ekuni
Maria Julia Hermida • Juan Valle-Lisboa
Editors

Cognitive Sciences and Education in Non-WEIRD Populations

A Latin American Perspective

Springer

Editors
Marcus Vinicius Alves
Sociocognitive Neuropsychology Laboratory (LINES)
Salvador, Bahia, Brazil

Maria Julia Hermida
Universidad Nacional de Hurlingham
Buenos Aires, Argentina

Roberta Ekuni
Universidade Estadual do Norte do Paraná
Bandeirantes, Paraná, Brazil

Juan Valle-Lisboa
University of the Republic
Montevideo, Uruguay

ISBN 978-3-031-06907-9 ISBN 978-3-031-06908-6 (eBook)
https://doi.org/10.1007/978-3-031-06908-6

This Springer imprint is published by the registered company Springer Nature Switzerland AG
The registered company address is: Gewerbestrasse 11, 6330 Cham, Switzerland

Foreword

The book you have in your hands is the expression of a necessity felt by an entire generation of researchers, in developing countries as well as developed ones: the urgent need to improve education for all people, across the cultural boundaries of class, race, gender, sexual orientation, and other schisms. Material inequality is produced by immaterial inequality, as reflected by the uneven distribution of opportunities for developing, learning, and maturing to the full extent of one's capability. If our species is to have a good future in this planet, it will be through the betterment and wide dissemination of education, in formal as well as informal settings. The increasing convergence of research in education, cognitive psychology, and neurobiology, well documented in this book, represents the active engagement of science in the effort to improve society, with a focus on the promotion of diversity and on the inclusion of multiple points of view from the developing world.

All this enthusiasm did not come out of the blue, but rather from the experience of the Latin American School for Education, Cognitive and Neural Sciences—or more warmly LaSchool—as an environment in South America for the close interaction of researchers from all career levels, with a formative focus on doctoral and postdoctoral students. Motivated by the support of the McDonnell Foundation, a group of scholars envisioned LaSchool as an initiative to foster a special kind of researcher, committed to the challenges of improving education while also being equipped by the methods and questions from psychology and biology. The comprehensive and in-depth set of chapters contributed by so many LASchool alumni shows that LaSchool's goal to foster education-oriented researchers has succeeded—and this is something to be cherished and hailed.

The chapters presented here deal with the expansion of the domains of cognitive science to populations that are underrepresented in the main literature. This is of course relevant to those who work with diverse populations in the fields of education, cognition, and human development. The book also provides challenging new data for classical theories that have often been developed with a quite restricted focus on the socially privileged cohorts of the planetary population. Undoubtedly, in the interest of all, the more the merrier.

Brain Institute, Universidade Federal do Rio Grande do Norte (UFRN)
Natal, Brazil

Sidarta Ribeiro

Contents

About the Editors

Marcus Vinicius Alves holds a PhD in psychobiology from Universidade Federal de São Paulo (Brazil) and has experience as a visiting researcher at Memorial University of Newfoundland (Canada) and as a research and development specialist at Université du Luxembourg (Luxembourg). At present, he is a full professor at Faculty of Technology and Sciences in Salvador, Brazil, and also the lead researcher of the Sociocognitive Neuropsychology Laboratory (LINES), registered at the National Council for Scientific and Technological Development (CNPQ). Alves' research interests include learning, memory, incidental and motivated forgetting, mental effort (processing systems, attention, and cognitive load theory), and social cognition, using mostly eye tracking and pupillometry.

Roberta Ekuni is an adjunct professor at Universidade Estadual do Norte do Paraná—graduate program in education. She has a PhD from Universidade Federal de São Paulo (Brazil), and she was a visiting student at Washington University in St. Louis. Also, she is one of the editors of Neuromyth's Busters book collection. She has interest in research that applies cognitive psychology to education, especially involving retrieval practice to improve student's memory.

Maria Julia Hermida is an adjunct professor at Universidad Nacional de Hurlingham and researcher at the National Council of Scientific and Technical Research in Argentina. She has obtained a PhD in psychology from Universidad Nacional de San Luis, Argentina. For more than 10, years she has been studying child cognitive development, how it is affected by poverty, and which interventions can help to prevent the negative cognitive consequences of poverty. Specifically, she has conducted different interventions combining cognitive neuroscience and educational approaches in diverse scholar contexts.

Juan Valle-Lisboa is Aggregate Professor of Biophysics and Neuroscience at Universidad de la República (UDELAR), sharing his time between the Schools of Science and Psychology. He is a member of the Program for the Development of Basic Science PEDECIBA Council (www.pedeciba.edu.uy). He has received a PhD in biology and a master's degree in biophysics from the PEDECIBA-UDELAR program. His main research interests are the study of learning and cognition, and the application of digital technologies to enhance education. In particular, in the last few years, he has directed projects aimed at measuring math and reading abilities as well as neuropsychological predictors of reading with the use of digital devices like tablets and cell phones. He is also interested in basic aspects of language processing, lexical representation and linguistic composition, and the use of neural network models to describe these processes.

About the Contributors

Jose Armando Aguasvivas is a Dominican-born Spanish citizen. He obtained his PhD in cognitive neuroscience in 2021, with his thesis primarily focusing on the influence of bilingualism on foreign language learning. Throughout his career, he has also worked on projects relating to reading acquisition in adult neo-literates and educational projects at the national scale with primary school children in the Dominican Republic. His main goal is to understand the mechanisms through which humans learn and use multiple languages in order to facilitate language instruction, particularly in the context of immigration and underrepresented populations.

Maria Soledad Assis is an MSc student in cognitive science at Universidad de la República (Uruguay. BSc in educational psychology, Universidad Nacional de Lomas de Zamora(Argentina). She is an educational psychologist and works clinically with children who have learning difficulties. Currently, she is finishing her postgraduate studies which focus on the adaptation of a game platform—*Kalulu*—for early literacy.

María Celeste Baiocchi is a researcher and instructor in Facultad de Psicología, Universidad Nacional de Rosario, Argentina. She was a CONICET fellow at Instituto de Investigaciones sobre Lenguaje, Sociedad y Territorio (INILSyT-UNaF). She received her PhD in psychology from Universidad Nacional de Córdoba, Argentina. Her doctoral research examined how Wichi children and adults categorize the animals native to their forest. Her work focuses on how conceptual organization and reasoning about nature vary throughout development across cultures.

Silmara Batistela is a psychologist and a professor at Centro Universitário São Camilo (São Paulo, Brazil) and has a PhD and a master's degree in psychobiology/neuropsychology from Universidade Federal de São Paulo (São Paulo, Brazil).

Guilherme Brockington is an associate professor at Universidade Federal do ABC, Brazil. He is a physicist and obtained his PhD in science education in 2012, developing a research on neuroscience and education. He stayed at the Brain and Creativity Institute in Los Angeles, California, led by the neuroscientist Antonio Damasio, for 2 years during his PhD. He got postdoc position in 2012 to work with neuroscience of emotion and education. Now, he is developing several researches in science education, physics education, neuroscience and education, neuroscience of consciousness, and storytelling.

Danilka Castro Cañizares BSch in Psychology (with honors). Holds a Master's Degree in Medical Psychology, and a PhD in Psychology, National Center for Scientific Research, Cuba. Postdoctoral fellow at the Center for Advanced Research in Education, Universidad de Chile, Chile. Teaching experience in undergraduate courses in Psychology and Education, in Cuba and Chile. Reader in Psychology, Neuropsychology and Neurosciences in Education, in postgraduate courses, bachelor's and Master's degree programs, Universidad Mayor, and Pontificia Universidad Católica de Chile. Her research focuses on cognitive psychology and neurosciences, particularly in the study of numerical mental representations and the early identification of developmental dyscalculia.

Alejandra Carboni is Professor of Cognitive Psychology at the University of Republic, Uruguay. She has a degree in psychology (UdelaR) and a PhD in neuroscience (Universidad Complutense de Madrid). Her research is focused on cognitive neuroscience, with the purpose of understanding attention mechanisms and its neural basis, as well as contributing to the understanding of the relation between cognitive development in early childhood and socio-economic status.

Manuel Carreiras is the scientific director of the BCBL (Basque Center on Cognition, Brain and Language, Donostia-San Sebastián, Spain) "Severo Ochoa" label of excellence. He is IKERBASQUE research professor, honorary professor of the University College of London (UCL), and distinguished researcher of the University of the Basque Country (UPV/EHU). His research focuses on reading, bilingualism, and second language learning. He has published more than 250 papers in high-impact journals. His research has been funded by national and international research agencies. He has been recipient of the Euskadi Research prize 2015, and of the Spanish National Research prize in Social Sciences Pascual Madoz, 2019.

Chrissie Ferreira de Carvalho Professor of Psychology Department, Universidade Federal de Santa Catarina. PhD in Developmental Psychology by Universidade Federal da Bahia, specialist in Neuropsychology, completed the postdoctoral studies at the Laboratory for Developmental Studies at Harvard University

Miladys Orraca Castillo is a medical doctor (with honors) and a second-grade specialist in clinical genetics, and holds a PhD in medical sciences. She is an associate professor at the University of Medical Sciences of Havana, Havana, Cuba, reader in 17 courses nationwide, and president of the Cuban Scientific Society for the Development of Families. She has won a National Health Award in the category of

Scientific Research in 2014 and a National Award of the Cuban Academy of Sciences in the category of Scientific Research in 2015. Her research focuses on the study of the neurocognitive phenotype of neurofibromatosis type 1 and the anatomical and functional brain networks underlying learning disorders.

Olalla Cutrín is an assistant professor and a member of the UNDERISK research group (research unit for research on risk behaviors and developmental disorders) at the University of Santiago de Compostela (Spain). Her research focuses mainly on understanding the risk factors for developing antisocial behavior during adolescence, such as rule-breaking or substance use, and specifically aims on how parenting practices and affiliations with antisocial peers influence adolescent behavior. Currently, she is involved in projects that focus on strengthening socio-emotional and life skills in children and youth to prevent early conduct problems as well as adolescent drug use.

Hernán Delgado is a researcher at the University of the Republic (UdelaR) in Uruguay. He has a degree in biology (UdelaR) and is a PhD student in the neuroscience program of the PEDECIBA. His research focuses on the development of self-regulation during early childhood. His main interest is in the analysis of the influences of poverty on self-regulatory development.

Nadir Díaz-Simón obtained a bachelor's degree in psychology from the University of Havana in Cuba and a master's degree in cognitive science from Universidad de la República in Uruguay. He has worked in the Department of Educational Neuroscience at the Cuban Neuroscience Center. He is currently working with the Mathematical Cognition Research Group in the Interdisciplinary Center in Cognition for Teaching and Learning at Universidad de la República, where he focuses on the development of basic skills training studies of approximate quantification in early education and school-aged children.

Guadalupe Rodriguez Ferrante is a PhD student at Universidad de Buenos Aires (UBA) working at Universidad Torcuato Di Tella (UTDT) and Universidad Nacional de Quilmes. She got her bachelor's/master's degree in biology from UBA. When she finished her degree, she got interested in computational tools, so now she is studying computer science at the UBA. She a teaching assistant in the Computer Sciences Department at UBA and teaches "Study of Human Behavior" at UTDT. Guadalupe's PhD focuses on the impact of the interaction between school timing and chronotype on adolescents' sleep, health, and academic performance.

Homègnon Antonin Ferréol Bah graduated in pharmacy, holds master's degree in pharmacy, and is a PhD candidate in public health—epidemiology at the Universidade Federal da Bahia.

Carolina Soledad Fracchia is a postdoctoral fellow at Unidad de Neurobiología Aplicada (UNA, CEMIC–CONICET). She is interested in the fields of developmental psychology. In particular, she is working in the association between executive resolution profiles and different individual and social environmental factors.

José Antonio Gómez-Fraguela is a PhD in psychology from the Universidade de Santiago de Compostela (USC), Since 2000, he works as associate professor and researcher in the USC's Clinical Psychology and Psychobiology. The investigation of him is aimed at the field of behavioral problems, antisocial behavior, and drug abuse in adolescence and youth. He has participated as a researcher in numerous projects and research contracts from which more than 100 scientific publications have been derived between books, book chapters and artists in scientific journals, as well as a large number of collaborations in international scientific congresses.

Melina Furman holds a PhD in science education from Columbia University. She is currently an associate professor at the University of San Andrés, Argentina, and a researcher at the National Scientific and Technical Research Council (CONICET). Melina is director of the science education research program in the School of Education at the University of San Andrés.

Juliana Perina Gândara is a speech-language therapist and holds a PhD in semiotics and general linguistics from the University of São Paulo (USP).

Adriana Soni García is a psychology lecturer at Arden University (UK). She is a fellow of the Higher Education Academy (UK), with a PhD (in Education) from the University of Bristol (UK) and experience as a visiting researcher at CICERO Learning at the University of Helsinki (Finland).

Federico Giovannetti is PhD fellow at Unidad de Neurobiología Aplicada (UNA, CEMIC–CONICET). He is interested and working in the fields of developmental psychology, social vulnerability, cognitive intervention design, and the study of developmental diversity. Federico is Adjunct Professor in Career of Psychopedagogy at Universidad Nacional de San Martín (UNSAM).

Andrea Paula Goldin is a junior research scientist at the Scientific and Technical Research National Council and the Universidad Torcuato Di Tella (UTDT) in Buenos Aires, Argentina. She got her biology degree and PhD from Universidad de Buenos Aires (UBA). She is an associated researcher at the Center of Evaluation of Evidence-Based Policies (UTDT). She is an enthusiastic educator and public communicator of science. Dr. Goldin specializes in cognitive neuroscience, and her main research topics are chronobiology of education, cognitive development and entrainment, and transfer of logical-mathematical learning and teaching of critical thinking.

Antônio Jaeger graduated in psychology (2000) and received his PhD from Universidade Federal do Rio Grande do Sul, Brazil, in 2007. He also held a 3-year postdoctoral position at Washington University in St Louis, USA (2008–2011). Currently, he holds a professor position at Universidade Federal de Minas Gerais, Brazil, where he is the principal investigator of the Research Group on Memory and Cognition. His main research focus is on human memory and learning from a cognitive psychology and cognitive neuroscience perspective.

Stephen Kulis (PhD, Columbia University) is Cowden Distinguished Professor of Sociology at Arizona State University (ASU), and director of research of the ASU Global Center for Applied Health Research. His research focuses on cultural processes in health disparities; cultural adaptation of prevention programs for ethnic minority youth and international populations; contextual neighborhood and school level influences on youth substance use; and the organizational sources of ethnic and gender discrimination. He has conducted studies to culturally adapt and test school-based substance use prevention programs for adolescents in urban American Indian communities, Mexico, Guatemala, Kenya, Spain, and Uruguay.

Dinorah de León obtained a bachelor's degree in psychology and a master's degree in cognitive sciences from Universidad de la República in Uruguay. She is currently a doctoral student in psychology at Universidad Católica, Uruguay, and a research assistant in the Mathematical Cognition Research Group at the Center for Interdisciplinary Cognition for Teaching and Learning. Her main research topic is the development of math skills and its relationship with the home learning environment.

María Juliana Leone is a junior research scientist in the Scientific and Technical Research National Council at Universidad Nacional de Quilmes (UNQ) and Universidad Torcuato Di Tella (UTDT) in Buenos Aires, Argentina. She got her biotechnology degree and her PhD at the UNQ. She is an associate researcher at the Center of Evaluation of Evidence-Based Policies (UTDT). Dr. Leone's research focuses on the impact of chronobiology and sleep on real life, including the effects of school timing on adolescents, the consequences associated with the pandemic-related lockdown, and the local characteristics of both sleep and circadian rhythms in Argentina.

Carlos Andrés Libisch is a graduate in psychology from Universiad de La República in Uruguay. Carlos is a doctoral student at the University of Santiago de Compostela, Spain, where he also earned a master's degree in clinical psychology and psychobiology. His research focuses on the prevention of drug use and other problem behaviors, active teaching, gamification, clinical psychology, virtual education, and social-emotional skills. He works as Head of the New Measurements team at the Global Learning Network of the CEIBAL program in Uruguay. https://redglobal.edu.uy/carloslibisch@gmail.com

Sebastián Javier Lipina is a researcher at the National Council of Research (CONICET); director of Unidad de Neurobiología Aplicada (UNA,

CEMIC-CONICET); principal investigator of the Academic Unit at Instituto Universitario CEMIC (IUC); co-director of the International Mind, Brain and Education School (Ettore Majorana Foundation); fellow of the Center for Neuroscience and Society (CNS, University of Pennsylvania); and member of the IRB at CEMIC.

Francisco Martín López obtained a bachelor's degree in physics and a master's degree in cognitive science from Universidad de la República, Uruguay. He recently joined the research group of Prof. Dr. Jochen Triesch at the Frankfurt Institute for Advanced Studies (FIAS), Germany, where he studies the development of multimodal sensorimotor coordination in artificial agents trained with intrinsically motivated reinforcement learning, with the intent of obtaining a PhD in physics.

Mariana Luzuriaga has a BA in education (PhD on-going, expected 2023) from the University of San Andrés, Argentina. She is currently a professor of graduate and postgraduate courses in the School of Education, as well as research assistant in the science education research program at the University of San Andrés, Argentina.

Alejandro Maiche is Full Professor of Cognitive Psychology in the School of Psychology at Universidad de la República. He obtained his PhD in perception, communication, and time from the Autonomous University of Barcelona (UAB) in 2002 and was lecturer professor at this University until 2010. Currently, he leads the Research Group in Mathematical Cognition of the Interdisciplinary Cognition Center for Teaching and Learning where educational interventions are developed aimed at stimulating cognitive abilities in the early stages of education (5- and 6-year-olds). His research interests include visual perception, number cognition, and education.

Flavio Marsiglia PhD is Regents' Professor in the School of Social Work at Arizona State University (ASU). As director of the Global Center for Applied Health Research (GCAHR), he conducts intervention health research in partnership with colleagues from around the world. His school-based "keepin'it REAL" program is widely used across the USA and in other countries, including in his native Uruguay. He is also the principal investigator in the specialized center on health disparities research at the Southwest Interdisciplinary Research Center (SIRC). Marsiglia is currently leading two COVID-19 projects, making testing and vaccination accessible to underserved and vulnerable communities of Arizona.

José Antonio Menezes-Filho is Professor of Toxicology in the Faculty of Pharmacy, Universidade Federal da Bahia. He is Doctor of Public Health and Environment and holds CNPq Research Productivity Scholarship 2.

Mônica Carolina Miranda is a psychologist, neuropsychologist, and an affiliated professor at Universidade Federal de São Paulo. She coordinates the Project for Early Childhood (Projeto Pela Primeira Infância—PPI) and is an associate researcher in the National Science Network for Education (Rede Nacional de Ciência para a Educação—RedeCpe) as well as Cnpq Research Productivity Fellow.

Ana Paula Moreira is a professor in the Psychology Department at the State Paulista University (UNESP). She is a psychologist with a PhD in psychology developing research in Scholar's Psychology and Child Development based on Vygotsky's theory. She is a specialist in Vygotsky's work and in cultural-historical psychology. Her area of interest is early child development, cultural-historical theory, and activity theory.

Migdalia Padilla works as a fellow at Instituto Rosario de Investigaciones en Ciencias de la Educación (IRICE, CONICET-UNR). She is a PhD student at Universidad Nacional de Córdoba (UNC). She studied psychology at the University of Puerto Rico, Rio Piedras precinct. Her research focuses on the relation between the mind and culture. As a PhD student, Migdalia is interested in maternal mental states attributions embedded in the mother-infant relation in the first 2 years of life.

Marcos Emanoel Pereira holds a PhD in social psychology from the Universidade Federal do Rio de Janeiro. He is a full professor at the Institute of Psychology, Universidade Federal da Bahia.

Nancy Estévez Pérez holds a BSc in psychology (with honors), Psychology Faculty, Havana University, Cuba; a master's degree in animal physiology from the Biology Faculty, Havana University; a master's degree in cognitive neuroscience from the Cuban Neurosciences Center (CNEURO); and a PhD in psychology, National Center of Scientific Research, Cuba. She is head of the Neurodevelopment Department, CNEURO, Cuba. She focuses on the study of anatomical and functional brain networks underlying typical and atypical development of numerical cognition and academic achievement. She is a permanent member of the Cuban National Examination Board for the Scientific Degree in Psychology. She became IBRO/IBE UNESCO Science of Learning Fellow in 2019.

Marcos Luis Pietto is a postdoctoral fellow at Unidad de Neurobiología Aplicada (UNA, CEMIC–CONICET). Marcos is interested and is working in the fields of developmental cognitive neuroscience and cognitive electrophysiology, and studying of the impact of childhood poverty on brain activity.

Marina Leite Puglisi is an adjunct professor in the Department of Communication Sciences at the Universidade Federal de São Paulo (UNIFESP). She has a PhD in communication sciences from the University of São Paulo (USP) and a postdoctoral degree from USP and the University of Oxford, and is former Newton International Fellow.

Mirela Cunha Cardoso Ramacciotti is an education researcher and consultant, is a PhD candidate in human communication disorders at the Universidade Federal de

São Paulo (UNIFESP), and holds an MSc in interdisciplinary education from Johns Hopkins University.

Juliana Reinais is a doctoral student in psychology, Universidade Federal da Bahia, and holds a Master of Experimental Psychology, Pontifical Catholic University of São Paulo, and Bachelor of Psychology, Salvador University.

Matías Fernández Ruiz works as a fellow at Instituto Rosario de Investigaciones en Ciencias de la Educación (IRICE, CONICET-UNR). He is currently is doing his PhD in anthropology at Universidad de Buenos Aires. He has an MS degree in cognitive psychology from FLACSO and Universidad Autónoma de Madrid, and a PGCE in higher education from Universidad de Buenos Aires. His work focuses on the interaction between culture, native categories, and cognition. Specifically, he studies how Wichi people represent and reason about causality of their ecosystem.

Paul Ruiz is a graduate and a Doctor of Psychology (PhD). Paul is a professor at the National University (UdelaR) and a researcher at the National Innovation Research Agency (SNI-ANII). His research areas are: factors that stimulate drug consumption in animals and humans.

Cassio Santos-Lima Psychologist, specialist in Neuropsychology, Master's degree, and PhD student in Developmental Psychology at the Universidade Federal da Bahia. Professor of Psychology at Escola Bahiana de Medicina e Saúde Pública.

Maria Soledad Segretin is a researcher in the National Council of Research (CONICET), codirector of Unidad de Neurobiología Aplicada (UNA, CEMIC-CONICET), and principal investigator of the Academic Unit of the Instituto Universitario CEMIC (IUC).

Diego Edgar Shalom is a physicist (Instituto Balseiro) and has a PhD in physics (Instituto Balseiro). He is a professor at Universidad de Buenos Aires and an independent researcher with the National Council of Scientific Research, both in Argentina. Diego has always been interesting in how things work; to explore that, he went from physics (he studied Josephson Junction Arrays) to neuroscience, analyzing a wide variety of issues like how humans read and learn. Currently, his main research line analyzes ocular movements during reading as a way to understand how we process, manipulate, and organize this information in the brain.

Heloisa Gonçalves da Silveira is a speech-language therapist, and master's and PhD candidate in human communication disorders at UNIFESP.

Helena Sousa is a speech-language therapist, specialist in child language, and master's candidate in human communication disorders at UNIFESP.

Andrea Taverna is a researcher member at Instituto Rosario de Investigaciones en Ciencias de la Educación (IRICE-CONICET-UNR) where she coordinates the *Hunhat lheley* lab. She obtained her PhD from Universidad Nacional de Córdoba, Argentina, and has made several postdoctoral visits to Northwestern University. Currently, she studies the acquisition of Wichi as primary language. This current work is complemented with her previous studies on how Wichí children represent their native environments. She has used this evidence to develop with the research team a series of bilingual Wichí-Spanish children's books.

Inés Taylor holds a BSc in biology (University of Bristol) and PGCE in science (Kings College, London). Currently, Inés is a research assistant in the science education research program at the University of San Andrés, Argentina. Limited Research Teacher Trainer Ltd Team Leader (Tanzania 2014, Ghana 2017 and Belize 2018).

Nikolay Veresov is an associate professor at Monash University, Australia. He has experience as a daycare center and kindergarten teacher (1987–1991) and secondary school teacher (1982–1987). He got his first PhD degree in Moscow in 1990 and started his academic career in Murmansk (Russia) as a senior lecturer (1991–1993) and the head of the Department of Early Childhood (1993–1997). The second PhD was obtained from the University of Oulu (Finland) in 1998. From 1999 to 2011, he had been affiliated to Kajaani Teacher Training Department (Finland) as a senior researcher and scientific director of international projects. He has published 5 books and over 160 articles/book chapters available in 9 languages. His area of interest is child development in the early years, cultural-historical theory, and research methodology.

Camila Zugarramurdi holds a PhD in linguistics from Basque Center on Cognition, Brain and Language, and University of the Basque Country (Spain); MSc in neuroscience, Universidad de la República-PEDECIBA (Uruguay); and BSc in human biology, Universidad de la República-PEDECIBA (Uruguay). She is an assistant professor in the School of Psychology at Universidad de la República, Uruguay. Her research focuses on language development and reading acquisition from an interdisciplinary perspective, integrating approaches from cognitive neuroscience, developmental psychology, and education. During her PhD, she developed a videogame—*Lexiland*—able to identify pre-reading children at risk of reading difficulties.

Abbreviations

AD	Arithmetic Dysfluency
ANEP	Administración Nacional de la Educación Pública Uruguay
ANS	Approximate Number System
ANT	Attentional Networks Task
BAU	Business-As-Usual
BIA+	Bilingual interactive Activation Plus
Ca^{2+}	Calcium ion
CA3	Cornu Ammonis 3
CABA	Ciudad Autónoma de Buenos Aires
CCPT	Conners' Continuous Performance Test
CEPAL	Comisión Económica para América Latina y el Caribe
CGT	Children's Gambling Task
CNS	Central Nervous System
CON	Control Group
CPMT	Coloured Progressive Matrices Test
DD	Developmental Dyscalculia
Dev-LEX	Developmental Lexicon
DL	Dyslexia
Dna	Naming discursive strategy
Dqst	Interrogative Discursive Strategy
Drsp	Respond speakers' questions discursive strategy
ECM	Educative Curriculum Materials
EEG	Electroencephalogram
EF	Executive Functions
ERN	Error Related Negativity
ERP	Evoked-Related Potentials
ET	Echo Time
FA	Fractional Anisotropy
FIST	Flexible Item Selection Task
fNRIS	Functional Near-Infrared Spectroscopy
FoMO	Fear of Missing Out

Hg	Mercury
HPG	High Performance Group
Hunhat lheley	Inhabitants of the Earth in Wichi language
Husek	Goodwill in Wichi language
ICTs	Information and Communication Technologies
IEA	International Association for the Evaluation of Educational Achievement
Iloy	Are alive in Wichi language
Inot	Water in Wichi language
INT	Intervention Group
IPS	Intraparietal Sulcus
IQ	Intelligence Quotient
KiREAL	Keepin it REAL
kir	Keepin' it Real
L1	Native language or first language
L2	Second language or foreign language
LAA	Low Arithmetic Achievement
LAC	Latin America and the Caribbean
LA	Latin America/Latin American
LAM	Low Academic Achievement in Math
LPG	Low Performance Group
MAT	Mathematical Attainment Test
MEG	Magnetoencephalography
MeHg	Methylmercury
MEQ	Morning-Eveningness Questionnaire
MESC	Morningness-Eveningness Scale for Children
MLD	Mathematical Learning Disability
MLU	Mean Length of Utterance
MM	Mind Mindedness
MMSA	Maternal Mental States Attributions
Mn	Manganese
MnTn	Manganese on Toenails
MP-RAGE	High resolution anatomic images
MSF	Midpoint of sleep on free days
MSFsc	Midpoint of sleep on free days sleep corrected
MSW	Midpoint of sleep on weekdays
NF1_LD	Children with Neurofibromatosis 1
NF1	Neurofibromatosis Type 1
Nin'ola	Invisible in Wichi language
OECD	Organization of Economic Co-operation and Development
OLPC	One Laptop Per Child
PAM	Plataforma Adaptativa de Matemáticas
PBA	Provincia de Buenos Aires
Pb	Lead

PCMC	Parents and Children Making Connections—Highlighting Attention
PISA	Programme for International Student Assessment
PUMA	Prueba Uruguaya de Matemáticas
RAN	Rapid Automatized Naming
RCT	Randomized Control Trial
RCTs	Randomized Controlled Trials
RDS	Relational Developmental Systems
REAL	Refuse, Explain, Avoid, and Leave
RFID	Radio-Frequency Identification
RT	Repetition Time
SDf	Sleep duration on free days
SDmean	Average sleep duration during a week
SD	Sleep Duration
SDw	Sleep duration on weekdays
SEP	Secretaría de Educación Pública
SES	Socioeconomic status
SITEAL	Sistema de Información de Tendencias Educativas en América Latina
SJL	Social jetlag
SLI	Socioeconomic Level Index
SNS	Social Networking Sites
SOff	Sleep Offset
SOMBIP	Self-Organizing Map of Bilingual Processing
SOM	Self-Organizing Map
SOn_f	Sleep onset on free days
SOn_w	Sleep onset on weekdays
SOn	Sleep onset
SPM	Statistic Parametric Mapping
STEM	Science, Technology, Engineering and Mathematics
TA	Typical Arithmetic Achievement
TIMSS	Trends in International Mathematics and Science Study
TOL	Tower of London
TSep	Temporo-spatial location
Tshotoy	Animals of the forest in Wichi language
TS	Turner Syndrome
UBN	Unsatisfied Basic Needs
UNESCO	United Nations Educational, Scientific and Cultural Organization
VBM	Voxel-Based Morphometry
VLA	Very Low Arithmetic Achievement
WEIRD	Western, Educated, Industrialized, Rich and Democratic
WM	Working Memory
Yilh	Can die in Wichi language

Part I
Cognitive Sciences and Education: Theories, Beliefs and Misconceptions

The Importance of Bringing the Latin American Perspective to Cognitive Sciences and Education

Marcus Vinicius Alves, Roberta Ekuni, Maria Julia Hermida, and Juan Valle-Lisboa

Introduction

Cognitive sciences and neuroscience have increased exponentially in their knowledge in the last decades. This increase eventually led to questions regarding the application of this knowledge to improve education in an integrated field known as Educational Neuroscience (Howard-Jones et al., 2016). Besides that, a wide variety of such attempts to integrate neuroscience into education resulted in neuromyths and misconceptions (Bruer, 1997; Ekuni & Pompéia, 2015; Hermida et al., 2016). Today, there is a corpus of knowledge about our central nervous system and its functioning that (adequately understood) has promising contributions to the education field (e.g., Sigman et al., 2014). Most of this knowledge, as can be seen in the reference list of Sigman et al. (2014), came from central countries (North America, Europe) and is based on studies conducted on what has been called WEIRD populations (i.e., western, educated, industrialized, rich, and democratic). However, much less is known about non-WEIRD societies, which are most of the world (Henrich

M. V. Alves (✉)
Sociocognitive Neuropsychology Laboratory (LINES), Salvador, Brazil

R. Ekuni
Universidade Estadual do Norte do Paraná, Graduate Program in Education, Jacarezinho, Paraná, Brazil

M. J. Hermida
Universidad Nacional de Hurlingham-Consejo Nacional de Investigaciones Científicas y Técnicas (UNAHUR-CONICET), Buenos Aires, Argentina

J. Valle-Lisboa
Centro de Investigación Básica en Psicología, Facultad de Psicología; Centro Interdisciplinario de Cognición para la Enseñanza y el Aprendizaje; Sección Biofísica y Biología de Sistemas, Facultad de Ciencias, Universidad de la República, Montevideo, Uruguay

M. V. Alves et al. (eds.), *Cognitive Sciences and Education in Non-WEIRD Populations*, https://doi.org/10.1007/978-3-031-06908-6_1

et al., 2010a). In sum, there is not much information about how the integration of cognitive sciences and neuroscience could impact education in populations that have these diverse cultural and social characteristics.

This book aims to present theoretical and practical innovations in the cognitive science and education field from a diversity view, focusing especially on studies and research conducted in Latin American populations. This field has been growing fast during the last decade in developing countries and, nowadays, they have valuable information to share with the world. Latin America made great advances in the area impulsed by different initiatives (for instance, the Latin American School on Education and the Cognitive and Neural Sciences; Bruer, 2014). This is the first book showing how Latin American researchers approach the fields of cognitive sciences, neurosciences, and education from their own perspective. With that, this book takes advantage of the varied communities that characterize developing countries to explore how underrepresented populations learn as well as what works and what does not for cognitive science and education, not only for the developing world but also for diversity in the whole world.

Cognitive Science, Education, and Diversity

Three concepts are fundamental in this book; thus, we'll start by defining them. Cognitive science is an attempt to unify views of thought developed by studies in psychology, linguistics, anthropology, philosophy, computer science, and neuroscience (Hunt, 1989). Broadly, it can be defined as the scientific study of minds and brains, be they real, artificial, human, or animal (Nadel & Piattelli-Palmarini, 2003). In other words, cognitive sciences are a group of disciplines that study the mind and the brain using a scientific approach.

Neuroscience is one of the disciplines included in the broader category of cognitive sciences. Neuroscience can be defined as the study of the central nervous system (Bergen, 1988). As the central nervous system cannot be seen as a separate entity, its study involves different disciplines (neurochemistry, neurophysiology, neurobiology, neuropharmacology, neuroanatomy, neuroimmunology, neurogenetics, neuroendocrinology, etc.). Also, as the central nervous system cannot be easily explored in an experimental way, disciplines such as computers or mathematics can contribute to modeling the central nervous system functioning. Thus, studies from neuroscientists, covering neural models, analytical and numerical techniques, neural simulation tools, data analysis tools, databases of models, and datasets are considered neuroscience (Jaeger & Jung, 2015). Finally, and maybe this neuroscience's face is the most important for educators, and our central nervous system generates mental representations: the activity of neural networks code information in the form of electrochemical activity, generating mental representations (Szűcs & Goswami, 2007). Thus, neuroscience overlapped at some point with disciplines such as cognitive psychology that focused on those mental representations.

Education has not a unified definition. However, in the context of this book, we will define it as socialization through knowledge publicly legitimate (Cullen, 1997). In other words, in the context of this book, education is linked with the learning and teaching of contents, habits, and procedures (generally included in the curriculum) that are given by trained and qualified professionals with pedagogical knowledge who are responsible for this socialization: teachers. Briefly, in the context of this book, education is understood as the process typically happening in schools. Of course, education and neuroscience work at different levels of analyses. However, the study of mental representation (including its development, training, and modulation) offers a level of analysis common both to cognitive sciences, neuroscience, and traditional behavioral approaches to learning and pedagogy (Szűcs & Goswami, 2007). This book will include chapters combining cognitive science (including behavioral methods) and the new methods and results of neuroscience to investigate aspects of mental representations important to educators. However, it has a particularity compared with previous books on this issue: a focus on diversity.

Diversity is another concept that has no unique definition. In education, diversity is linked to racial and ethnic diversity and linked to arguments for social justice and equal access to education (Haring-Smith, 2012). This concept strongly overlaps with what anthropologists and psychologists consider to be the least studied groups, that is, those who do not belong to WEIRD societies or people. Regarding cognitive sciences and neuroscience, in the last years there has been a call for scientists to move their focus from the typical population of study to a more diverse population in order to increase the generalization of results (Henrich et al., 2010a). Evidence in different areas of cognitive sciences showed considerable variation among human populations, and actually, WEIRD people are some of the most psychologically unusual people on the planet (Henrich et al., 2010b). However, today, most cognitive sciences, neuroscience, and education literature came from WEIRD societies, and the evidence generated in non-WEIRD populations remains highly unknown.

The main objective of this book is to bring evidence in the interface of cognitive sciences, neuroscience, and education produced in or applied to non-WEIRD communities. Each chapter describes experiences from scientists in non-weird societies (mostly Latin American researchers) and/or applications or worldwide current evidence to diverse groups (indigenous population, low-income children, children with disabilities, etc.). By doing that, we aim to provide useful information to scientists and educators interested in the area and working in non-WEIRD populations.

An Outline of the Book

The book is organized into five parts. The first one exposes the general approach of the book through two main claims: neuroscience and education are not opposite; but for that, neuroscience should be adequately understood. The first claim is developed in the second chapter, "Neuroscience and Vygotsky: Putting Together Contemporary Evidence and Cultural-Historical Psychology." There, authors showed how the most classical educational approaches, like those stemming from Vygotskian and related theories, fit nicely with what we know from neuroscience and cognitive science, while providing necessary updates and details not present in these foundational theories. Authors review studies from neuroscience and describe how their results can be seen as evidence for Vygotsky's main conceptions. This idea is ubiquitous in all the book: approaches from education, neuroscience, and cognitive science are not contradictory, and they have a lot in common and provide evidence for the same phenomena from different levels of analyses.

The third chapter presents another crucial point: for a genuine integration among neuroscience, cognitive science, and education, it is necessary to eradicate misconceptions and neuromyths. Thus, the chapter "Dialogue with Latin American Teachers: Beliefs About Neuroscience and Education" shows data on myths and correct beliefs about neurosciences, drawing attention to the importance of the connection between neuroscience and education in the construction of this knowledge. Thus, it points out important aspects, such as whether teaching people basic functions (i.e. sleep, nutrition, cognition and others) and how they are actually related to learning may end up affecting how education as a whole is viewed, consequently also influencing populations in Latin America and in situations of extreme poverty.

The second part of the book refers to child and adolescent cognitive development and factors influencing it. The six chapters of this first part are ordered going from general issues to more specific ones. First, the authors will review attention and executive functions' relations with academic performance as well as the factors that modulate the development of these and other cognitive processes (i.e., home environment, exposure to toxics, and peer pressure in adolescence). Second, chapters will go further on diversity, describing research experiences and results in an indigenous community and exploring stereotypes of women.

In the chapter "Attention and Academic Performance: From Early Childhood to Adolescence," as attentional mechanisms are essential for activities that involve cognitive and motor functions, especially in schooling contexts, the authors highlight studies that demonstrate the importance of attention for the acquisition of reading, writing, and mathematics. Their chapter focuses on the development of attention from early childhood to adolescence.

In the chapter "Child Cognitive Development in Latin American Rural Poverty: What Should Researchers Consider for Conducting Fieldwork?" the authors describe Latin American rural contexts and showed that poverty has a greater impact on rural than on urban non-WEIRD populations and support that claim with evidence from Argentina. Then, they summarize important practical information for conducting studies in rural poverty across Latin America.

The authors of chapter "Concepts, Language, and Early Socialization in the Indigenous Wichi Perspective: Toward a Relational-Ecological Paradigm" presents the cognitive and language development patterns of this population, using an ecological-relational paradigm, which they propose as a comprehensive conceptual framework in developmental science. In particular, they focus on categorization, causal reasoning, and their development during language acquisition.

In the chapter "The Role of Local Violence on Children's Affective Decision-Making," the authors showed the results of a study analyzing the effects of violence on children's decision-making at different socioeconomic status levels. Since violence in non-WEIRD countries is approximately 40% higher than in developed WEIRD countries and is higher in Latin America than in any other region, local evidence regarding its impact on child development is needed and this chapter provides an insight into the issue.

The chapter "The Role of Peer Pressure in Adolescents' Risky Behaviors" describes the results from a prevention curriculum that could mitigate the peer pressure for risky behaviors. Importantly, the authors describe how that curriculum, originally designed for the US population, was carefully adapted for Uruguay, the first Latin American nation to decriminalize the use and commercialization of cannabis, which also regulates the application of effective drug prevention programs for adolescents.

The chapter "The Implications of Exposure to Neurotoxic Metals for Cognitive Development of Children and Adolescents" highlights the cognitive consequences of exposure during development to the neurotoxic metals more present in Latin America and explains why those metals can affect specially non-WEIRD low socioeconomic status populations.

The third part of the book focuses on mathematics and language academic performance. Chapters of this part will address the development of mathematical cognition and educational experiences with developmental dyscalculia in Latin America. Also, chapters will revise evidence regarding bilingualism and education. Finally, this part will describe evidence in Latin American countries regarding chronotype and education in adolescence.

In the chapter "Bilingualism, Foreign Language Learning, and Cognition: Insights into Education," the authors show the linguist and cognitive views of learning two languages, as opposed to learning only their native one, on the experience of foreign language learning. The chapter also highlights the Latin American perspective on bilingualism, as it is ubiquitous in Latin America. The chapter ends with some insights into education showing how the research in this field could bring some ideas to educational practice.

In the chapter "Development of Mathematical Cognition: The Role of Technology in low SES Populations," the authors describe how technology could be used to reduce the gap between high and low SES children in mathematical cognition and in non-WEIRD countries. They do it by describing their own experience designing computerized mathematical cognition tests and intervention in Uruguayan school-aged children. Importantly, they also address the problems that emerged while doing that in non-WEIRD countries and how they could be overcome.

In the chapter "Number Processing and Low Arithmetic Achievement in Cuban and Chilean Children: From Neurocognitive Theories to Educational Practice," the authors present a series of studies conducted in Chile and Cuba addressing the neural basis of mathematical skills and the contribution of domain-specific and general-domain cognitive processes to math achievement. Then, they describe the mathematical skills that are most demanded by school curricula in Latin American countries and neural correlations involved in those abilities, followed by educational implications.

In the chapter "The Perfect Hurricane in Latin America: School Start Time, Chronotype, Sleep, and Academic Performance During Adolescence," Goldin and co-workers explore the consequences of the mismatch between chronotypes and social pressures on children and adolescents, like those involving the educational system. One noteworthy characteristic in the southern cone is the extreme social jet-lag adolescents exhibit in these areas, something which is not apparent anywhere else.

In the chapter "Stereotype Threat and Professional Motivation: Assessing Career Expectations of Undergraduate and Graduate Female Students," the authors describe the results of experimental manipulation and changed the career expectations of female university students.

The fourth part of the book will point to educational practices and interventions. Here, the authors will describe teaching strategies effective in non-WEIRD populations to develop higher order thinking and to improve retention of content in school. Also, this part will inform how diversity influences learning (i.e., how individual differences—in terms of socioeconomic status, place of residence, and level of language—can make an intervention succeed or not). A final chapter in this section will present an example of how technology could be used to bridge cognitive science and education.

In the chapter "Retrieval Practice as a Learning Strategy for Diverse Populations," the authors reviewed the literature regarding the benefits of retrieval practice focused on diverse populations, as most studies in this field involved college students from WEIRD populations. Although retrieval practice is a robust strategy to enhance learning, studies show that both students and educators prefer less effective techniques. This chapter shows that the benefits of retrieval are also found in a diverse population and should be taken into account in evidence-based education.

In the chapter "Consideration of Individual Differences in Cognitive Interventions for Children at Risk for Poverty," the authors open the "poverty" label in order to analyze the weight that individual factors have on the effects of poverty. They analyze how each cognitive construct has both a global, general aspect and a particular aspect, and both can be affected differently by poverty, thus enriching the usual outlook.

In the chapter "Technology on Our Side: Using Technology for Transferring Cognitive Science to Education," the authors show how the low-income population can be benefited from using technology for learning mathematics and writing. All the research shown here is based on technology development that was based on cognitive research, and it could be implemented in educational practice. Furthermore, the authors highlight differences in the results of research from Spanish readers and other language readers, such as English. This also highlights the need to consider cultural underpinnings when devising technological applications.

In the chapter "Early Language Intervention in School Settings: What Works for Whom?" the authors addressed the factors that can make or not succeed language interventions in WEIRD and non-WEIRD populations. They reviewed early language interventions in school settings and concluded that there are some intervention aspects (who delivered the intervention, which abilities were stimulated, and intensity of the intervention) that can make it succeed, besides children's individual factors. However, they note the gap in the number and quality of studies conducted in WEIRD and non-WEIRD populations and the lack of data about the last ones.

In the chapter "Developing Higher Order Thinking in Elementary School Science: A Narrative Around a Research Group Quest," Furman and co-workers present their results around the attempt to develop scientific thinking skills among schoolchildren. This is particularly relevant in the Latin American region, as it is well known that science education is one of the recognized problems of education. What these authors find is that it is useful, but insufficient to provide teachers with educative curriculum materials. Teacher training is also needed. As the authors mention, this is relevant worldwide but critical in Latin America and other non-WEIRD countries.

The final part of the book is dedicated to future perspectives. This part presents a Latin American view on the addressed issues. Particularly, the chapter in this part will express Latin American visions of childhood poverty, as well as an overview of teachers' professional development. The last chapter, "Latin American Perspectives in the Study of Childhood and Adolescence Poverty Through the Lenses of Neural Sciences," brings us a breaking idea in terms of diversity: beyond sharing human neural mechanisms, there are some qualitative aspects of them that can be modulated by individual and contextual differences. Those variations should lead to broader conceptions of the categories of childhood and development and Latin America could constitute a propitious setting for the construction of notions that honor the variety of childhood experiences.

The theoretical and practical innovations in the cognitive science and education field have been growing fast during the last decade in developing countries and, nowadays, they have valuable information to communicate to the whole world. In this book, we take advantage of the diverse population that characterizes developing countries to sum up what the research in these countries can tell the world. The chapters, written by authors from different countries, take the diversity of the developing world as a strength to explore how underrepresented populations learn and what works and what does not for cognitive science and education in these diverse regions.

Conclusions

Specifically, our book will take advantage of developing world diversity to explore cognitive science and education interaction through the following lines: (1) basic and applied research on cognitive sciences of education related to new experiences and contexts; (2) an integrative vision of cognitive science that is social, affective, moral, political, and interpersonal; and (3) containing specific chapters of theoretical and methodological foundation, including directions to young researchers—from all around the globe—in their possible studies about education and cognition. Our book will have a specific focus that will broaden scientists, teachers, and students' views about education. Therefore, our book brings a perspective of updating the areas concomitant with cognition and educational science, addresses aspects that allow it to be used both in undergraduate and graduate courses, as well as by consolidated researchers, and addresses the field of cognitive and educational sciences in a holistic way.

References

Bergen, D. (1988). Encyclopedia of neuroscience. *JAMA: The Journal of the American Medical Association, 260*(1), 104. https://doi.org/10.1001/jama.1988.03410010112050

Bruer, J. T. (1997). Education and the brain: A bridge too far. *Educational Researcher, 26*(8), 4–16.

Bruer, J. T. (2014). The Latin American School on Education and the cognitive and neural sciences: Goals and challenges. *Trends in Neuroscience and Education, 3*(1), 1–3.

Cullen, C. (1997). *Crítica de las razones de educar*. Buenos Aires.

Ekuni, R., & Pompéia, S. (2015). The impact of scientific dissemination in the perpetuation of neuromyths in education. *Revista da Biologia, 15*(1), 21–28.

Haring-Smith, T. (2012). Broadening our definition of diversity. *Liberal Education, 98*(2), 6–13.

Henrich, J., Heine, S. J., & Norenzayan, A. (2010a). Most people are not WEIRD. *Nature, 466*(7302), 29–29.

Henrich, J., Heine, S. J., & Norenzayan, A. (2010b). The weirdest people in the world? *Behavioral and Brain Sciences, 33*(2–3), 61–83.

Hermida, M. J., Segretin, M. S., Soni García, A., & Lipina, S. J. (2016). Conceptions and misconceptions about neuroscience in preschool teachers: A study from Argentina. *Educational Research, 58*(4), 457–472.

Howard-Jones, P. A., Varma, S., Ansari, D., Butterworth, B., De Smedt, B., Goswami, U., et al. (2016). The principles and practices of educational neuroscience: Comment on Bowers (2016).

Hunt, E. (1989). Cognitive science: Definition, status, and questions. *Annual Review of Psychology, 40*(1), 603–629.

Jaeger, D., & Jung, R. (Eds.). (2015). *Encyclopedia of computational neuroscience*. Springer.

Nadel, L., & Piattelli-Palmarini, M. (2003). *What is cognitive science? Encyclopedia of cognitive science*. Macmillan.

Sigman, M., Peña, M., Goldin, A. P., & Ribeiro, S. (2014). Neuroscience and education: Prime time to build the bridge. *Nature Neuroscience, 17*(4), 497–502.

Szűcs, D., & Goswami, U. (2007). Educational neuroscience: Defining a new discipline for the study of mental representations. *Mind, Brain, and Education, 1*(3), 114–127.

Neuroscience and Vygotsky: Putting Together Contemporary Evidence and Cultural-Historical Psychology

Guilherme Brockington, Ana Paula Moreira, and Nikolay Veresov

Introduction

The stage of existence on which the human development processes has been observed, questioned, and provoked by several characters who seem to dispute the sovereignty of a kind of explanatory principle capable of defining the bitterness and beauties of this spectacle. In the voices of these characters, despite locating them in the poetic understanding of the world, in Greek theater, in artistic expressions, in Science, in Philosophy, or in everyday life, they reverberate as possible definitions for what this peculiar humanity is.

That is why, for example, Walsh et al. (2014) present the risks we take when we want to study the movements of history. Now, we do not extract ourselves from history to talk about it, to compose historiographies or to dwell on historical analyses. Thus, considering ourselves as participants in this history, we propose here a possible relationship between the elaborations made by Vygotsky and his collaborators in the field of Cultural-Historical Psychology and Neuroscience and Contemporary Psychology. In this endeavor, history demands that the questions that have long been of interest, mainly, to Philosophy, almost always split between idealist and materialist positions.

G. Brockington (✉)
Center of Natural and Humans Sciences, Universidade Federal do ABC, São Paulo, Brazil
e-mail: brockington@ufabc.edu.br

A. P. Moreira
Psychology Department, Universidade Estadual Paulista (Unesp), São Paulo, Brazil

N. Veresov
Faculty of Education, Monash University, Monash, Australia
e-mail: nikolai.veresov@monash.edu

M. V. Alves et al. (eds.), *Cognitive Sciences and Education in Non-WEIRD Populations*, https://doi.org/10.1007/978-3-031-06908-6_2

In that regard, the studies by Walsh and colleagues point to the genesis of Psychology in the midst of a dichotomy that sometimes understood it from idealistic bases, other times emphasized it through materialist principles. If we consider that the contours of this duality define human existence itself, it is reasonable to assume that this is also a founding structure of what we now call Neuroscience. Between the material limits and subjectivity, consciousness, personality, thought, and volition are established. We have long glimpsed the dynamics of this existence in the synthesis between culture and biology. There is less that we understand that science must not do without the characteristics of this synthesis while it is dedicated to studies on human specificity.

We can say that these two forces, as biology and culture, correspond to an old clash that slithers into the definition of which scientific procedures are consistent. The argument that equalizes Science with the observation of facts is usually followed by its opposite, which asserts any human activity as Science, both careless and expressively mischievous. Now, Science, as a practice, is one of the human activities, which produce and reproduce existence. However, two statements are necessary: not all daily practice is scientific and Science fits (and it does) specific refinements given its nature within the economic and social powers division (De Waal, 1973).

Having an intention of overcoming these misunderstandings, this work aims to unveil materiality as the genesis of subjectivity, thus announcing a monist perspective and, from there, seeking evidence of what seems to be a ballast since the experiments proposed by Vygotsky and what is currently organized as a research field for Neuroscience.

Vygotsky and Cultural-Historical Psychology: Possible Relations Between Contemporary Psychology and Neuroscience

Lev Semionovitch Vygotsky was a Soviet thinker who lived between 1896 and 1934. His precocious 37 years of life coincided with the period of the Socialist Revolution in 1917 and went through the circumstances of an overwhelming civil war that affected millions of children's lives. He studied law, also humanities at the Faculty of History and Philosophy at the Chaniavski People's University. He got involved in the practical work of teaching people, wrote more than 70 theatrical reviews, organized the Psychology office at the School of Pedagogy and several psychological research laboratories and institutes, and, from there, expanded his work in the area, carrying out numerous experimental research programs on the development of human psychological processes (Van der Veer & Valsiner, 1991; Veresov, 1999; Prestes, 2010; Dafermos, 2018). The magnitude of his life, work, and career represented, according to Kozulin (1994, p.13), "a remarkable and suggestive literary quality, which often recalls the lives of the literary heroes of Thomas Mann, Herman Hesse or Boris Pasternak."

Vygotsky produced a solid and broad theoretical foundation on the development process of higher psychological functions, emphasizing the role of culture on the evolutionarily established biological substrate. He studied with a great wealth of experimental content the organization of language, the occurrence of gestures, the appearance of the word, the development of personality, and the specificity of play in the constitution of human beings' consciousness, especially studying the development of babies and children. Its theoretical set substantially deviates from other theoretical elaborations organized by other authors within Developmental Psychology.

Thus, it is inevitable that anyone who knows the expanse of this theoretical set, which we refer to as Cultural-Historical Psychology, does not feel a great surprise when faced with the most recent research results produced by Neuroscience about the processes of child development. It is impossible not to be affected by a strange feeling that some of the experiments and conclusions drawn up by Vygotsky are being reproduced. However, the amazement would be less disturbing if it were not accompanied by a surprising scarcity of references to his prolific previous works.

By limelighting the higher psychological functions' development from their elementary bases, Vygotsky (1987) highlights the role of speech and thought and, more precisely, the meaning of the word, as a constitutive force of a type of system that involves memory, attention, and imagination in the organization of self-control of conduct. Since the work called Prehistory of the Development of Written Language,[1] he has devoted himself to a scrutiny of the paths that emerge from vocalizations and gestures for the elaboration of abstractions produced by what he called thought by concepts. The process of signification mediated by using signs, that is, symbolic representations of knowable objects, advances from the inter-psychological dimensions to the intra-psychological ones. Along these lines, from the materiality of social relations, within specific economic and political structures, constitute the most sophisticated functions of human consciousness or, in Vygotsky's words, "each higher function was thus originally shared between two persons" (Vygotsky, 1995).

To study the establishment of the mastery of behavior, in other words, the possibility that a person is able to use abstract thought to deliberate on situations and direct their own behavior in everyday life, Vygotsky (1995) proposes the free-choice experiments. In this scenario, the child was led to impute a new meaning to a certain set of neutral objects. Here, the force of stimuli such as dice or letters submits to a type of criterion artificially created by the movement of thought. So, for example, a child, while conducting this kind of experiment, could agree with himself that by throwing a dice and getting a certain number; as a result, he/she would act in a certain way. What we have here is the delineation of mediated actions due to the meanings attributed to external objects. An election or a choice always carries what is

[1] It is important to highlight that the correct title of this seminal Vygotsky's work is in Russian original "The pre-history of development of written speech" (речь) (Vol 3, 1983, Russian edition), but in English edition, Vol. 4, 1997 the title was translated as written language—one of the major translation inaccuracies.

called a conflict of motives. Those reasons go through previous experiences and are constituted as affective-cognitive dimensions.

From this synthesis, Vygotsky (1995) discusses the emergence of human choice, stating that our freedom does not concern any independence from the laws of nature, their determinations or circumstances, but, on the other hand, consists in the knowledge of these laws and their manipulation through thought.

Thus, we emphasize the overwhelming materiality of the meaning process that arises in the daily relations established between people through language and, this way, the internalization of signs from vocals, through gesture to abstract thought; it is reasonable to assume that this dynamic is also interesting to the contemporary field of study about the neuroscience of language. There is currently very significant evidence that language learning occurs much earlier than previously assumed, appearing to happen from the fetus.

Language from the Fetus

The impacts of culture on the brain seem to occur much earlier than most child development experts believed. It was already known that newborn babies can distinguish between their mother tongue and a foreign language. A classic experiment used an adapted pacifier connected to a computer that was able to record the baby's sucking rate. In this way, it was possible to measure the reaction of 40 French babies, 3 days old, when they heard a speech in French or Russian. Each time the babies were exposed to the foreign language, their sucking pattern changed, indicating an ability to discriminate between the two languages (Mehler et al., 1988). This study has been replicated several times, and this pacifier technique has become a standard methodology for studies about early language development (Moon et al., 1993; Byers-Heinlein et al., 2010; Butler et al., 2014). However, more recent research has brought evidence that this ability of infants to recognize their mother's language occurs even earlier, within a few hours of birth, enabling the hypothesis that language learning occurs in utero.

The research group led by Patricia Kuhl, using Swedish and American babies, had already shown that phoneme discrimination occurs first for vowels than for consonants and such ability can be measured at around 6 months of age (Kuhl et al., 1992). In 2013, research by the same group conducted the first study showing evidence that language learning already occurs in the fetus, probably 6 months before birth. Again, using the sucking rate measure, 40 Swedish and 40 American babies, now only 30 h of birth, were divided into two groups. Each half-listened to a recording of 17 sounds in the mother tongue and 17 in the foreign language. Each time the baby sucked on the pacifier, a vowel was produced until he stopped sucking, and with a new suck, a new vowel was heard by him. All babies, when they heard the foreign language vowel, sucked the pacifier much more often. For the authors, this result reveals not only the ability to discriminate between languages but also that the baby is much more interested in the vowels of the language that is new to him, which he had not heard yet, suggesting a learning of features of the mother tongue still in the womb (Moon et al., 2013).

Interestingly, the authors state that, until then, it was the consensus view that learning in language was innate, universal, and would take place only after birth. They conclude that the results they found are due to the immersion of babies, still in the womb, to the mother tongue. As they have never been exposed to a foreign language, they perceive the vowels of this new language as something different from what they have already learned.

Still in the intrauterine phase, research using biomagnetometers, equipment that is placed on the mother's belly, is able to detect small magnetic fields generated from electrical currents in the fetuses' body, thus measuring the variation of its heart rate, corroborated the results of the research described above. Twenty-four pregnant women, all American, around 35 months of pregnancy, listened to excerpts of stories sometimes in English and sometimes in Japanese, while a fetal magnetocardiography was collected. The results showed that the fetuses' heart rates varied when they listened to the passages in Japanese, an unfamiliar language and rhythmically quite distinct from English. For the authors, this would be further evidence that language development begins before the baby is born (Minai et al., 2017).

A recently published study in 2021 revealed that the length of intrauterine life affects babies' ability to discriminate prosody. The researchers used functional near-infrared spectroscopy to measure the neural activity of 62 neonates born between 23 and 41 weeks. The results revealed that babies born before 32 weeks had hemodynamic response patterns to speech perception that were distinct from babies born after that time. There was a positive correlation between length of intrauterine life and neuronal activation during language discrimination, suggesting that language development must begin in utero (Alexopoulos et al., 2021).

Perhaps one of the most striking and interesting pieces of evidence about language learning before birth comes from studies of babies' crying. Published in 2009, research revealed that babies cry with an accent, reflecting the rhythm and melody of the language they heard while they were inside the womb. Through acoustic analysis of the cries of 60 newborn babies, 30 French and 30 German, all from monolingual families, the researchers showed that the babies' cries correspond to the specific melodic qualities of each mother tongue. Thus, German babies produce a cry that goes from a higher to a lower tone—very similar to the intonation of the German language, while French babies have a cry that is completely different from the German one, now reproducing the typical intonation of the French language. Since this phenomenon occurs as early as the third day of life, evidence suggests that this learning occurred before birth, already in the womb (Mampe et al., 2009).

Of course, babies are not able to understand language at such an early age, but they can perceive that some sounds have their own rhythm that distinguishes them from others. Around the third trimester, a fetus can hear the rhythm and melody of its mother's voice, an acoustic characteristic known as "prosody." This characteristic of the sound waves is the most important for intrauterine language learning since tissue and amniotic fluid abolish the sound of individual words. Thus, the first cry is marked by the maternal language. But after they are born, babies can imitate several different sounds. However, prosody, already marked by what they hear in the womb, allows them not only to distinguish the infinite sounds coming from the environment, but aids in the development of language by grouping the flow of sound into words and sentences, producing speech, through melodic variations present in speech. And it is these characteristics of intonation and the rhythm behind each word that enable the production of meaning when communicating. Thus, when analyzing and quantitatively comparing the melodic structure of the crying of 52 Swedish newborns and 79 German newborns, researchers revealed that the prosody of the mother's language shapes the melody of the babies' crying (Prochnow et al., 2019).

Revealing that language acquisition begins well before babies speak their first words, there is an increasingly solid body of evidence of the influence of culture while he is still in the womb. This becomes even more evident when studying tonal language peoples, that is, any language in which intonation is a fundamental part of its semantic structure. Thus, the same word can have very different meanings, determined by the pitch with which a syllable is pronounced. Thanks to this, newborns whose mothers speak tonal languages tend to produce more complex crying melodies. This was revealed by research on babies born in Cameroon, which has a tonal language, and babies born in Germany, which speaks a non-tonal language. They were between 37 and 42 weeks of birth and had their cries recorded and analyzed, showing that the cries of Cameroonian neonates were higher in variability than those of German neonates (Wermke et al., 2016).

Another study, from this same group, researched the crying of Chinese babies, such as Mandarin being a tonal language, and German babies, also during the first days of life. The results are consistent, showing that Chinese babies, immersed in Mandarin, produce more complex cry melodies (Wermke et al., 2017). Interestingly, the authors assess that there is a degree of universality in these cry results. Even though the babies come from very different cultural and economic conditions (Nos, a rural community, and no access to technology in Cameroon, and Beijing, an urban and technological community in China), the babies born in these countries, who speak a tonal language, differ from the German babies, accentuating how much culture does indeed act on the brain.

To emphasize the nuances of crying from the understanding of the materiality on which language learning is based is to unveil the consistency of the method proposed by Vygotsky for the study of the development of higher psychological functions. It is not wrong to assume that contemporary research in psychology and neuroscience will not indulge in what is usually called the biologization of human phenomena, in the same way that the field of Cultural-Historical Psychology will have the possibility to free itself from postmodern or idealistic conceptions that insist, for example, in rejecting the importance of experimental studies for the appropriation of the directions of psychological development.

In 1934, Vygotsky elected the meaning of the word as the constitutive unit of the intimate relationship between thought and speech. Everyday activities, in the materiality of their conditions and circumstances, organize themselves as the source or sustain what we call the subjective reflection of objective reality. Sensations and perceptions advance from biology from substances that are cultural always mediated by language. This force, which conducts social experiences, drives the processes of attention and memory, offering ballast for thought that increasingly becomes capable of abstracting and grounding our decisions, attitudes, and feelings.

The Inner Speech

Childhood, in its various expressions, always contextual and historical, assumes, then, unquestionable importance for the study of the initial processes of these developments. This conjuncture appears in the work entitled "Vygotsky Meets Neuroscience: The Cerebellum and the Rise of Culture Through Play," when Vandervert (2017) proposes the hypothesis that the activities related to childhood play have driven the improvement of the cerebellum, which, in turn, would have constituted itself as the determining structure for what we more sophisticatedly do, such as fantasy.

While studying these early periods, Vygotsky (1987) debates the elaboration made by Piaget about what we know as egocentric speech. Vygotsky's statement that thought is not expressed, but realized in the word, makes explicit his understanding of the opposite directions of development of the sound and semantic aspects of language. Here, it is evident in the synthesis that the functions that are internalized as subjectively and uniquely psychic were, before, social relations between people. Development is, therefore, the process of how the social becomes the individual. And thus, what, in the realm of speech expression, Piaget considered egocentric, opened the doors of investigation to what Vygotsky called inner speech. This peculiar mode of speech would preserve the most crucial and delicate dimensions of the transformation of thought into word. Therefore, it is a blunt mistake to conceive it as a kind of thought devoid of vocalization.

Thus, what Piaget called egocentric speech would be, for Vygotsky (1987), a moment of vocalized expression of inner speech, that is, a type of speech external in its manifestation, but internal as to its functions and structures. Such understanding removes from this period of development any possible association with something that compares to selfishness or absence of link with what is essentially social.

What we here call the delicate and sophisticated constitution of humanity, from the revolutionary relationship that is established between thought and speech by means of the word, emerges consistently within the genetic experiments proposed by Vygotsky, his partners, and collaborators. The specificity of inner speech as a crucial expression of this relationship has also been the object of study in contemporary science. Thus, it is now considered that internal speech is deeply involved in important psychological functions such as memory, cognition, emotional regulation, and consciousness (Alderson-Day & Fernyhough, 2015). Several research findings support the Vygotskyan theory about inner speech and its importance for the development of the psyche (Winsler et al., 2009).

In the last three decades, hundreds of researches have been produced to investigate internal speech, both in psychology and neuroscience. Thus, as pointed out by Vygotsky, empirical results reveal that it is at the preschool age, between 4 and 7 years, that children's internal speech reaches its peak, reducing its frequency during infancy (Winsler et al., 2009). Furthermore, it is known, for example, that there is intense activation in the auditory cortex even when the subject is at rest within the scanner, not performing any task and not exposed to any stimuli, which suggests the spontaneous emergence of internal speech (Hunter et al., 2006). The use of functional magnetic resonance imaging (fMRI) has enabled advances in understanding and recognizing the neural correlates that support internal speech (Owen et al., 2004; Shuster & Lemieux, 2005; Frings et al., 2006; Hoeft et al., 2007). From a wide variety of tasks and experimental designs, the evidence points out that internal speech recruits activation of the left inferior frontal gyrus, and the left angular and supramarginal gyri (Geva & Fernyhough, 2019), revealing how much progress has already been made in the neuronal understanding of this phenomenon.

In this scenario, we will highlight here the works that explicitly take the Vygotskyan theory as a basis, revealing how fruitful the approximation with the Russian psychologist's works can be. Researcher Charles Fernyhough and his collaborators are based on Vygostky to argue that, in infancy, the maturation of a whole imbricated neuronal system that supports language occurs together with the development of internal speech, pointing to an intimate relationship between neuroanatomical and psychological developments (Fernyhough, 2016). One of the significant results produced by the researcher and his group concerns the first neuroimaging study investigating neural differences between two varieties of internal speech: dialogic and monological internal speech.

To this end, 21 college students had their brains scanned while performing internal speech tasks. They received a written description of a scenario, for example, a visit to their old school. The task then proposed two situations in which the subjects had to generate internal speech until they were shown a signal indicating that they should stop. The first involved a monologue: you are giving a speech to the students at school. The second involved dialogues with other people: you are talking to a teacher. In addition, the subjects, still inside the scanner, also performed Theory of Mind (ToM) tasks. In this way, it was possible to investigate whether different neural substrates were involved in the different types of inner speech, and more than that, whether such differences in activation had any correlation with the activation of the brain circuitry that supports ToM.

The results revealed that when comparing the brain activations involved in monologue inner speech, dialogic speech showed significantly greater activation, compared to monologue, in the precuneus, posterior cingulate, the right superior temporal gyrus, the left insula, inferior frontal gyrus, and cerebellum (Alderson-Day et al., 2016). In addition, the researchers found an overlap in the brain activations of dialogic internal speech with those of ToM. Such activations occurred near sections of the Temporal Parietal Junction, an area strongly associated with empathy, the ability to infer the internal state of others.

Therefore, such results rightly suggest the involvement of social and cognitive processes in dialogic inner speech events. Corroborating perhaps the most profound dimension of Vygostky's theory of inner speech production, these data bring evidence of the biological support for the dialogical, and therefore social, process of internal speech development and, consequently, language. The work conducted by Fernyhough brings the perspective that this kind of speech is a type of dialogue that unveils, at one and the same time, the substances of thought and speech, in the synthesis of what constitutes our most ordinary singularity: astonishment, joy, fright, or indignation.

With respect to internal speech, an extensive review discusses the theoretical approaches and methodological challenges for studying internal speech, presenting diverse empirical evidence in both children and adults (Alderson-Day & Fernyhough, 2015). The authors end, once again, by reinforcing the Vygotskyan proposal, which points to the emergence of internal speech as a social process and playing an important role in human cognition. It should be noted that already in 1996, Fernyhough affirmed the importance of recognizing Vygotsky's work for the advancement not only of the compression about language but of the relevance of his theory as a theory of psychic development (Fernyhough, 1996).

Thus, in more than two decades of research, Fernyhough combines findings from neuroscience with Vygotsky's psychology precisely so that the understanding of issues dear to contemporary psychology can be advanced. However, the author states that such evidence about the parallel development of neural substrates and psychological processes cannot be interpreted in an uncritical way. To avoid the biologization of psychism, it is necessary, first of all, to consider that the development of language is not only the result of the maturation of brain structures. "Following Vygotsky, Luria argued for bidirectional causation between biological maturation and sociocultural experience, fitting with the view that the internalization of social exchanges creates a new functional system of inner speech" (Alderson-Day et al., 2016). In this sense, here is the understanding that the psychological functions or cognitive functions, as studies in the field of neuroscience seem to refer to them, are formed from everyday materiality. And, as Smirnov et al. (1960) say, there is the evolution, the brain and the reflex that organize the bases on which the conquest of culture is established, the mediation of language, and, therefore, the tone of higher development. In Vygotsky, and for contemporary science that is positioned this way, lie the marks for understanding the most subtle and sophisticated subjectivity without having to slip into idealistic perspectives.

Culture in the Brain: Possible Redemptions Since Vygotsky

It is beyond the scope of this chapter to define what culture is, something that has been a matter of discussion in the humanities since the eighteenth century. However, it should be noted that for a little more than 10 years cultural aspects have been investigated by the theoretical and methodological framework of neuroscience

(Chiao & Ambady, 2007). Since then, the number of research studies investigating the interactions between culture, social practices, psychological processes, brain, and genes has been growing annually. The importance of presenting some results of this research is precisely to emphasize the determining role of culture in all of Vygotsky's developmental theory. Far from trying to biologize culture, a real risk that should be combated by researchers in the field (Choudhury & Kirmayer, 2009), recent research on the fundamental relationship between culture and biology shows how bold and plausible Vygotsky's theoretical propositions were.

Instead of considering development as a universal process, as reflected in the stages proposed by Piaget, Vygostky announced that the development of higher psychological functions varied profoundly as the contexts and contents of human activity varied. Unlike most theorists of the time, the Soviet psychologist's propositions emphasized the social dynamics of personality production. Although, obviously, Piaget and many other theorists of the period did not disregard culture, virtually none other than Vygotsky placed it at the center of their theory. The idea that culture conquers biology in the process of construction of the human being has been, nowadays, increasingly evidenced.

Thus, investigating the relationship between culture and self-representation, researchers (Zhu et al., 2007) analyzed the brains of Chinese and Westerners as they judged trait adjectives characteristic of their own personality, of a public person, and of their mothers, such as honesty, for example. For both groups, an activation of the medial prefrontal cortex (mPFC) and anterior cingulate cortex was observed when the subjects thought about themselves and not when they thought about public people or each other. However, only for the Chinese, there was activation of the medial prefrontal cortex when they thought about their mothers. This suggests that Western subjects, when thinking about whether they were honest, had brain activities quite distinct from when they were thinking about whether another person was honest, be it their mother or the distant celebrity. This did not happen with the Chinese. When they thought about whether they were honest, their brain activity was quite similar to that displayed when they thought about whether their mothers were honest. According to the authors, Westerners use the mPFC exclusively to represent the self, and this would be evidence of the impacts of culture on neuroanatomy. This and other results suggest that North Americans and Europeans, immersed in a strongly individualistic culture, tend to view the self as independent, separate from others, while Chinese, immersed in a collectivist culture, value interdependence and interconnectedness more, and these cultural differences impact both the psyche and the brain (Markus & Kitayama, 1991; Chiao et al., 2009; Park & Huang, 2010; Liew et al., 2011; Sul et al., 2012; Huang et al., 2019; Medina et al., 2019).

Still corroborating the hypotheses made by Vygotsky in 1920, a research study using fMRI revealed how culture makes the same stimulus activate completly different neural circuitries. Based on historical data that point to a social valuation for submission in Asian countries while Western culture tends to value domination more, researchers analyzed the brain activity of Japanese and Americans viewing silhouettes of bodies in submissive postures (head and arms hanging down) and in dominant postures (head up and arms crossed). All subjects had activated the medial prefrontal cortex and the caudate nucleus, an area of the brain strongly associated with reward detection and the associative learning of rewards. However, in the Americans, this reward neuronal circuitry was activated only when they saw the dominating postures, while the Japanese had the same circuitry activated only for the submissive postures. Furthermore, the intensity of activation was positively correlated with self-report of how much the participants valued dominance and submission. For example, the more the Japanese placed importance on submission, the more intense was the activation in the caudate and mPFC when he saw a submissive posture. The same was true for the American, but only with dominant postures. For the authors, "The findings provide a first demonstration that culture can flexibly shape functional activity in the mesolimbic reward system, which in turn may guide behavior" (Freeman et al., 2009).

Studies about the impacts of culture on the brain and the psyche become even more important when we pay attention to the fact that the overwhelming majority of scientific evidence we have today about what makes us human comes from research conducted mostly in Western countries, from a group of participants that is, as we have seen, largely immersed in an individualistic culture. Using the acronym WEIRD to designate these research subjects (Western, Educated, Industrialized, Rich, Democratic) in the last decade, the academic community has announced the problems of scientific practice that takes the singular as universal. Over 95% of research in psychology uses subjects with these characteristics, and they represent only 12% of the world's diverse cultural populations (Henrich et al., 2010). Thus, it is impressive that it took almost 100 years for the academic community to come into line with Vygotsky's propositions and to consider the extreme relevance of culture in human development.

This and other research seek to investigate the cultural impacts on the brain and behavior. It is obvious that emphasizing social and cultural processes does not mean to disregard genetic, biological conditioning. However, to believe in a universal cognitive development, free or with little influence from culture, contradicts what contemporary neuroscience reveals nowadays and what Vygotsky already said in the 1920s.

Concluding Remarks

Vygotsky's works were ahead of their time and now we already understand that this is not an exaggeration or a metaphor. Contemporary research in the most complex areas of neuropsychology confirms his ideas about two lines of development—natural and cultural—and about their dialectical relationship. There is a significant amount of modern research confirming that the social and cultural environment is not a development factor, but a source of psychological development. The cultural neuropsychology that emerged with the work of Alexander Luria, who relied on Vygotsky's cultural-historical theory, finds more and more research-based evidence. The child's ability to socially interact and learn does depend on the development of the brain—but the development of the brain is determined by the kind of culture environment the child is a part of and in what collective forms of behavior and activity she is involved. Such a view opens up new possibilities for the development of training programs and new social practices.

Bibliography

Alderson-Day, B., & Fernyhough, C. (2015). Inner speech: Development, cognitive functions, phenomenology, and neurobiology. *Psychological Bulletin, 141*(5), 931.

Alderson-Day, B., et al. (2016). The brain's conversation with itself: Neural substrates of dialogic inner speech. *Social Cognitive and Affective Neuroscience, 11*(1), 110–120.

Alexopoulos, J., et al. (2021). The duration of intrauterine development influences discrimination of speech prosody in infants. *Developmental Science, 24*(5), e13110.

Butler, S. C., et al. (2014). Preference for infant-directed speech in preterm infants. *Infant Behavior and Development, 37*(4), 505–511.

Byers-Heinlein, K., et al. (2010). The roots of bilingualism in newborns. *Psychological Science, 21*(3), 343–348.

Chiao, J. Y., & Ambady, N. (2007). Cultural neuroscience: Parsing universality and diversity across levels of analysis.

Chiao, J. Y., et al. (2009). Neural basis of individualistic and collectivistic views of self. *Human Brain Mapping, 30*(9), 2813–2820.

Choudhury, S., & Kirmayer, L. J. (2009). Cultural neuroscience and psychopathology: Prospects for cultural psychiatry. *Progress in Brain Research, 178*, 263–283.

Dafermos, M. (2018). *Rethinking cultural-historical theory: A dialectical perspective to Vygotsky*. Springer.

De Waal, F. (1973). The bonobo and the atheist: In search of humanism among the primates., WW Norton & Company.

Fernyhough, C. (1996). The dialogic mind: A dialogic approach to the higher mental functions. *New Ideas in Psychology, 14*(1), 47–62.

Fernyhough, C. (2016). *The voices within: The history and science of how we talk to ourselves*. Basic Books.

Freeman, J. B., et al. (2009). Culture shapes a mesolimbic response to signals of dominance and subordination that associates with behavior. *NeuroImage, 47*(1), 353–359.

Frings, M., et al. (2006). Cerebellar involvement in verb generation: an fMRI study. *Neuroscience Letters, 409*(1), 19–23.

Geva, S., & Fernyhough, C. (2019). A penny for your thoughts: Children's inner speech and its neuro-development. *Frontiers in Psychology, 10*, 1708.

Henrich, J., et al. (2010). The weirdest people in the world? *Behavioral and Brain Sciences, 33*(2–3), 61–83.

Hoeft, F., et al. (2007). Functional and morphometric brain dissociation between dyslexia and reading ability. *Proceedings of the National Academy of Sciences, 104*(10), 4234–4239.

Huang, C.-M., et al. (2019). Culture-related and individual differences in regional brain volumes: A cross-cultural voxel-based morphometry study. *Frontiers in Human Neuroscience, 13*, 313.

Hunter, M., et al. (2006). Neural activity in speech-sensitive auditory cortex during silence. *Proceedings of the National Academy of Sciences, 103*(1), 189–194.

Kozulin, A. (1994). *La psicología de Vygotski; biografía de unas ideas* (p. 294). La psicología de Vygotski; biografía de unas ideas.

Kuhl, P. K., et al. (1992). Linguistic experience alters phonetic perception in infants by 6 months of age. *Science, 255*(5044), 606–608.

Liew, S.-L., et al. (2011). Who's afraid of the boss: Cultural differences in social hierarchies modulate self-face recognition in Chinese and Americans. *PLoS One, 6*(2), e16901.
Mampe, B., et al. (2009). Newborns' cry melody is shaped by their native language. *Current Biology, 19*(23), 1994–1997.
Markus, H. R., & Kitayama, S. (1991). Culture and the self: Implications for cognition, emotion, and motivation. *Psychological Review, 98*(2), 224.
Medina, L. D., et al. (2019). Collectivism is associated with greater neurocognitive fluency in older adults. *Frontiers in Human Neuroscience, 13*, 122.
Mehler, J., et al. (1988). A precursor of language acquisition in young infants. *Cognition, 29*(2), 143–178.
Minai, U., et al. (2017). Fetal rhythm-based language discrimination: A biomagnetometry study. *Neuroreport, 28*(10), 561.
Moon, C., et al. (1993). Two-day-olds prefer their native language. *Infant Behavior and Development, 16*(4), 495–500.
Moon, C., et al. (2013). Language experienced in utero affects vowel perception after birth: A two-country study. *Acta Paediatrica, 102*(2), 156–160.
Owen, W. J., et al. (2004). FMRI of two measures of phonological processing in visual word recognition: Ecological validity matters. *Brain and Language, 90*(1–3), 40–46.
Park, D. C., & Huang, C.-M. (2010). Culture wires the brain: A cognitive neuroscience perspective. *Perspectives on Psychological Science, 5*(4), 391–400.
Prestes, Z. R. (2010). Quando não é a mesma coisa: análise de traduções de Lev Semionovitch Vygotsky no Brasil: repercussões no campo educacional.
Prochnow, A., et al. (2019). Does a 'musical' mother tongue influence cry melodies? A comparative study of Swedish and German newborns. *Musicae Scientiae, 23*(2), 143–156.
Shuster, L. I., & Lemieux, S. K. (2005). An fMRI investigation of covertly and overtly produced mono-and multisyllabic words. *Brain and Language, 93*(1), 20–31.
Smirnov, A. A., et al. (1960). Psicología, Grijalbo.
Sul, S., et al. (2012). Cultural modulation of self-referential brain activity for personality traits and social identities. *Social Neuroscience, 7*(3), 280–291.
Van der Veer, R., & Valsiner, J. (1991). *Understanding Vygotsky: A quest for synthesis*. Blackwell Publishing.
Vandervert, L. (2017). Vygotsky meets neuroscience: The cerebellum and the rise of culture through play. *American Journal of Play, 9*(2), 202–227.
Veresov, N. (1999). *Undiscovered Vygotsky: Etudes on the pre-history of cultural-historical psychology*. Lang Frankfurt.
Vygotsky, L. S. (1987). The collected works of LS Vygotsky: Problems of general psychology.
Vygotsky, L. S. (1995). *The collected works of LS Vygotsky: Problems of the theory and history of psychology*. Springer Science & Business Media.
Walsh, R. T., et al. (2014). *A critical history and philosophy of psychology: Diversity of context, thought, and practice*. Cambridge University Press.
Wermke, K., et al. (2016). Fundamental frequency variation within neonatal crying: Does ambient language matter? *Speech, Language and Hearing, 19*(4), 211–217.
Wermke, K., et al. (2017). Fundamental frequency variation in crying of mandarin and German neonates. *Journal of Voice, 31*(2), 255. e225–255. e230.
Winsler, A. E., et al. (2009). *Private speech, executive functioning, and the development of verbal self-regulation*. Cambridge University Press.
Zhu, Y., et al. (2007). Neural basis of cultural influence on self-representation. *NeuroImage, 34*(3), 1310–1316.

Dialogue with Latin American Teachers: Beliefs About Neuroscience and Education

Adriana Soni García

Abundant international research in the field of neuroscience and education has largely focused on identifying what misconceptions about the brain are held by educators (Cole, 2013; Dekker et al., 2012; Deligiannidi, 2009; Geake, 2008; Howard-Jones, 2010; Pasquinelli, 2012; Rato et al., 2010; Tardif et al., 2015; Torrijos-Muelas et al., 2021) and such a trend appears to be similar in Latin America (Cosme et al., 2018; de Sá et al., 2020; Gleichgerrcht et al., 2015; Hermida et al., 2016; Lopes et al., 2020). Being able to identify these misconceptions is an important step toward pinpointing issues that should be priorities toward further developing the field of neuroscience of education—however, most evidence available appears to stay at that stage.

The present chapter focuses on Latin America—a region with a great diversity of educational contexts—where many teachers face challenges beyond those encountered in educational practices itself. Admittedly, the data that will be discussed in this chapter comes from a doctoral project (Soni Garcia, 2016) that—originally—had very similar objectives to the available evidence reviewed at the time and described just above. Some of those original objectives aimed to explore the knowledge about mind and brain that Latin American educators had and to better understand how such knowledge was formed. However, through the process of conducting such a doctoral project (Soni Garcia, 2016), it became evident that more than simply testing teachers on their knowledge about mind and brain, it was fundamental to understand what education looks like in Latin America. This was in order to better understand what a Latin American teacher may need from neuroscience-informed research and why they may hold certain beliefs about the brain. For this reason, specific results from this doctoral project will be discussed—a mixed-methods research project completed in 2016 with qualitative data derived from semi-structured interviews carried out with 19 teachers and quantitative data obtained

A. S. García (✉)
Arden University, London, UK

M. V. Alves et al. (eds.), *Cognitive Sciences and Education in Non-WEIRD Populations*, https://doi.org/10.1007/978-3-031-06908-6_3

from an online questionnaire administered to 314 participants from Latin American countries. Next, the chapter will explore Latin American teachers' perceptions and descriptions of their educational context—as these should be powerful influences that should continuously be informing neuroscience and education. The final section of this chapter will take a closer look at the available evidence relating to the collaboration between neuroscience and educators—has it moved to the next step or does it continue to identify what Latin American teachers *do not know* about the brain?

Neuroscience and Education: Views from Latin American Teachers

The following section will discuss findings derived from Soni García's (2016) doctoral project, which ran from 2012 to 2016. This project consisted of several objectives and stages (see Fig. 1) because it was a mixed-methods project, which started with a qualitative stage to identify, among other issues, which ideas about the brain were most prevalent in Latin America. This qualitative stage helped design a pilot questionnaire (stage 2), which ultimately was refined into a final online questionnaire composed from items used from other available studies (e.g., Dekker et al., 2012; Herculano-Houzel, 2002; Howard-Jones, 2010) and those ideas that were identified as prevalent during stage 1. Here, only findings relating to the following research question will be discussed: What knowledge about the brain do teachers in Latin America have? Most particularly, quantitative findings that were indicative of Latin American teachers' neuroscience literacy and neuromyth scores in that period of time are discussed. However, qualitative data will be presented, too, as it helps provide a fuller picture of the results.

Operationalizing "Neuroscience Literacy" and "Neuromyths"

Neuroscience literacy, in part, can be defined as the ability to "understand basic concepts of neuroscience and its research methods and to use this knowledge to separate science from science fiction in the media" (Bergmann et al., 2017 p. 236). For the purposes of that project, neuroscience literacy was operationalized as the accurate knowledge Latin American teachers held regarding the brain (being able to agree with the factual information provided in the questionnaire and disagree with the inaccurate items). In parallel, inspired by the prevailing available evidence, the project was also concerned with identifying neuromyths. This term has been around for decades, with "neuromyth" first being used in the 1980s (Howard-Jones, 2010). However, its most conventional definition was provided by the UK's Organization of Economic Co-Operation and Development (OECD): "a misconception generated by a misunderstanding, a misreading or a misquoting of facts scientifically

Fig. 1 Research stages

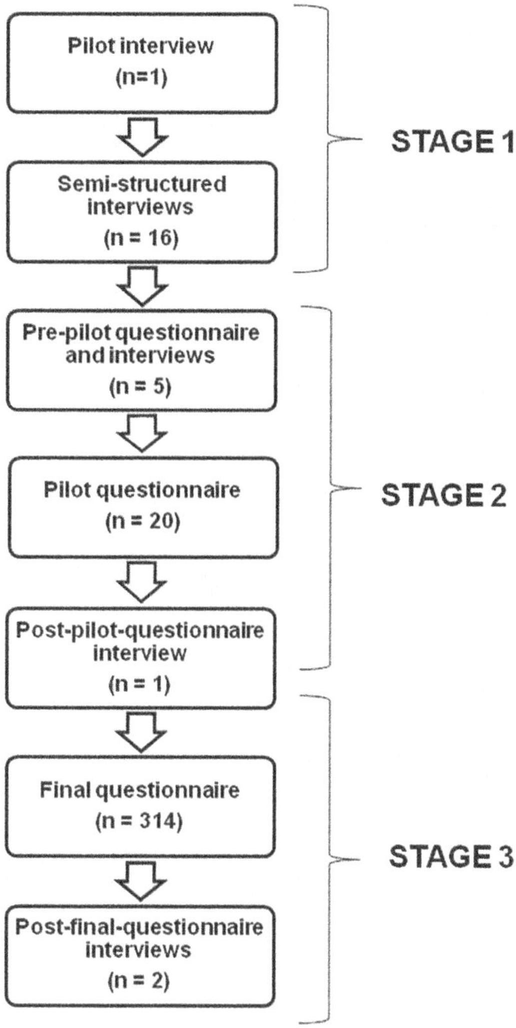

established (by brain research) to make a case for the use of brain research in education and other contexts" (OECD, 2002 p. 258). Thus, not only was it relevant to the project to identify neuroscience literacy, but the prevalence of neuromyths as well.

*Stage 3: Online Questionnaire (*N *= 314) – Participants*

The final online questionnaire (stage 3) was administered to 314 participants from 12 Latin American countries (with most participation from Mexico, Argentina, Chile, and Colombia, see Table 1). Within this sample, the mean age of the participants was 41.33 years (SD 11.42); only 21% of them were male; the average year of

Table 1 Neuroscience literacy score items ($N = 314$)

Item	Don't know %	Disagree %	Agree %	Mean (SD)
1 "'Mental abilities are inherited and education cannot modify them."	5.4	90.4	4.1	1.47 (0.85)
4 "'There are sensitive periods in childhood when it is easier to learn certain things."	7.3	7.0	87.5	4.11 (1.52)
5 "Education cannot remediate learning difficulties associated with developmental differences amongst students."	9.9	79.9	10.2	1.54 (1.08)
7 "Memory is stored in networks of cells distributed throughout the brain."	12.1	11.5	76.5	3.68 (1.76)
8 "'Skipping breakfast affects students' cognitive function."	10.8	11.2	78.0	3.83 (1.74)
9 "'The use of technology affects children's brain development."	17.5	60.9	21.7	1.89 (1.49)
11 "'There are changes in our neural connections when we learn something."	6.7	4.1	89.2	4.29 (1.43)
14 "Poverty can impact children's brain development."	12.4	24.2	63.4	3.29 (1.83)
15 "'Extensive practice of some mental processes can change the shape and structure of some parts of the brain."	16.9	10.2	72.9	3.47 (1.86)
16 "'Not sleeping properly affects the consolidation of the day's learning in long-term memory."	9.2	6.0	84.8	4.04 (1.57)
20 "'Teenage behaviour can be due to the brain still being in development during that stage."	21.7	19.8	58.6	2.91 (1.93)
22 "'Vigorous exercise cannot improve mental function."	15.6	75.5	8.9	1.45 (1.12)
23 "'Stress can be detrimental for learning."	10.8	11.5	77.7	3.81 (1.70)
28 "'Hormones do not influence personality."	14	74.2	11.8	1.58 (1.19)
30 "'The change in sleep patterns during adolescence can diminish students' mental function."	18.5	10.8	70.7	3.32 (1.85)
31 "'It is easier to learn a topic in several sessions distributed throughout time than it is in a single session."	11.8	10.8	77.4	3.71 (1.69)
33 "IQ is inherited and cannot be modified by education."	6.1	87.3	6.7	1.44 (0.91)
35 "'During old age, there is no production of new connections in the brain."	12.1	76.6	11.1	1.59 (1.18)

Note. In the original questionnaire (Soni Garcia, 2016), participants were able to select from the following answering categories: "don't know," "completely disagree," "disagree," "agree," and "completely agree." Here, agreement and disagreement options have been condensed into single categories.

teaching experience was 14.03 years (SD 10.77); 87% of teachers worked in urban settings; there was an even representation of sectors with 43% of teachers working in the private sector and 47% working in the public sector (9.6% reported working on both sectors); 36.9% of teachers reported having zero days of neuroscience training in the 2 years previous to the administration of the questionnaire.

Stage 3: Online Questionnaire (N = 314) – Neuroscience Literacy and Neuromyth Findings

Tables 1 and 2 contain the list of items that were used to identify teachers' neuroscience literacy score and neuromyth score, respectively.

Table 2 Neuromyth score items ($N = 314$)

Item	Don't know %	Disagree %	Agree %	Mean (SD)
10 "When we work in a pleasant environment, the brain releases dopamine."	14.6	5.7	79.7	3.79 (1.81)
12 "Caffeine consumption favours the reduction of mental abilities."	36.0	50.3	13.7	1.38 (1.40)
13 "When we sleep, the brain shuts down."	3.2	90.2	6.6	1.31 (0.91)
18 "For the most part, we only use 10% of our brain."	21.0	37.6	41.4	2.37 (1.86)
19 "Learning more than one language from childhood reduces mental development."	4.1	90.4	5.4	1.27 (0.87)
21 "The intake of sugary drinks and snacks reduces children's attention."	23.9	30.9	45.3	2.52 (1.93)
24 "Without a brain it is not possible to have consciousness."	14.6	15.9	69.5	3.51 (1.87)
25 "It is easier to learn when it is through our preferred learning style."	10.2	9.0	80.8	3.98 (1.68)
27 "Individual differences in students can be explained by the dominance of their left or right brain hemisphere."	22.3	23.6	54.1	2.77 (1.95)
34 "Emotional processes in the brain interrupt processes associated with learning."	14.0	21.4	64.7	3.18 (1.82)
36 "Teachers cannot develop students' creativity, they are born with it."	3.8	92.0	4.1	1.30 (0.83)
37 "Brain Gym does not improve the integration of the brain hemispheres."	8.9	83.1	7.1	1.39 (1.04)

Note. In the original questionnaire (Soni Garcia, 2016), participants were able to select from the following answering categories: "don't know," "completely disagree," "disagree," "agree," and "completely agree." Here, agreement and disagreement options have been condensed into single categories.

It is beyond the scope of this chapter to run an in-depth analysis of each neuromyth in order to debunk it (for such purposes, please consult Grospietsch & Mayer, 2020, and Howard-Jones, 2014a). Rather, three main points will be highlighted: only 63.4% of the sample agreed with the statement that poverty can impact children's brain development; some very popular neuromyths around the world remained prevalent in Latin America; and the qualitative interviews helped identify a new neuromyth ("dopamine is released when we work in a pleasant environment") which 79.7% of the sample agreed with.

Popular Neuromyths Around the World That Are Prevalent in Latin America

As it can be seen in Table 2, similar to what has been found in other regions (Grospietsch & Mayer, 2020; Howard-Jones, 2014a), there seems to be several persistent neuromyths present in Latin America. Close to half of the sample (41.4%) still appears to believe that for the most part we use 10% of the brain. There was great agreement with the "it is easier to learn when it is through our preferred learning styles" neuromyth as 80.8% of the sample seems to believe this is accurate. Similarly, 57% of participants still appear to believe that individual differences in their students are explicable by the dominance of their left or right brain. And, not surprisingly, Brain Gym remains a rooted neuromyth as 83.1% of the sample disagreed with the idea that it does not improve integration of brain hemispheres. These neuromyths, and others, are not exclusive to Latin America—they are found among educators internationally (Ferrero et al., 2016; Grospietsch & Mayer, 2020; Howard-Jones, 2014a; Idrissi et al., 2020; Painemil et al., 2021; Papadatou-Pastou et al., 2017) teaching at different levels of education (Betts et al., 2019).

While it is quite important to keep track of the dissemination of these neuromyths, there is a seeming emphasis on testing teachers on their knowledge about the brain. As stated at the beginning of this chapter, that was the starting point of the doctoral project (Soni Garcia, 2016) that produced this data. However, through the process of conducting qualitative research (and having in-depth conversations with the Latin American teachers that took part in the study), it was possible to see that more than testing teachers on what they do and do not know about the brain—a conversation with them was more collaborative. Though there are now many reliable instruments (e.g., Dekker et al., 2012; Herculano-Houzel, 2002; Howard-Jones, 2010) to score teachers on neuroscience literacy and beliefs in neuromyths, it is worth questioning if using such materials miss out on picking up on the nuances of teachers' beliefs. Most importantly, if such instruments fail to identify ideas that have not been included as an item but are held by teachers regardless. It was thanks to qualitative data that a new belief was identified in Soni García's (2016) study.

New Neuromyth: When We Work in a Pleasant Environment, the Brain Releases Dopamine

Dopamine appears to be the go-to explanation for a number of behaviors that range from cupcake consumption to obsessively checking one's emails (Howard-Jones, 2014a). It has been labeled the media's neurotransmitter of choice (Bell, 2013); thus, it is not surprising that some Latin American teachers appear to use it to add a neuroscientific explanation to some of their practices. As seen in Table 2, 79.7% of the sample agreed with the neuromyth stating that "when we work in a pleasant environment the brain releases dopamine." Interestingly, this idea was included in the online questionnaire as some participants mentioned the following during the qualitative stage (see Fig. 1):

> **Participant 12 (32 years of age; 2 years of teaching experience)**: "*... the fact that a good working environment influences how the student will be prepared, or not, to have a good experience. How it will be recorded in the brain and it will be more inclined to having a maths lesson, how it will predispose him and remain in his memory. If it was negative or positive, how it will influence in the future. That is what we try to emphasize the most, a good working environment to release dopamine, so that they feel good and in such a way it is recorded as something positive.*"

> **Participant 10 (58 years of age)**: "*...but I use joy and surprise, and the fact that we are here to enjoy and have fun and we are always laughing. And with this, this is neuroscience, right? It becomes more bearable and the kid has this dopamine that is released in the brain so he is more open to learn and enjoying what he is learning and he knows it.*"

As with any neuromyth, there is some truth to these ideas, but they are overly simplified. Dopamine does play a role in experiencing a pleasurable state (Willis, 2008), and in memory consolidation (Stern et al., 2005; Wise, 2004), but there are certain conditions that promote a greater release of dopamine. One of the areas of research in education that involves dopamine is the use of learning games in the classroom (Howard-Jones & Demetriou, 2009; Howard-Jones et al., 2011). Dopamine is part of the reward system, and uncertain rewards are more stimulating than predictable ones (the dopamine levels after a reward-response peaks at 50% certainty in games); cognitive neuroscience research has shown that the formation of declarative memory in the brain is strongly linked to its response to reward (Howard-Jones, 2014b).

The last part of this section will focus on discussing an alarming finding—the need to raise awareness of the impact of poverty on children's development in a region plagued with high poverty rates (Bellei et al., 2013).

Neuroscience Literacy: Poverty Can Impact Children's Brain Development

Quantitative findings revealed that 12.4% of the sample did not know that poverty can impact children's brain development. And, more worryingly, 24.2% disagreed with this statement. This is concerning as, indeed, neuroscience research has shown that poverty can impact children's brain development in a number of ways that ultimately may impact their academic performance (as cognitive skills needed for education are negatively affected by the result). For example, low socioeconomic status has been found to be linked with changes in the structure of the brain, particularly in the areas of memory, executive control, and emotion (Brito & Noble, 2014). A study in the state of Oregon, United States, where children were recruited from 11 public preschools, revealed that the acute stress to which some children living in poverty are exposed during their development can affect the prefrontal cortex and the hippocampus, which are central to many aspects of working memory and execute function (Neville et al., 2013); chronic stress associated with poverty was also found to be related to a greater amygdala volume and reduced prefrontal activity during cognitive tasks with children recruited from public schools and anti-poverty programs in rural counties in the Northeastern United States (Kim et al., 2013). Another study found evidence that children from low-socioeconomic-status families living in Boston and St. Louis's metropolitan regions have significantly lower average total gray matter volumes in comparison to children from high-socioeconomic-status families (Hanson et al., 2013).

The prevalence of poverty in Latin America reported among countries varied a lot during the time this project took place (2012–2016), but countries such as Guatemala, Honduras, and Nicaragua had more than 50% of their population experiencing low-socioeconomic status then (SITEAL, 2015). Unfortunately, these trends have been worsened by the effects of COVID-19 in the entire Latin American region, and poverty has reached levels in 2020 that were not seen since 2008 whereas rates of extreme poverty have gone back to what they used to look like in the early 2000s (CEPAL, 2021). This is the sort of issue that should be a priority when trying to move the field of neuroscience and education forward. To better understand the challenges faced by these teachers (and students!) so that specific messages and valuable neuroscientific research can be created with and for them.

The next section will focus on highlighting the educational side of the "neuroscience and education" field. More specifically, it will present some of the perspectives shared by some of the 19 Latin American teachers that were interviewed for Soni García's (2016) doctoral project.

Latin American Teachers' Perceptions of Their Educational Context

As concluded in the last section, it is important to remember that messages and research coming from neuroscience (for the purpose of enriching education) should be relevant for their audiences. How can neuroscientific messages and neuroscience research findings be meaningful if they are not ready to be used (or valuable) for every kind of classroom or setting? In parallel, as established throughout this chapter, testing teachers on their knowledge about the brain is certainly informative.

However, it does not seem to lead to collaboration between educators and neuroscientists—nor does it seem to do much toward helping each field better understand one another. What should neuroscientists know about educational settings to better understand educators? This is a difficult question to answer in such a large and diverse region as Latin America is.

Generally, the quality of teaching in Latin America ranks low, and the region has proved to have a deficient teaching system, as suggested by the results of student achievement obtained in the TIMSS of the IEA and the PISA of the OECD; in the 2003 PISA, the three (out of 40 participant countries) Latin American countries that participated, Brazil, Mexico, and Uruguay, ranked 40th, 37th, and 35th, respectively (de Moura Castro & Ioschpe, 2007). Even if the available information is not complete, it indicates that teaching in this region does not match the description of a profession with a high status, as professional development opportunities are limited and the initial training quality is low (Bellei et al., 2013).

Latin American teachers' level of initial training varies across the region, but close to 90% of the teachers participating in the SERCE study (conducted in 2006) were reported to have teaching training. About 60% of this sample completed a university-level education, 32% attended nonuniversity technical or pedagogical training programs, but there were still 11% working just having completed secondary education level or lower (Duarte et al., 2010). The countries where teachers have little to no teaching formation are Guatemala, where more than a third of the teachers had no pedagogical studies or higher education, Nicaragua, and to a less extent, Panama and Ecuador (Falus & Goldberg, 2011). In Mexico, one in four teachers have higher education levels without any teaching training, and in El Salvador, this happens in one out of ten teachers (Falus & Goldberg, 2011). In general, in Mexico and other countries, teachers with a postgraduate level of education are the ones that engage more in professional development activities (SEP, 2009).

A condition that could influence Latin American teachers' attitude toward becoming informed about neuroscience (or not) is the little amount of time that they seem to have available to dedicate to professional development. Two of the obstacles found in a Mexican survey with secondary school teachers ($n = 62{,}650$ participants) were that more than 40% of them said that they had conflicts with their schedule and that some of the professional development activities were too costly (SEP, 2009). In relation to this point, in some countries (such as Argentina, Brazil, and Uruguay), there is a high proportion of teachers (in a sample of 1816 teachers 44% of Brazilian, 22% of Uruguayan, and 23% of Argentine sixth-grade teachers) with a job in more than one school, or outside the educational sector (Duarte et al., 2010), which means it could limit their free time and funds to invest in professional development.

With regard to the differences between the rural and urban areas, there is great diversity evident in Latin America. A regional study showed the distribution that each sector has in urban and rural areas in 12 countries in Latin America (Argentina, Bolivia, Brazil, Honduras, Mexico, Nicaragua, and Paraguay) in the different levels that are considered as formal education: in rural areas, the distribution of the private sector only reached 4% whereas it was 24.1% in urban areas (Pereyra, 2008). The distribution of teachers across urban and rural areas is different and reflects the slant in resources distribution (both physical and human): in Uruguay, rural teachers make for 12% of the total population of teachers; in Colombia, rural teachers make up 25% of the total teacher population. However, in countries like Nicaragua and Honduras, the proportion of rural teachers reaches more than 50% of the total (Vaillant & Rossel, 2006).

Some of the participants in Soni García's (2016) study shared the following regarding their educational settings:

Participant 1 (48 years of age; 23 years of teaching experience): "*I work in a private school that is dedicated to help people experiencing low socioeconomic status…it's a marginal zone…*"

"*…they live in conflict areas, and there may be instances in which people start to drink alcohol, or are heavy users of tobacco and some may even be doing drugs…*"

"*…Their family environment, their social environment, of course, ends up being detrimental for them. Poverty, and the lack of books in the home, newspapers, access to TV shows, computers and all of that, it ends up causing the students to be… well that their level of development is not comparable with students who have both parents taking care of them…*"

Participant 5 (34 years of age; 16 years of teaching experience): "*It is one of the areas where you see a lot of poverty. So, we are surrounded by the context, socially speaking, the context is full of neighbourhood gangs and our school is at the centre of where they all hang out*".

"*…[students] when they go back home, they go back to what is traditional, to monotony, to hearing ´no, no don't do that, instead get to work, right now I do not need you to do your homework, you, better start mixing the cement with me because we are going to build a wall' or something*".

"*There are many parents that are embarrassed to reach out and say 'I have no money for this'.*"

"*We do have many children that go without eating. I have given away my lunch, I have asked for credit in the school cafeteria to feed some students…*"

"*I ask my students 'what did you have for dinner at home' and they say 'well sometimes beans, sometimes tortillas' and sometimes they don't eat…that's how poverty affects them. They are thinking 'I'm hungry, my stomach is rumbling, my head hurts, I want to throw up'. We have to fight against all of that.*"

Participant 4 (30 years of age; 15 years of teaching experience): "*People survive, basically, working the field and they work very scarcely., As in, a paid job is only for people that can go to the nearest communities to work as pawns or construction workers, or women who go to work in houses or cities.*"

"*Obviously, there are no libraries in rural communities. So, we start to, along with the rural teacher, organise the place and ask for permission to whomever we have to and we start building the library.*"

"...parents prefer to take them to work in the field, because they will be able to produce food for self-consumption. You have to work to eat. Many kids do that, they go to school in the morning and go work in the field in the afternoon."

"...the teacher has a very hard experience in Mexico. The teacher is a hero, almost ... he has to deal with lack of nutrition, sometimes domestic violence, and with child labour."

"The rural teacher has many other worries before being a good academic teacher"

It can be learned from these selected extracts that there are important challenges existing in the Latin American educational setting. Teachers who kindly donated their time to share their views were passionate about trying their best to do their jobs (as participant 5 shared, going as far as to giving away their lunch so that their students do not have to learn on an empty stomach). Better understanding their working conditions and the great inequalities they experience on a daily basis can help form a picture regarding their knowledge of neuroscience. With little time available to invest in professional development, with little available resources (physical and human), and other social issues that impact their working conditions, it is not surprising that learning about neuroscience is not a priority.

Educational Contexts as Powerful Influences That Should Guide Neuroscience and Education Research

As established in the earlier section, there are many different kinds of classrooms (urban, rural, private, public) around Latin America (and the world as well). The complexity of an educational setting directly influences the available time and interests a teacher has in order to invest in professional development. Thus, in the hopes of evolving the field of neuroscience and education in Latin America, context-sensitive neuroscientific research is needed. For example, international reports on neuroscience and education (Howard-Jones, 2014b) have attempted to estimate the distances of neuroscience research available to how close (or distant) they are to classroom applications (based on educational settings that are not as complex as the Latin American one), but such distances are not true for all kinds of classrooms. Within this report (see Howard-Jones, 2014b), there were two examples provided that may prove problematic in certain Latin American contexts.

First, focusing on the topic of "reading" the report (Howard-Jones, 2014b) stated that the intervention proposed would require computer-based training. This may work in a school that is equipped with enough computers and staff that are trained to solve any technological issues that may come up during its course. However, if we imagine a rural school in any country in Latin America, where classrooms may lack the sort of equipment needed for this intervention, the distance to application suddenly doesn't seem so near. The region is still behind in the use of information technologies to teach; from a sample of 91,000 sixth grade students in Latin

America, it was reported that more than half of them have never used a computer in school (Román & Murillo, 2014).

Then, there is the issue of the physiological aspects of learning discussed in the report (sleep, nutrition, and hydration) (Howard-Jones, 2014b). Latin American teachers may already be aware of the importance of a good night's sleep, a balanced nutrition for their students, and that they should stay hydrated to improve their learning outcomes, but what is it that they could do to help students living in extreme poverty? Are these really issues that can be solved by teachers telling their students to eat breakfast in the morning and have a good night's sleep? These scenarios in Latin America emphasize the possibilities for collaboration between researchers and educators. Instances like these can help re-shape and grow the field of neuroscience and education by making neuroscience researchers sensitive to the fact that there are many children who live in conditions that are less than optimal and that teachers have to find ways to educate them with the resources that are available to them. They present the opportunity for neuroscience and education to find ways of adapting, or editing, existing research for it to be meaningful and useful in every context: from the most remote and challenging indigenous rural classrooms in a country like Guatemala to an urban public school in metropolitan Argentina (Soni Garcia, 2016).

The final portion of this chapter will take a look at examples that help understand the current state of neuroscience and education research in Latin America.

Neuroscience and Education: Current Trends in Latin America

There seems to still be a predominant trend that focuses on identifying prevalence of neuromyths in Latin American teachers (for example, Añazco Hermosilla et al., 2016; de Sá et al., 2020; Jiménez Pérez & Calzadilla-Pérez, 2021; Lopes et al., 2020; Maureira et al., 2021; Mendoza Fox et al., 2018; Torres & Alvarado, 2018). However, there is also evidence of more collaborative research exploring professional development (da Silva, 2020; Grossi et al., 2014) and including qualitative stages to identify messages that are a priority (Crespi, 2020; Crespi et al., 2020).

In 2011, Brazil invested in a series of pilot projects ("Projecto Plural," "NeuroEduca," and more) that are aimed at consolidating training contents for teachers linked to neuroscience and education (Gonchoroski et al., 2014). In parallel, there seems to be an interest in empirically exploring the value of neuroscience-informed interventions in Brazilian educational settings (for example, Brockington et al., 2018; Muñoz et al., 2019).

There appears to be a mixture of different research trends happening in Latin America, which suggests that there is still emphasis on monitoring what teachers know about the brain (and their beliefs in neuromyths). However, many other neuroscience-informed initiatives that seem to be created for the Latin American context and in communication with Latin American teachers appear to be taking hold, which suggests that collaboration between neuroscience and education may be growing stronger.

Conclusion

As explored in this chapter, a lot can be learned about what teachers know about the brain through quantitative questionnaires. These are helpful in identifying the prevalence of neuromyths and the status of their neuroscience literacy in regard to how these may impact their educational practices. However, only using those questionnaires as a means of information may result in missing out on important detail such as beliefs that are held but are not presented as items.

Conversations, through the means of qualitative research, are helpful toward generating a more collaborative field where not only are teachers prompted to share their knowledge regarding the brain, but neuroscientists can better understand diverse educational settings and the working conditions experienced by teachers so that more meaningful messages can be generated. Context-sensitive, collaborative neuroscience and education research should be the main trend in Latin America.

References

Añazco Hermosilla, L., Contreras Novoa, T., Millafilo Antilef, C., & Rodríguez Rodríguez, A. (2016). *Prevalence of neuromyths amongst student-teachers from Chile* (Doctoral dissertation, Universidad Católica de la Santísima Concepción).

Bell, V. (2013). The unsexy truth about dopamine. *The Observer*. Retrieved November 27, 2014.

Bellei, C., Poblete, X., Sepúlveda, P., Orellana, V., & Abarca, G. (2013). *Situación educativa de América Latina y el Caribe. Hacia la educación para todos 2015*. UNESCO.

Bergmann, A., Biehl, A., & Zabel, J. (2017). Toward neuroscience literacy?—theoretical and practical considerations. In *The human sciences after the decade of the brain* (pp. 231–248). Academic Press.

Betts, K., Miller, M., Tokuhama-Espinosa, T., Shewokis, P. A., Anderson, A., Borja, C., ... & Dekker, S. (2019). International report: Neuromyths and evidence-based practices in higher education. *Online Learning Consortium*.

Brito, N. H., & Noble, K. G. (2014). Socioeconomic status and structural brain development. *Frontiers in Neuroscience, 8*(276), 1–12.

Brockington, G., Balardin, J. B., Zimeo Morais, G. A., Malheiros, A., Lent, R., Moura, L. M., & Sato, J. R. (2018). From the laboratory to the classroom: The potential of functional near-infrared spectroscopy in educational neuroscience. *Frontiers in Psychology, 9*, 1840.

CEPAL, N. (2021). Social Panorama of Latin America 2020. See: http://hdl.handle.net/11362/46688

Cole, L. (2013). *Neuromyths in education: An empirical study of trainee teachers' understanding of the brain*. (Unpublished MEd. Dissertation). University of Bristol, Bristol, United Kingdom.

Cosme, F. A. D. O., Agra, B. C. N., & Ximenes-da-Silva, A. (2018). NEUROMITOS: MITOS E VERDADES SOBRE O CÉREBRO. [Paper presentation]. 70ª Reunião Anual da SBPC 22 a 28 de julho de 2018 – UFAL, Maceió, Brazil. http://www.sbpcnet.org.br/livro/70ra/trabalhos/resumos/3098_1ea0dfda042563ea544883c3b6bdab555.pdf

Crespi, L. R. S. (2020). *Neurociências na formação docente continuada: Valorizando o desenvolvimento e a aprendizagem na primeira infância*. Tese (Doutorado em Educação) - Universidade Federal do Rio Grande do Sul.

Crespi, L., Noro, D., Nóbile, M. F., & Peruzzo, V. (2020). Neurociências na formação continuada de docentes da pré-escola: Lacunas e diálogos. *EDUCA-Revista Multidisciplinar em Educação, 7*(17), 62–81.

Dekker, S., Lee, N. C., Howard-Jones, P., & Jolles, J. (2012). Neuromyths in education: Prevalence and predictors of misconceptions among teachers. *Frontiers in Psychology, 3*, 429.

Deligiannidi K., (2009). *What do you know about the brain? The Neuroscience Literacy of Greek and British teachers, and the Greek Public*. [Unpublished master's thesis]. University of Bristol.

De Moura Castro, C., & Ioschpe, G. (2007). *La remuneración de los maestros en América Latina: ¿Es baja? ¿Afecta la calidad de la enseñanza?* Documento de PREAL, No. 37.

de Sá, A. L., do Carmo Narciso, A. L., & Fumiã, H. F. (2020). Neurociência cognitiva e Educação: Análise sobre a prevalência de neuromitos entre os docentes de Matemática e das demais áreas do conhecimento atuantes na SRE de Carangola-MG. *Educação (UFSM), 45*, 58–51.

da Silva, M. D. C. G. (2020). Plasticidade neural, neurociência e educação: As bases do aprendizado. *Arquivos do Mudi, 24*(2), 30–41.

Duarte, J., Bos, M. S., & Moreno, M. (2010). *Los docentes, las escuelas y los aprendizajes escolares en América Latina: Un estudio regional usando la base de datos del SERCE Banco Interamericano de Desarrollo*. Nota Técnica # 8.

Falus, L., & Goldberg, M. (2011). *Perfil de los docentes en América Latina. Cuaderno 09 de SITEAL*. IIPE-UNESCO-OEI.

Ferrero, M., Garaizar, P., & Vadillo, M. A. (2016). Neuromyths in education: Prevalence among Spanish teachers and an exploration of cross-cultural variation. *Frontiers in Human Neuroscience, 10*, 496.

Geake, J. (2008). Neuromythologies in education. *Educational Research, 50*(2), 123–133.

Gleichgerrcht, E., Lira Luttges, B., Salvarezza, F., & Campos, A. L. (2015). Educational neuromyths among teachers in Latin America. *Mind, Brain, and Education, 9*(3), 170–178.

Gonchoroski, T., Garcia da Silva, T. L., & Isaia Kindel, E. A. (2014). *Neurociências na educação: Conhecimento e opiniões de professores*. See: https://www.lume.ufrgs.br/bitstream/handle/10183/117634/000967593.pdf?sequence=1

Grospietsch, F., & Mayer, J. (2020). Misconceptions about neuroscience–prevalence and persistence of neuromyths in education. *Neuroforum, 26*(2), 63–71.

Grossi, M. G. R., Lopes, A. M., & Couto, P. A. (2014). A neurociência na formação de professores: Um estudo da realidade brasileira. *Revista da FAEEBA - Educação e Contemporaneidade, 23*, 27–40.

Hanson, J. L., Hair, N., Shen, D. G., Shi, F., Gilmore, J. H., Wolfe, B. L., & Pollak, S. D. (2013). Family poverty affects the rate of human infant brain growth. *PLoS One, 8*(12), e80954.

Herculano-Houzel, S. (2002). Do you know your brain? A survey on public neuroscience literacy at the closing of the decade of the brain. *The Neuroscientist, 8*(2), 98–110.

Hermida, M. J., Segretin, M. S., Soni García, A., & Lipina, S. J. (2016). Conceptions and misconceptions about neuroscience in preschool teachers: A study from Argentina. *Educational Research, 58*(4), 457–472.

Howard-Jones, P. (2010). Neuromyths. In *Introducing neuroeducational research: Neuroscience, education and the brain from contexts to practice*. Routledge.

Howard-Jones, P. (2014a). Neuroscience and education: Myths and messages. *Nature Reviews Neuroscience, 15*(12), 817–824.

Howard-Jones, P. (2014b). *Neuroscience and education: A review of educational interventions and approaches informed by neuroscience*. Education Endowment Foundation (EEF).

Howard-Jones, P., & Demetriou, S. (2009). Uncertainty and engagement with learning games. *Instructional Science, 37*(6), 519–536.

Howard-Jones, P., Demetriou, S., Bogacz, R., & Yoo, J. H. (2011). Toward a science of learning games. *Mind, Brain, and Education, 5*(1), 33–41.

Idrissi, A. J., Alami, M., Lamkaddem, A., & Souirti, Z. (2020). Brain knowledge and predictors of neuromyths among teachers in Morocco. *Trends in Neuroscience and Education, 20*, 100135.

Jiménez Pérez, E. H., & Calzadilla-Pérez, O. O. (2021). Neuromyth prevalence in teachers at the University of Cienfuegos. *Ciencias Psicológicas, 15*(1), e2358. Epub 01 de junio de 2021. https://doi.org/10.22235/cp.v15i1.2358

Kim, P., Evans, G. W., Angstadt, M., Ho, S., Sripada, C. S., Swain, J. E., Liberzon, I., & Luan Phan, K. (2013). Effects of childhood poverty and chronic stress on emotion regulatory brain function in adulthood. *Proceedings of the National Academy of Sciences, 110*(46), 18442–18447.

Lopes, F. M., Dias, N. M., Mendonça, B. T., Coelho, D. M. V., Andrade, A. L. M., & Micheli, D. D. (2020). O que sabemos sobre neurociências?: Conceitos e equívocos entre o público geral e entre educadores. *Revista Psicopedagogia, 37*(113), 129–143.

Maureira, F., Ferro, E. F., Retamal, F. C., Cortés, M. C., Troncoso, S. P., Acevedo, V. B., … Escaffi, B. C. (2021). Prevalencia de neuromitos en estudiantes de Pedagogía en Educación Física de Chile (Prevalence of neuromyths in students of Physical Education Pedagogy of Chile). *Retos, 42*, 426–433.

Mendoza Fox, R. M., Bringas, N., Riquelme-Uribe, D., & Roosevelt, C. F. (2018). Neuroscience and education: Teachers' perceptions and misconceptions regarding the biological basis of learning. See: https://www.aassa.net/blog/?p=2737

Muñoz, J. E., Montoya, M. F., Lopez, D. S., & Henao, O. A. (2019, September). Embodying the homunculus in virtual reality: Creating an interactive laboratory procedure for brain mapping and visualization. In *Proceedings of the 5th Workshop on ICTs for improving Patients Rehabilitation Research Techniques* (pp. 144–147).

Neville, H. J., Stevens, C., Palulak, E., Bell, T. A., Fanning, J., Klein, S., & Isbell, E. (2013). Family-based training program improves brain function, cognition, and behaviour in lower socioeconomic status preschoolers. *Proceedings of the National Academy of Sciences of the United States of America, 110*(29), 12138–12413.

Organisation for Economic Co-operation and Development. (2002). *Understanding the brain: Towards a new learning science*. OECD.

Painemil, M., Manquenahuel, S., Biso, P., & Muñoz, C. (2021). Creencias versus conocimiento en futuro profesorado. Un estudio comparado sobre neuromitos a nivel internacional. *Revista Electrónica Educare, 25*(1), 246–267.

Papadatou-Pastou, M., Haliou, E., & Vlachos, F. (2017). Brain knowledge and the prevalence of neuromyths among prospective teachers in Greece. *Frontiers in Psychology, 8*, 804.

Pasquinelli, E. (2012). Neuromyths: Why do they exist and persist? *Mind, Brain, and Education, 6*(2), 89–96.

Pereyra, A. (2008). La fragmentación de la oferta educativa en América Latina: La educación pública vs la educación privada. *Perfiles educativos, 8*, 132–146.

Rato, J. R., Abreu, A. M., & Castro-Caldas, A. (2010). Neuromyths in education: What is fact and what is fiction for teachers. In *Conference Abstract: EARLI SIG22-Neuroscience and Education*.

Román, M., & Murillo, F. (2014). Disponibilidad y uso de TIC en escuelas latinoamericanas: Incidencia en el rendimiento escolar. *Educação e Pesquisa, 40*(4), 869–895.

Secretaría de Educación Pública (SEP). (2009). *Estudio Internacional sobre la Enseñanza y el Aprendizaje (TALIS) Resultados de México*. México-SEP Unidad de Planeación y Evaluación de Políticas.

SITEAL. (2015). 01 *Resumen Estadistico Comentado: Desarrollo en America Latina 2000–2013*.

Soni Garcia, A. L. (2016). *Latin American teachers' perceptions about the mind-brain relationship, brain function and neuroeducation* (Doctoral dissertation, University of Bristol).

Stern, E., Grabner, R., Schumacher, R., Neuper, C., & Saalbach, H. (2005). *Educational research and neurosciences—expectations, evidence, research prospects*. Federal Ministry of Education and Research.

Tardif, E., Doudin, P. A., & Meylan, N. (2015). Neuromyths among teachers and student teachers. *Mind, Brain, and Education, 9*, 50–59.

Torres, J. F. F., & Alvarado, J. C. O. (2018). Del conocimiento científico al malentendido. Prevalencia de neuromitos en estudiantes ecuatorianos. *Revista Iberoamericana de Educación, 78*(1), 87–106.

Torrijos-Muelas, M., González-Víllora, S., & Bodoque-Osma, A. R. (2021). The persistence of neuromyths in the educational settings: A systematic review. *Frontiers in Psychology, 11*, 3658.

Vaillant, D., & Rossel, C. (2006). *Maestros de Educación básica en América latina: Hacia una radiografía de la profesión*. PREAL.

Willis, B. (2008). Building a bridge from neuroscience to the classroom. *Phi Delta Kappan, 89*(6), 424–427.

Wise, R. (2004). Dopamine, learning and motivation. *Nature Review Neuroscience, 5*(6), 483–494.

Part II
Factors Influencing Child and Adolescent Development

Attention and Academic Performance: From Early Childhood to Adolescence

Mônica Carolina Miranda, Silmara Batistela, and Marcus Vinicius Alves

Introduction

It is now a common knowledge among scientists that most of the published research in the world's leading scientific journals on human behavior and cognition is primarily based on samples that represent little of the global diversity. These research studies use mainly WEIRD populations (W = western; E = educated; I = industrialized; R = rich; and D = democratic) (Henrich et al., 2010; Rad et al., 2018). This would already be a misrepresentation of the world population in scientific databases, but there is yet another aggravating factor: studies conducted in populations that are not representative of the global context tend to be extremely generalized, with researchers assuming that their WEIRD findings are universal (Henrich et al., 2010). Moreover, considering that studies in cognitive science carried out in large universities take place primarily with undergraduates of this discipline, it is also possible to notice a wide experimentation with college students who, categorically speaking, represent an almost irrelevant portion of the global population spectrum. Thus, in many cases, when they set out to study the human mind and human behavior, many scientists are actually presenting results related to a small portion of an already peculiar WEIRD population (Henrich et al., 2010). It is no stranger to the notion that much of the psychology and neuroscience of the last century actually represents how this group thinks, behaves, learns, and remembers, stipulating an unattainable threshold for the true mix of ethnic, contextual, environmental,

M. C. Miranda (✉)
Associate Researcher at the National Science Network for Education (Rede Nacional de Ciência para Educação - Rede CpE), São Paulo, Brazil

S. Batistela
Centro Universitário São Camilo, São Paulo, Brazil

M. V. Alves
Sociocognitive Neuropsychology Laboratory (LINES), Salvador, Brazil

M. V. Alves et al. (eds.), *Cognitive Sciences and Education in Non-WEIRD Populations*, https://doi.org/10.1007/978-3-031-06908-6_4

nutritional, and psychological conditions of the real world. With this, in order to truly understand the human vicissitudes, it is pertinent to have works in non-WEIRD populations, which present relevant data to the most diverse groups and, thus, can contribute to a greater understanding of individuals in their various contexts.

An example of this proposal can be the study by Begus et al. (2016), which assessed the working memory of infants aged 12–16 months in rural Africa, specifically in Gambia, one of the smallest countries on the continent. This study's results showed that the neural activity when infants observe objects being hidden for 3 seconds compared to 6 seconds is different, and it is possible to assume that this activation reflects the babies holding in their minds the representation of objects as naturally occurs in child development, but when these results are compared to British babies, it is possible to see a difference between the two samples, this result implies a significant difference between the two populations, and such difference can be related to the socioeconomic context present in the development of these groups, also suggesting that neurodevelopment in at-risk and impoverished regions is less prominent (Katus et al., 2020). Furthermore, similar studies in marginalized groups in India have also shown impaired performance (Wijeakumar et al., 2019). Considering this example, it is possible to assume that results considering only children born in the United Kingdom would invariably be biased and, more broadly, could not be functional for cognitive educational practices applied to populations in extreme poverty contexts, but when one takes only this WEIRD sample as representative of so-called ideal child development, any and all disadvantaged contexts will a priori fall short. Therefore, a qualitatively different look at the contextual differences implies an understanding of the developmental potentials in certain situations and, with this, a perspective that allows for more effective practices in an educational environment, as an example.

Growing up in a family with a low socioeconomic status can be associated with a wide range of negative consequences throughout life, affecting cognition and physical and mental health (Hackman et al., 2010; McEwen & Gianaros, 2010; Ursache & Noble, 2016), but it is still not known exactly which elements influence development in the face of socioeconomic risk factors, and even studies with different socioeconomic perspectives need to be observed in relation to context. As an aggravating factor, situations of violence and chronic stress can happen during childhood and are inexorably present in less developed countries with high social inequality. Nevertheless, even in industrialized countries, the gap found between individual income is considerable, influencing developmental stimulation due to the social inequalities (Farah, 2018). Studies that only consider individuals within the same country, yet still in a particular portion of it, similarly limit the very representativeness of their data.

In view of what has been presented so far, in this chapter we will discuss the development of attention, relating it to academic performance in children and adolescents from non-WEIRD populations. As a primary cognitive function, attention is one of the most vulnerable mechanisms to adversities in early life and, despite this, it is involved in a number of fundamental mental skills needed for various daily applications, including those related to academic performance (Blair & Raver, 2016;

Posner et al., 2006). In view of such importance, we will examine how this cognitive function can be impacted in various non-WEIRD contexts, aiming at an understanding that may allow future educators and researchers to better prepare their practice in these contexts, considering populations that live and develop in emerging countries, with diverse ethnicities and/or unique social contexts, but no less important, such as countries with volatile political scenarios, few educational resources, and/or extreme poverty.

Conceptualization of Attention

According to William James (1890),

> Every one knows what attention is. It is the taking possession by the mind, in clear and vivid form, of one out of what seem several simultaneously possible objects or trains [p. 404] of thought. (p. 256).

However, despite being a process known by everyone (at least in James' view), attention is difficult to characterize. As a prerequisite for satisfactory performance in many cognitive and behavioral tasks, attention has a complex functioning and is not related to a single brain region, neither can it be characterized by a singular functioning—contrary to the assumption of much of the common sense, attention is not a unitary entity, but one that is divided into different domains (Cohen, 1993; Fiebelkorn & Kastner, 2020; Parasuraman, 1998; Posner et al., 2019).

Attention involves the allocation of mental resources, that is, the selective directing of these resources toward sensory stimuli, actions, or cognition, such as memory and thoughts (Cohen, 1993; Raz & Buhle, 2006). Thus, this cognitive function, despite its limited capacity—an aspect that is also sometimes ignored by popular knowledge—allows for the careful use of cognitive resources (Cohen, 1993; Krauzlis et al., 2018; Treisman, 2004). One of the most cited and discussed attention models in the literature is the one proposed by Posner and Petersen (1990), which differentiates at least three attentional systems: (1) initiation and maintenance of awareness and vigilance—important for identifying and responding to stimuli; (2) orientation to the appropriate sensory stimulus to process (i.e., the so-called target stimulus); (3) executive control to identify and focus on the stimulus, i.e., to volitionally control responses to stimuli (Petersen & Posner, 2012). Although these are considered independent attentional networks, as well as functionally and neuroanatomically distinct, they still operate in close relation for the intentional regulation of behavior and cognition (Fan et al., 2002; Posner & Boies, 1971; Raz & Buhle, 2006). In this sense, executive attention has also been related to concepts such as emotional regulation and self-control, i.e., behavioral self-regulation (Rueda et al., 2005). Posner's proposed model is widely used in experimental models, serving as a basis for understanding neurobiological aspects of attention and providing clinical applications. However, there are other taxonomies, which will also be discussed below.

Since the 1980's, an effort has been made to develop behavioral tests to assess the different components of attention. As mentioned previously, several authors have emphasized that attention is not a unitary construct but consists of distinct and often complementary mechanisms; hence, some operational definitions of attention have been established. Attention can also be classified as selective, divided, and concentrated, and although there is no consensus in the literature about this taxonomy, everyone agrees that it involves multi-components (see Fig. 1), different processes that can even influence each other and cooperate together for cognitive or behavioral actions (Parasuraman, 1998; Raz & Buhle, 2006).

The definition of selective attention is intertwined with the definition of attention itself. It is a system responsible for selection and inhibition: activating cognitive resources and prioritizing the processing of only one (or some) stimulus while voluntarily ignoring others available at the moment (Cohen, 1993; Krauzlis et al., 2018; Pashler, 1998). Therefore, this attentional system has the function of selective orientation, being influenced by stimulus characteristics such as novelty and relevance and by internal factors such as motivation (see Fig. 2). The existence of this phenomenon in other animals—such as amphibians, fish, and birds—indicates its highly adaptive function, evolutionary importance, and complexity (Krauzlis et al., 2018).

On the other hand, sustained attention is characterized by the ability to maintain attentional focus for a prolonged period on a specific, monotonous task (such as responding to a stimulus that arises infrequently and unpredictably), but which requires fast and accurate responses to stimuli (Fortenbaugh et al., 2017; Raz & Bohle, 2006). Thus, an important differentiation of this type of attention from the others is the task duration, and although some authors have proposed specific delineation of this duration, there is no consensus regarding its extent (Fortenbaugh et al., 2017) (Fig. 1a). There are different hypotheses for performance impairment in this type of task over time, such as (1) decline in attentional resources over time; (2) its repetitive and/or monotonous aspect, which would lead to task disengagement; (3) mind-wandering, which is experienced as time goes by while performing the task, consuming attentional resources that were initially focused on the activity in progress; and (4) motivation or reward involved in the task (Helton et al., 2005;

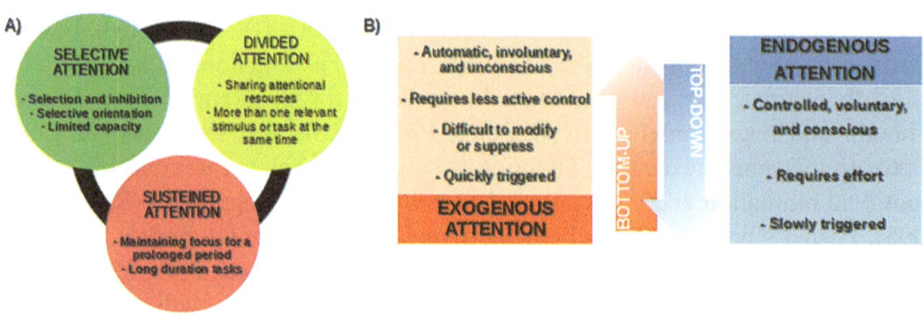

Fig. 1 (**a**) Different types of attention; (**b**) two mental processes related to the attention function

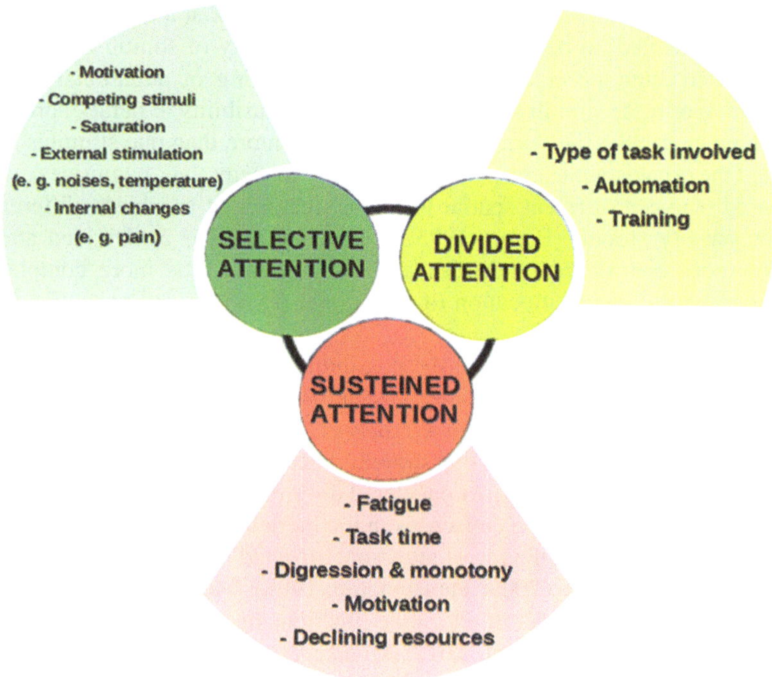

Fig. 2 Factors that affect different types of attention

Helton & Russel, 2015; Manly et al., 1999; Parasuraman et al., 2009; Robertson et al., 1997; Smallwood & Schooler, 2006; Stuus et al., 1995) (see Fig. 2).

This concept, sustained attention, is often used as a synonym for vigilance; other authors do not consider them synonymous, but rather that it depends on the task. For example, sustained attention tasks require a continuous response over a long period of time. An example is Conners' Continuous Performance Test (CCPT), where the subject must respond to every letter that appears on a computer screen, except for the letter X, and is thus used to obtain quantitative information regarding an individual's ability to sustain attention over time (Conners et al., 2018). Vigilance tasks, on the other hand, have been considered those that require responses to rare and non-predictive events, for example, radar operators waiting for a nonfrequent signal. In these situations, the quality of attention is fragile and declines over time—this is called vigilance decrement. Vigilance decrement is defined as a gradual decline in the detection rate of target signals over time, or an increase in the response time to detect something. In other words, vigilance decrement also depends on perceptual and decision-making factors because the key is to determine whether vigilance declines due to a loss in perceptual sensitivity to detect the signal, or whether it is due to changes in decision criteria, which has been studied as signal detection theory (Parasuraman, 1998).

However, it is not always possible to perform one task at a time, so attention is continuously subject to being divided between a variety of stimuli and activities. Thus, divided attention is characterized by the sharing of attentional resources between two or more stimuli (or tasks) due to the infeasibility of parallel processing (Parasuraman, 1998) (Fig. 1a). In these situations, more than one stimulus is relevant, and the organism must respond concurrently to them; thus, cognitive resources are shared between different spatial locations, features of an object, different targets, or sensory stimuli (Braun, 1998; Cohen, 1993) (Fig. 2). Divided attention requires more control than selective attention because it is more complex and requires more voluntary allocation of attentional resources and executive control (Hahn et al., 2008).

There is also a discussion about the possibility of another subtype of attention: alternating attention. In this modality, instead of the simultaneous processing of stimuli (or tasks), there would be an alternation of attentional resources, that is, the individual would disengage from one stimulus to engage in another (Fiebelkorner & Kastner, 2020). Although they are considered different subsystems by some authors (Hirst et al., 1980), there is no consensus in the literature in this regard, and studies consider it to be the same subtype of attention, involved in the so-called "dual-task" (Franz, 2012; Hahn et al., 2008; Parasumaran, 1998). Many authors still research the topic within the construct of executive functions under the name of alternation (shifting) (Miyake & Friedman, 2012; Miyake et al., 2000; Petersen & Posner, 2012). When executing two or more tasks simultaneously—as has possibly been observed at some point by everyone—there may be interference of an activity in another, an event that will depend on the intrinsic characteristics of each task, including the automatization of one of them, and, according to Shneider and Shiffrin (1977), it is only possible to execute two actions if one has undergone the process of automatization (Hirst et al., 1980).

Mental processes can be controlled or automatic (Fig. 2), despite some controversies about this dual division (for a review, see Melnikoff & Bargh, 2018). Controlled processes are intentional, involve active control by the subject, and have more limited capacity; thus, they can be easily modified (Moors, 2016; Schneider & Shiffrin, 1977). In contrast, automatic processes are rapid, involuntarily triggered, modifying controlled actions that are in progress, directing the attentional focus, and independent of the subject's active control (Moors, 2016; Schneider & Shiffrin, 1977). Some tasks involve automatic processes from their initial execution, while others may become automated after successive training, and even complex tasks may undergo the automatization process and then become actions performed involuntarily, without effort (Bago & De Neys, 2017; Pennycook, 2017; Schneider & Shiffrin, 1977; Spelke et al., 1976). Once automatized, such processes become difficult to suppress, ignore, modify, or the ongoing actions interrupted (Moors, 2016; Schneider & Shiffrin, 1977). Such processes are also called endogenous and exogenous attentional modalities, or top-down and bottom-up, respectively. The former (top-down) can also be called voluntary attention, is triggered at a slower rate, but is maintained for longer periods of time than exogenous (bottom-up, or involuntary) attention. On the other hand, involuntary attention ensures more prompt allocation of resources and is maintained for a shorter period of time and in a less flexible manner than endogenous attention (Bowling et al., 2019; Chica et al., 2013).

One can notice, then, that despite different classifications, several mechanisms are acting together to perform a certain task. For example, a student during a class should direct his attentional resources to the subject taught by the teacher, while ignoring competing stimuli (for example, the conversation of a classmate next to them, which requires them to self-regulate their behavior). This is a scenario that involves endogenous attention, that is, top-down processing—which will require effort, voluntary direction. Considering the length of a regular class, the student must maintain this attentional focus, and sometimes divide (or alternate) it with the task of writing what the teacher says, for example. Also, according to the assumptions elucidated above, for a student who is already proficient in writing, this divided task will be less demanding—after all, the writing process has already been automatized. Hence, one can observe the heterogeneity involved in this phenomenon, which is often not treated with due complexity, being disregarded, for example, the social variability inherent to its development, which will be discussed below.

Childhood and Adolescence Development in Non-WEIRD Contexts and Academic Achievement

Over the years, the development of attention has been studied from different points of view, either to understand other cognitive processes such as memory, perception, regulation of attention, and behavior related to brain development. In the clinical population, there is also an important impact of these studies, where the predictive value of early attention problems for future dysfunctions in both attentional processes and other areas of development has been sought to be understood (van de Weijer-Bergsma et al., 2008). One factor to be considered is that temporal dissociations of attention processes are evident throughout development (Amso & Scerif, 2015), a model known as attentional networks that divides attention into networks of functions consisting of alertness, orientation, and executive control, which seem less independent in childhood (Rueda et al., 2004; Suades-González et al., 2017), as already explained in the previous topic.

The development of attentional orientation is important because efficient attentional orientation to environmental stimuli is essential in activities of daily living and is a prerequisite for the acquisition of skills such as reading comprehension, operational memory, and executive control. The period between 6 and 10 years old is crucial for the development of endogenous attention and inhibitory control, although the preschool period is also important (Leclercq & Siéroff, 2013).

Reynolds and Romano (2016) point out that the general arousal/attention system shows significant developmental changes throughout early childhood, characterized by gains in both the magnitude and duration of sustained attention periods, and these developmental gains in sustained attention are related to improved performance on working memory tasks. Studies of infants in Western countries have shown that the basic attentional capacities (orienting, anticipating, and processing) emerge gradually during the first year (Pyykkö et al., 2019). With regard to selective attention, there is a prolonged development during childhood, with studies pointing out that younger children fail to filter out irrelevant information, distributing their attention between what is relevant and what is not, which results in processing more information than necessary and therefore less efficiently (Plebanek & Sloutsky, 2019). At school age, the influences of attention on the maintenance of working memory are less efficient in 6- and 11-year-old children in comparison to older adults and adolescents (Amso & Scerif, 2015). The abilities to select among competing stimuli and to preferentially process more relevant information are essentially available in very young children, but the speed and efficiency of these behaviors improve with development (Stevens & Bavelier, 2012), and so there are functional changes in selective attention that occur during childhood and adolescence.

For most developmental theories, maturation of selective and sustained attention processes occurs during adolescence (Smith et al., 2011), but there is a paucity of studies in the period between 12 and 20 years old. For many authors, attention would not constitute a simple construct with a fixed ontogenetic trajectory, but a multidimensional construct with different developmental trajectories in its various components (Plude et al., 1994). Another relevant factor to be considered in studies of attention development in adolescence is the pubertal stage, a criterion of physiological/sexual maturity with neuroendocrine events that influence cognition and that explain this developmental trajectory better than chronological age (Steinberg, 2005; Zanini et al., 2021). Furthermore, the scarcity of systematic research in other cultural contexts is a major impediment to theoretical progress in this area (Nielsen et al., 2017), as cultural differences may in part reflect differences in socioeconomic status (Zanini et al., 2021), which has been strongly supported by many researchers (Nisbett & Masuda, 2003).

In a literature review study, van de Weijer-Bergsma et al. (2008) showed that prematurity seems to be related to lower efficiency of attentional orientation in infants (first year of life), and problems with sustained attention become more evident at preschool age as age increases. Other environmental factors such as maternal IQ, educational level, and socioeconomic status did not show consistent results and the authors point out the lack of studies in this age group (infants and preschool children). Regarding cross-cultural studies, Köster et al. (2018) evaluated 144 5-year-old children from three different cultural contexts (urban area of Germany, rural area of Cameroon Republic, urban area of Japan) to investigate different aspects of children's visual attention processes. The results showed that children's visual attention in rural Cameroon differs from both urban contexts, characterized by a high object focus across tasks. According to the authors, these differences may be based on different cultural and environmental learning experiences, such as children's basic familiarity with materials and stimuli, rather than culturally transmitted attentional styles. In this same line of studies, Jurkat et al. (2020) evaluated the visual attention of different age groups from middle-class urban families in Germany and from a rural area in the Republic of Cameroon. The results were similar; i.e., differences in visual attention between cultures were contingent on the familiarity of the corresponding culture-specific stimuli, thus suggesting that the familiarity of a stimulus strongly affects individuals' visual attention, meaning that stimulus familiarity needs to be considered when investigating culture-specific differences in attention styles.

Pyykkö et al. (2019) analyzed the development of visual attention abilities in a sample of infants in rural Malawi, focusing on three aspects: (a) visual search, (b) anticipatory responses, and (c) allocation of attention by measures known to be sensitive for assessing the development of visual attention processes. The authors showed that compared to studies in Western populations there was a similar pattern of group-level results across tasks; they also showed that there are changes in most of these measures between 7 and 9 months. Individual variations in infants' attention abilities were moderately stable across tests, but not related to socioeconomic factors such as prematurity, nutritional status, and psychosocial stress. First, it must be considered that socioeconomic status corresponds to a complex set of social and economic factors such as educational level, income, living conditions, and family purchasing power, which should be considered together in studies that analyze this variable in cross-cultural studies (Zanini et al., 2021). Second, Pyykkö et al. (2019) consider that infants' early attention abilities may emerge relatively independently of variations in the early environment; the development of these functions is more dependent on "optimal" physical growth and environmental support or stimulation. A recent paper assessed selective attention in 4-year-old children from low and high socioeconomic backgrounds over the course of a year. This study used an auditory selective attention paradigm, in which children listened to two different stories at the same time to check their ability to retain attention on only one of them. The results of both the behavioral test and electroencephalography measurements showed that the development of selective attention is influenced by economic background; that is, individuals from groups with high socioeconomic standards showed more efficient responses to stimuli than their peers of the same age but from a poorer socioeconomic background (Hampton Wray et al., 2017).

Because of this, it is important to analyze studies that are related to academic performance, as this is determined by a variety of factors, including educational opportunity, socioeconomic status, social skills, personality traits, and cognitive abilities (Stevens & Bavelier, 2012). These authors reviewed studies that evaluated the relationship between auditory attention and aspects relevant to academic performance, such as language, literacy, and mathematics. They found that deficits in selective auditory attention are related to language processing difficulties, that selective attention may be important for establishing the neural circuits important for efficient reading, and finally that there is a relationship between attention and math skills, mediated by the effect of selective attention on working memory. However, most of the studies reviewed were conducted in a WEIRD population. Due to this, the study by Alavi et al. (2019) stands out, which evaluated typically developing Malaysian children, aged 7–12 years, to determine whether attention or impulse control could predict overall academic performance, as well as gender difference. Teacher-returned questionnaires were used, which showed that the higher the teacher's perception of attention and impulse control skills, the better the academic performance. Girls were rated with higher levels of attention and impulse control than boys; the gender of the children did not moderate the relationships between attention or impulse control and academic performance or impulse control. This reinforces previous studies, i.e., that the data is similar when studying typically developing children from far eastern countries living in Malaysia compared to western children.

According to Resett (2021), there was an increase in the levels of schooling achievement by children and adolescents throughout the twentieth century in industrialized and post-industrialized nations. This is a psychosocial variable of great relevance, and academic performance is linked to educational institutions, i.e., an estimated measure of skills, aptitudes, or knowledge that a student learned as a result of a formal instructional process, and that these educational institutions may vary in countries with non-WEIRD populations. Resset (2021) examined whether the attention performance of children and adolescents would be a predictor of school performance, and for this, they evaluated 155 children and adolescents (9–15 years old) from a province in Argentina. It was observed that the reaction time and the number of correct answers in the D2 test, which evaluates selective attention, were predictors of math, language, and basic science scores in children. In adolescents, on the other hand, they predicted scores in math, language, English, and biology, corroborating previous studies in the American population. Similar results were found by Abreu et al. (2017) on selective attention tasks, which found correlations with a test of performance in reading, writing, and arithmetic, although these correlations were weak. In none of these studies, from Argentina and Brazil, did the authors control for socioeconomic variables. What is important for teachers and professionals who work with children and adolescents? The literature generally points out that sustained attention is especially relevant in the school context because complex problem solving is an important prerequisite for school performance; there is a causal relationship between visual attention and reading acquisition; attention and visual perception predict math scores (Anobile et al., 2013; Steinmayr et al., 2010), and that the literature in non-WEIRD populations is especially sparse to conclude how cultural factors, and here we emphasize socioeconomic level, influence attention development and academic performance.

According to Henrich et al. (2017, p. 79),

> perhaps there are some domains in which researchers could expect phenomena to be more universal than they are in other domains. We believe that the degree of universality does likely vary across domains, although this has yet to be demonstrated.

According to this author, some cognitive functions, such as attention, have a low interpopulation variability and this may be due to physiological or even genetic factors. In fact, in some studies conducted by our research group, we compared the performance of Brazilian children and adolescents in the CCPT, which assesses sustained attention, with American performance norms. Both young children (4 and 5 years old) and older children (6–11 years old) had better scores in the Brazilian sample on some measures than in the American sample. We attribute these differences to sample selection, as the Brazilian sample excluded participants with attention deficit disorder, screened by the Conners' scale, unlike the American sample (Miranda et al., 2008, 2009).

Conclusion

Attention is a prerequisite for good performance in other cognitive functions (such as memory and executive functions) and is involved in the performance of motor or mental activities. Thus, interest in the functioning of attention (in its healthy or pathological state) permeates the study of several fields, such as cognitive, learning, and developmental models. However, despite a large number of studies on attention, most of them have samples from the WEIRD population, raising the question whether the findings of these studies can be extended to the non-WEIRD population. Thus, this chapter sought to present the main attentional models, different taxonomies proposed, and the explanation of their non-unitary functioning. We also presented studies with a WEIRD population, which were contrasted with those with non-WEIRD populations. Finally, it is noted that there is still a gap to be filled in this field, requiring more studies, and especially, with methods that allow more robust evidence regarding possible divergences between the development and functioning of attention in WEIRD and non-WEIRD populations.

References

Alavi, M., Seng, J. H., Mustaffa, M. S., Ninggal, M. T., Amini, M., & Latif, A. A. (2019). Attention, impulsiveness, and gender in academic achievement among typically developing children. *Perceptual and Motor Skills, 126*(1), 5–24. https://doi.org/10.1177/0031512518809163

Amso, D., & Scerif, G. (2015). The attentive brain: Insights from developmental cognitive neuroscience. *Nature Reviews Neuroscience, 16*(10), 606–619. https://doi.org/10.1038/nrn4025

Anobile, G., Stievano, P., & Burr, D. C. (2013). Visual sustained attention and numerosity sensitivity correlate with math achievement in children. *Journal of Experimental Child Psychology, 116*(2), 380–391. https://doi.org/10.1016/j.jecp.2013.06.006

Bago, B., & De Neys, W. (2017). Fast logic?: Examining the time course assumption of dual process theory. *Cognition, 158*, 90–109. https://doi.org/10.1016/j.cognition.2016.10.014

Begus, K., Lloyd-Fox, S., Halliday, D., Papademetriou, M., Darboe, M. K., Prentice, A. M., Moore, S. E., & Elwell, C. E. (2016). Using fNIRS to study working memory of infants in rural Africa. *Advances in Experimental Medicine and Biology, 876*, 273–279. https://doi.org/10.1007/978-1-4939-3023-4_34

Blair, C., & Raver, C. C. (2016). Poverty, stress, and brain development: New directions for prevention and intervention. *Academic Pediatrics, 16*(3), S30–S36. https://doi.org/10.1016/j.acap.2016.01.010

Bowling, J. T., Friston, K. J., & Hopfinger, J. B. (2019). Top-down versus bottom-up attention differentially modulate frontal–parietal connectivity. *Human Brain Mapping, 9*(1116), 928–946. https://doi.org/10.1002/hbm.24850

Braun, J. (1998). Divided attention: Narrowing the gap between brain and behavior. The attentive brain: Issues and prospects. In R. Parasuraman (Ed.), *The attentive brain* (pp. 327–351). The MIT Press. https://mitpress.mit.edu/books/attentive-brain

Chica, A. B., Bartolomeo, P., & Lupiáñez, J. (2013). Two cognitive and neural systems for endogenous and exogenous spatial attention. *Behavioural Brain Research, 237*, 107–123. https://doi.org/10.1016/j.bbr.2012.09.027

Cohen, R. A. (1993). *The neuropsychology of attention*. Plenum Press. https://doi.org/10.1007/978-0-387-72639-7_19

Conners, K. C., Sitarenios, G., & Ayearst, L. E. (2018). Conners' continuous performance test third edition. In J. S. Kreutzer, J. DeLuca, & B. Caplan (Eds.), *Encyclopedia of clinical neuropsychology*. Springer. https://doi.org/10.1007/978-3-319-57111-9_1535

de Abreu, E. S., Viapiana, V. F., Hess, A. R. B., Gonçalves, H. A., Sartori, M. S., Giacomoni, C. H., … Fonseca, R. P. (2017). Relação entre atenção e desempenho em leitura, escrita e aritmética em crianças. *Revista Avaliação Psicológica, 16*(4), 458–467. https://doi.org/10.15689/ap.2017.1604.12989

Fan, J., McCandliss, B. D., Sommer, T., Raz, M., & Posner, M. I. (2002). Testing the efficiency and independence of attentional networks. *Journal of Cognitive Neuroscience, 3*, 340–347. https://doi.org/10.1162/089892902317361886

Farah, M. J. (2018). Socioeconomic status and the brain: Prospects for neuroscience-informed policy. *Nature Reviews Neuroscience, 19*(7), 428–438. https://doi.org/10.1038/s41583-018-0023-2

Fiebelkorn, I. C., & Kastner, S. (2020). Functional specialization in the attention network. *Annual Review of Psychology, 71*, 221–249. https://doi.org/10.1146/annurev-psych-010418-103429

Fortenbaugh, F. C., DeGutis, J., & Esterman, M. (2017). Recent theoretical, neural, and clinical advances in sustained attention research. *Annals of the New York Academy of Sciences, 1396*, 70–91. https://doi.org/10.1111/nyas.13318

Franz, E. A. (2012). The allocation of attention to learning of goal-directed actions: A cognitive neuroscience framework focusing on the basal ganglia. *Frontiers in Psychology, 535*(3), 1–15. https://doi.org/10.3389/fpsyg.2012.00535

Hackman, D. A., Farah, M. J., & Meaney, M. J. (2010). 5. How does the brain change after birth? Socioeconomic status and the brain: Mechanistic insights from human and animal research. *Neuroscience, 11*(9), 651–659. https://doi.org/10.1038/nrn2897

Hahn, B., Wolkenberg, F. A., Ross, T. J., Myers, C. S., Heishman, S. J., Stein, D. J., Kurup, P. K., & Stein, E. A. (2008). Divided versus selective attention: Evidence for common processing mechanisms. *Brain Research, 1215*, 137–146. https://doi.org/10.1016/j.brainres.2008.03.058

Hampton Wray, A., Stevens, C., Pakulak, E., Isbell, E., Bell, T., & Neville, H. (2017). Development of selective attention in preschool-age children from lower socioeconomic status backgrounds. *Developmental Cognitive Neuroscience, 26*, 101–111. https://doi.org/10.1016/j.dcn.2017.06.006

Helton, W. S., & Russell, P. N. (2015). Rest is best: The role of rest and task interruptions on vigilance. *Cognition, 134*, 165–173. https://doi.org/10.1016/j.cognition.2014.10.001

Helton, W. S., Hollander, T. D., Warm, J. S., Matthews, G., Dember, W. N., Wallart, M., Beauchamp, G., Parasuraman, R., & Hancock, P. A. (2005). Signal regularity and the mindlessness model of vigilance. *British Journal of Psychology, 96*, 249–261. https://doi.org/10.1348/000712605X38369

Henrich, J., Heine, S. J., & Norenzayan, A. (2010). The weirdest people in the world? *Behavioral Brain Science, 33*(2–3), 61–83. https://doi.org/10.1017/S0140525X0999152X

Hirst, W., Spelke, E. S., Reaves, C. C., Caharack, G., & Neisser, U. (1980). Dividing attention without alternation or automaticity. *Journal of Experimental Psychology: General, 109*(1), 98–117. https://doi.org/10.1037/0096-3445.109.1.98

James, W. (1890). *The principles of psychology* (Vol. 1). Dover Publication. https://doi.org/10.1037/10538-000

Jurkat, S., Köster, M., Yovsi, R., & Kärtner, J. (2020). The development of context-sensitive attention across cultures: The impact of stimulus familiarity. *Frontiers in Psychology, 11*, 1526. https://doi.org/10.3389/fpsyg.2020.01526

Katus, L., Mason, L., Milosavljevic, B., McCann, S., Rozhko, M., Moore, S. E., Elwell, C. E., Lloyd-Fox, S., de Haan, M., Drammeh, S., Mbye, E., Touray, E., Ceesay, M., Jobarteh, B., Darboe, M. K., Austin, T., & Prentice, A. (2020). ERP markers are associated with neurodevelopmental outcomes in 1–5 month old infants in rural Africa and the UK. *NeuroImage, 210*, 116591. https://doi.org/10.1016/j.neuroimage.2020.116591

Köster, M., Itakura, S., Yovsi, R., & Kärtner, J. (2018). Visual attention in 5-year-olds from three different cultures. *PLoS One, 13*(7), 1–21. https://doi.org/10.1371/journal.pone.0200239

Krauzlis, R. J., Bogadhi, A. R., Herman, J. P., & Bollimunta, A. (2018). Selective attention without a neocortex. *Cortex, 102*, 161–175. https://doi.org/10.1016/j.cortex.2017.08.026

Leclercq, V., & Siéroff, E. (2013). Development of endogenous orienting of attention in school-age children. *Child Neuropsychology, 19*(4), 400–419. https://doi.org/10.1080/09297049.2012.682568

Manly, T., Robertson, I. H., Galloway, M., & Hawkins, K. (1999). The absent mind: Further investigations of sustained attention to response. *Neuropsychologia, 37*, 661–670. https://doi.org/10.1016/s0028-3932(98)00127-4

Mcewen, B. S., & Gianaros, P. J. (2010). Brain stress adaptation SES health. *Annals of the New York AcademyofSciences, 1186*,190–222.https://doi.org/10.1111/j.1749-6632.2009.05331.x.Central

Melnikoff, D. E., & Bargh, J. A. (2018). The mythical number two. *Trends in Cognitive Sciences, 22*(4), 280–293. https://doi.org/10.1016/j.tics.2018.02.001

Miranda, M. C., Sinnes, E. G., Pompéia, S., & Bueno, O. F. A. (2008). A comparative study of performance in the conners' continuous performance test between Brazilian and North American children. *Journal of Attention Disorders, 11*(5), 588–598. https://doi.org/10.1177/1087054707299412

Miranda, M. C., Sinnes, E. G., Pompeia, S., & Bueno, O. F. A. (2009). O K-CPT em uma amostra brasileira: Descrição do desempenho e comparação com as normas norte-americanas. *Revista de Psiquiatria Do Rio Grande Do Sul, 31*(1), 60–66. https://doi.org/10.1590/S0101-81082009000100011

Miyake, A., & Friedman, N. (2012). The nature and organization of individual differences in executive functions: Four general vonclusions. *Current Directions in Psychological Science, 21*(1), 8–14. https://doi.org/10.1177/0963721411429458

Miyake, A., Friedmana, N. P., Emersona, M. J., Witzkia, A. H., Howerter, A., & Wagerb, T. D. (2000). The unity and diversity of executive functions and their contributions to complex "frontal lobe" tasks: A latent variable analysis. *Cognitive Psychology, 41*(1), 49–100. https://doi.org/10.1006/cogp.1999.0734

Moors, A. (2016). Automaticity: Componential, causal, and mechanistic explanations. *Annual Review of Psychology, 67*, 263–287. https://doi.org/10.1146/annurev-psych-122414-033550

Nielsen, M., Haun, D., Kärtner, J., & Legare, C. H. (2017). The persistent sampling bias in developmental psychology: A call to action. *Journal of Experimental Child Psychology, 162*, 31–38. https://doi.org/10.1016/j.jecp.2017.04.017

Nisbett, R. E., & Masuda, T. (2003). Culture and point of view. *Proceedings of the National Academy of Sciences of the United States of America, 100*(19), 11163–11170. https://doi.org/10.1073/pnas.1934527100

Parasuraman, R. (1998). The attentive brain: Issues and prospects. In R. Parasuraman (Ed.), *The attentive brain* (pp. 3–18). The MIT Press. https://mitpress.mit.edu/books/attentive-brain

Parasuraman, R., de Visser, E., Clarke, E., McGarry, W. R., Russey, E., Shaw, T., & Thompson, J. C. (2009). Detecting threat-related intentional actions of others: Effects of image quality, response mode, and target cuing on vigilance. *Journal of Experimental Psychology: Applied, 15*, 275–290. https://doi.org/10.1037/a0017132

Pashler, H. E. (1998). *The psychology of attention*. The MIT Press. https://mitpress.mit.edu/books/psychology-attention

Pennycook, G. (2017). A perspective on the theoretical foundation of dual-process models. In W. De Neys (Ed.), *Dual process theory 2.0* (Vol. 2, pp. 1–34). Psychology PressDual Process Theory.

Petersen, S. E., & Posner, M. I. (2012). The attention system of the human brain: 20 years after. *Annual Review of Neuroscience, 35*(21), 73.89. https://doi.org/10.1146/annurev-neuro-062111-150525

Plebanek, D. J., & Sloutsky, V. M. (2019). Selective attention, filtering, and the development of working memory. *Developmental Science, 22*(1), 1–12. https://doi.org/10.1111/desc.12727

Plude, D. J., Enns, J. T., & Brodeur, D. (1994). The development of selective attention: a life-span overview. *Acta Psychologica (Amst), 86*(2-3), 227–272. https://doi.org/10.1016/0001-6918(94)90004-3

Posner, M. I., & Boies, S. J. (1971). Components of attention. *Psychological Review, 78*, 391–408. https://doi.org/10.1037/h0031333

Posner, M. I., & Petersen, S. E. (1990). The attention system of the human brain. *Annual Review of Neuroscience, 13*, 25–42. https://doi.org/10.1146/annurev.ne.13.030190.000325

Posner, M. I., Sheese, B. E., Odludaş, Y., & Tang, Y. Y. (2006). Analyzing and shaping human attentional networks. *Neural Networks, 19*(9), 1422–1429. https://doi.org/10.1016/j.neunet.2006.08.004

Posner, M. I., Rothbart, M. K., & Ghassemzadeh, H. (2019). Restoring attention networks. *Yale Journal of Biology and Medicine, 92*, 139–143. PMC6430178.

Pyykkö, J., Forssman, L., Maleta, K., Ashorn, P., Ashorn, U., & Leppänen, J. M. (2019). Early development of visual attention in infants in rural Malawi. *Developmental Science, 22*, e12761. https://doi.org/10.1111/desc.12761

Rad, M. S., Martingano, A. J., & Ginges, J. (2018). Toward a psychology of Homo sapiens: Making psychological science more representative of the human population. *Proceedings of the National Academy of Sciences of the United States of America, 115*(45), 11401–11405. https://doi.org/10.1073/pnas.1721165115

Raz, A., & Buhle, J. (2006). Typologies of attentional networks. *Nature Reviews Neuroscience, 7*, 367–379. https://doi.org/10.1038/nrn1903

Resett, S. (2021). Relación entre la atención y el rendimiento escolar en niños y adolescentes. *Revista Costarricense de Psicología, 40*(1), 3–22. https://doi.org/10.22544/rcps.v40i01.01

Reynolds, G. D., & Romano, A. C. (2016). The development of attention systems and working memory in infancy. *Frontiers in Systems Neuroscience, 10*, 1–12. https://doi.org/10.3389/fnsys.2016.00015

Robertson, I. H., Manly, T., Andrade, J., Baddeley, B. T., & Yend, J. (1997). 'Oops! ': Performance correlates of everyday attentional failures in traumatic brain injured and normal subjects. *Neuropsychologia, 35*(6), 747–758. https://doi.org/10.1016/s0028-3932(97)00015-8

Rueda, M. R., Fan, J., McCandliss, B. D., Halparin, J. D., Gruber, D. B., Lercari, L. P., & Posner, M. I. (2004). Development of attentional networks in childhood. *Neuropsychologia, 42*(8), 1029–1040. https://doi.org/10.1016/j.neuropsychologia.2003.12.012

Rueda, M. R., Posner, M. I., & Rothbart, M. K. (2005). The development of executive attention: Contributions to the emergence of self-regulation. *Developmental Neuropsychology, 28*, 573–594. https://doi.org/10.1207/s15326942dn2802_2

Schneider, W., & Shiffrin, R. M. (1977). Controlled and automatic human information processing: I. Detection, search, and attention. *Psychological Review, 84*(1), 1–66. https://doi.org/10.1037/0033-295X.84.1.1

Smallwood, J., & Schooler, J. W. (2006). The restless mind. *Psychological Bulletin, 132*, 946–958. https://doi.org/10.1037/0033-2909.132.6.946

Smith, A. B., Halari, R., Giampetro, V., Brammer, M., & Rubia, K. (2011). Developmental effects of reward on sustained attention networks. *NeuroImage, 56*(3), 1693–1704. https://doi.org/10.1016/j.neuroimage.2011.01.072

Spelke, E., Hirst, W., & Neisser, U. (1976). Skills of divided attention. *Cognition, 4*, 215–230.

Steinberg, L. (2005). Cognitive and affective development in adolescence. *Trends in Cognitive Sciences, 9*(2), 69–74. https://doi.org/10.1016/j.tics.2004.12.005

Steinmayr, R., Ziegler, M., & Träuble, B. (2010). Do intelligence and sustained attention interact in predicting academic achievement? *Learning and Individual Differences, 20*(1), 14–18. https://doi.org/10.1016/j.lindif.2009.10.009

Stevens, C., & Bavelier, D. (2012). The role of selective attention on academic foundations: A cognitive neuroscience perspective. *Developmental Cognitive Neuroscience, 2*(SUPPL. 1), 30–48. https://doi.org/10.1016/j.dcn.2011.11.001

Stuss, D., Shallice, T., Alexander, M., Picton, T., & W. (1995). A multidisciplinary approach to anterior attentional functions. *Annals of the New York Academy of Sciences, 769*, 191–212. https://doi.org/10.1111/j.1749-6632.1995.tb38140.x

Suades-González, E., Forns, J., García-Esteban, R., López-Vicente, M., Esnaola, M., Álvarez-Pedrerol, M., … Sunyer, J. (2017). A longitudinal study on attention development in primary school children with and without teacher-reported symptoms of ADHD. *Frontiers in Psychology, 8*, 1–10. https://doi.org/10.3389/fpsyg.2017.00655

Treisman, A. (2004). Psychological issues in selective attention. In M. S. Gazzaniga (Ed.), *The cognitive neurosciences III* (pp. 529–544). MIT Press. https://mitpress.mit.edu/books/cognitive-neurosciences-iii-third-edition

Ursache, A., & Noble, K. G. (2016). Socioeconomic status, white matter, and executive function in children. *Brain and Behavior, 6*(10), e00531. https://doi.org/10.1002/brb3.531

van de Weijer-Bergsma, E., Wijnroks, L., & Jongmans, M. J. (2008). Attention development in infants and preschool children born preterm: A review. *Infant Behavior and Development, 31*(3), 333–351. https://doi.org/10.1016/j.infbeh.2007.12.003

Wijeakumar, S., Kumar, A., Delgado Reyes, L. M., Tiwari, M., & Spencer, J. P. (2019). Early adversity in rural India impacts the brain networks underlying visual working memory. *Developmental Science, 22*(5), 1–15. https://doi.org/10.1111/desc.12822

Zanini, G. A. V., Miranda, M. C., Cogo-Moreira, H., Nouri, A., Fernández, A. L., & Pompéia, S. (2021). An adaptable, open-access test battery to study the fractionation of executive-functions in diverse populations. *Frontiers in Psychology, 12*, 467. https://doi.org/10.3389/fpsyg.2021.627219

Child Cognitive Development in Latin American Rural Poverty: What Should Researchers Consider for Conducting Fieldwork?

María Julia Hermida and Diego Edgar Shalom

Introduction

It is well established that living in poverty limits child cognitive-developmental opportunities. Several decades of studies conducted by developmental psychologists and cognitive scientists have shown how those limitations are associated with lower levels of intelligence, executive functions, and other cognitive functions (Farah et al., 2006) as well as lower academic achievement (Bradley & Corwyn, 2002). Further recent research has also shown differences in brain structure according to income (Noble et al., 2015). Given this evidence, a number of interventions have been implemented aiming to diminish poverty effects on cognition, and many of them showed positive impact (Sianes et al., 2021).

However, most of the studies analyzing cognitive development in poverty have been carried out in urban populations. A rather remarkable tension arises between real life and academic research, since the incidence of extreme poverty is much higher in non-urban contexts. Furthermore, the majority of inhabitants living in extreme poverty are children (Lichter & Johnson, 2007; Olinto et al., 2013; Schreuder, 2010). The importance of filling this void through the study of child cognitive development in rural contexts relies at least upon four facts:

M. J. Hermida (✉)
Universidad Nacional de Hurlingham-Consejo Nacional de Investigaciones Científicas y Técnicas (UNAHUR-CONICET), Buenos Aires, Argentina
e-mail: julia.hermida@gmail.com

D. E. Shalom
Instituto de Física de Buenos Aires (IFIBA) CONICET, Facultad de Ciencias Exactas y Naturales, Departamento de Física, Buenos Aires, Argentina

Universidad de Buenos Aires, Facultad de Ciencias Exactas y Naturales, Departamento de Física, Buenos Aires, Argentina

Universidad Torcuato Di Tella, Laboratorio de Neurociencia, Buenos Aires, Argentina

M. V. Alves et al. (eds.), *Cognitive Sciences and Education in Non-WEIRD Populations*, https://doi.org/10.1007/978-3-031-06908-6_5

(a) Almost half of all people (45% of the worldwide population) live in rural areas (United Nations International Children's Emergency Fund [UNICEF], 2012). Also, most low-income populations (68%) are rural (World Bank, 2018).
(b) In those areas, families tend to have more children (Food and Agriculture Organization [FAO], 2012). According to recent reports (UNICEF, 2016), at least four out of every five children living in extreme poverty are in rural areas worldwide.
(c) Mainly pushed by unemployment (Lyu et al., 2019), migration to urban cities is increasing worldwide. Thus, each year a growing number of children reared in rural contexts become part of urban low socioeconomic status (SES) children, yet the family gets at greater vulnerability as it lacks its original social network.
(d) Rural places account for more than 90% of land on earth (Joint Research Centre [JRC], 2008), contain the major number of natural resources, and produce the majority of food the world needs. Hence, making rural places a healthy habitat is of note for the society as a whole.

By and large, very little is known about the cognitive development of children that grow up in rural settings, even less in Latin American (LA) rural contexts. Ergo, this chapter is aimed at addressing this issue in two parts: the first one will describe differences between LA urban and rural poverty regarding factors affecting child cognitive development; the second one will give recommendations for conducting studies and interventions in rural child cognition.

Part 1: Characterization of LA Rural Contexts

As we mentioned, extreme poverty is concentrated in rural areas across all world regions. In South Asia, the percentage of extremely poor people living in rural areas is 83%; in Sub-Saharan Africa, 82%; in East Asia and the Pacific, 74%; in Europe and Central Asia, 66%; and in LA and the Caribbean, 53% (Castaneda et al., 2016). However, "extreme poverty" is different across regions. According to World Bank (2018), being extremely poor in Oriental Asia, the Pacific, Middle East, and North Africa means a lack of access to sanitation; in Europe and Central Asia is associated with no schooling enrollment; in South Asia and LA implies lack of adequate sanitation or electricity; and in Sub-Saharan Africa means having all those needs together. At this point is important to point out that this chapter will refer only to rural populations that do not identify themselves as indigenous population.[1]

[1]Various rural LA contexts' characteristics associated with poverty are shared with indigenous contexts: extreme poverty is concentrated in both of them (extreme poverty is 2.7 times greater in indigenous than in the rest of the population; 17% of people living in extreme poverty in LA are indigenous). However, there are important particularities: first, 48% of indigenous people live in an urban region. Also, they have specific vulnerabilities (e.g., have lower access to health and education than any other group) and each group has its particular cultural identity. Thus, research with indigenous populations require specific approaches that are out of the scope of this chapter, but that will be addressed in others (see chapter "Concepts, Language, and Early Socialization in the Indigenous Wichi Perspective: Toward a Relational–Ecological Paradigm").

Rural LA contexts differ from WEIRD (Western, Educated, Industrialized, Rich, and Democratic) rurality in several aspects, this being the core rationale for this chapter: categories to explain child cognitive development in rural WEIRD societies might be inappropriate to explain how cognition develops in rural LA. The deeper scarcity of LA rural contexts justifies revising the adequation of results from studies in the WEIRD world.

LA rurality is disproportionately poor even compared to urban contexts. Although rural territories were home to only 18% of the population in LA in 2016, they accounted for 29% of the total population living in poverty and 41% in extreme poverty (Economic Commission for Latin America and the Caribbean [ECLAC], 2018). Therefore, although the rural population does not exceed one-fifth of the total population, there is a disproportionate number of people living in extreme poverty in rural territories (FAO, 2019).

For the vast majority of these people, agriculture is the main economic activity. Current characteristics of this activity in LA have a significant role in reproducing LA rural poverty (Macias & Cueto Calderón, 2020): (a) land concentration in the hands of the few, (b) transference of land to foreign hands, (c) the ownership of the main local commerce into foreign hands, and (d) scarce industrializations relying on imported inputs or foreign-owned suppliers. As a result, not only employment in rural LA is vulnerable, characterized by lower wages and little social protection, including child work and temporal work, but also unemployment is increasing year by year (International Labour Organization [ILO], 2016). Thus, most people abandon agricultural activities to migrate to big cities looking for more opportunities. While in 1950 60% of the LA population was rural, in 2015 only 20% of the total population in LA was rural, and migration to big LA cities continued increasing (FAO, 2019). In this scenario, the people who emigrate are not "the poorest of the poor," but sectors that have some assets and that can finance high travel costs. Although the poorest are the ones with the greatest incentives to migrate, they are also the ones having the greatest limitations to do so (FAO, 2019). In sum, besides the urban–rural worldwide gap, the LA rural scenario (extremely poor) is very different from rural families in WEIRD societies, which mostly still can get a stable life quality at their places. Those contextual differences are likely meaning diverse opportunities for child development.

Child Cognitive Development in Rural Low Socioeconomic Status Contexts

Evidence from WEIRD Societies

Most of the little evidence on child development in rural contexts came from studies conducted in WEIRD societies. According to this literature, living in a rural low SES setting has been reported to be negative, neutral, or positive for child cognition, depending on the study. For example, in France, Gouin and collaborators (2015) showed lower cognitive performances for poor urban born preterm children than for rural children. In the USA, Tine (2014) found that low-income rural children exhibited asymmetrical deficits when compared with their high-income rural counterparts, with more extreme visuospatial working memory deficits than verbal working memory deficits; conversely, urban low-income children showed symmetrical working memory differences. Additionally, in the USA, Miller and collaborators (2013) found that the relationship between income and early academic skills was at least three times stronger in large cities when compared with rural areas.

Besides the fact that it is difficult to compare studies because they used different cognitive measures, at different ages and different definitions of "rural" contexts, overall, the results are inconclusive. In other words, the few studies analyzing cognition in rural areas of WEIRD societies showed differences between urban and rural children's achievement. Notwithstanding those results, it is not clear whether living in a low SES rural place presents more or less risk to children's cognitive development than living in a poor urban setting. Probably, following what others have argued (Miller et al., 2013), it might depend, mostly, on the living condition of those places.

Evidence from LA Rural Contexts

Considering only rural children in LA, there are a number of studies analyzing the relationship between SES and cognition. For instance, Macours and collaborators (2012) found a positive SES gradient of child development on the Peabody Picture Vocabulary Test, among other tests, for children aged 3–7 living in highly disadvantaged communities in rural Nicaragua. Also, Schady and collaborators (2011) administered three cognitive tests to different SES rural children at 3 and 5 years old. They found a strong association between child cognitive achievement and vocabulary and schooling of their mothers. The vocabulary levels of mothers and children were more strongly correlated among older children in the sample, suggesting that the effects of a richer maternal vocabulary are cumulative. In another study (Schady et al., 2014), results showed that rural children in Nicaragua and Perú obtained vocabulary scores more than 2 SD lower than the reference population, while that difference was 1.5 SD in Colombia and less than 1 SD in Chile. In all cases, rural children tend to have lower scores than in urban areas. However, those differences should be interpreted considering that in LA access to public resources, salaries, and general economic development is higher in urban areas than in rural areas. Not to mention the impact of malnutrition, which is also more present in rural children and has also been associated not only with cognitive development (Carrasco Quintero et al., 2016) but also with school absenteeism (Vargascruz et al., 2016), which, in turn, affects child cognition. The flip side of the association between living conditions and child cognition is the studies scrutinizing the positive impact of cash transfer programs in child cognitive development. For instance, Macours and collaborators (2012) in a randomized controlled trial in rural Nicaragua showed that cash increased intake of nutrient-rich foods, early stimulation, and use of preventive health care and had significant effects on cognitive outcomes, especially language.

Generally, this evidence suggests that the context of extreme poverty that characterizes the LA rural population generates a child cognitive gap with their urban counterparts and that mere cash transfer is associated with increments in cognitive scores. However, it is challenging to compare those results because of the variability in the cognitive measures used across studies. The next section will explore what happens when comparing urban and rural populations with the same tests.

The Example of Argentina

Living in Rural and Urban Contexts

In a study conducted in 2015 (Hermida et al., 2019a), poverty and rural/urban contexts were investigated to determine how they affect children's cognitive performance in Argentina. Executive functions and nonverbal intelligence performance, as well as individual and environmental information, were obtained from 131 five-year-old children living in low SES rural and urban contexts. Additionally, some contextual and individual variables that are typically associated with child cognitive development (i.e., parental education and occupation, school assistance, family size, and other socioeconomic information of the families) were measured in rural and urban places.

Table 1 shows variables that differ significantly between rural and urban Argentinean contexts. Rural children have significantly more siblings (to sustain a family with various children is associated with higher levels of poverty), and their families receive significantly more government subsidies (which is also associated with higher levels of vulnerability, as subsidies are given to families reaching certain vulnerabilities). Also, rural children have significantly fewer time of preschool education than their urban counterparts, and their fathers (usually, the person in charge of family income) have lower education and occupational levels than the others. All this data goes in one direction: risks for child cognitive development are more present in rural than in urban contexts.

Other variables, although did not reach statistical significance, also are in line with that claim (see details in Hermida et al., 2019a, 2021). For instance, on average, children living in rural places have mothers with uncompleted elementary school and no occupation at all, while their counterparts in urban places have mothers with uncompleted college and some occupations (although of low range).

In conclusion, living conditions in rural and urban settings inside the same country are different in terms of family size, education and occupation levels, number of government subsidies, and past preschool attendance, with poorer conditions in rural contexts. Now the question is, how do these different contexts influence child cognitive development?

Table 1 Descriptive statistics and ANOVA results of the variables that differ according to context

	Urban			Rural		ANOVA results			
Continuous variables	*n*	Media	SD	*n*	Media	SD	*F*	*p*	Eta[a]
Number of siblings of the child	83	1.34	0.99	47	2.53	1.36	20.151	1.59E-05	0.178
Father's completed level of education[b]	72	7.39	2.84	41	2.41	1.72	46.741	4.74E-10	0.314
Father's occupation[a]	78	3.73	1.79	45	1.51	1.01	24.377	2.59E-6	0.165
Number of government subsidies received by the family	77	0.51	0.62	47	1.02	0.49	13.334	3.86E-4	0.099
Past preschool attendance (in months)	77	21.97	9.62	47	7.91	5.75	50.798	8.06E-11	0.311

Note. [a]is a code used by the INDEC (2010) to measure occupation in function of salary where 0 = unemployed; 1 is, e.g., peddler; 2 is. e.g., a street sweeper; 4 is, e.g., taxi driver; etc.
[b]is a code used by the INDEC (2010) to measure educational level where 0 = no studied; 1 = primary school uncompleted; 3 = primary school completed; 6 = high school uncompleted; 9 = high school completed/college uncompleted; 10 = college completed/graduate school uncompleted; 12 = graduate school completed

Comparing Child Cognitive Development in Rural and Urban Contexts

In the mentioned study (Hermida et al., 2019a), we measured executive function (EF) and nonverbal intelligence, using the same tests and evaluations in preschoolers of the same age and SES, both in the rural and urban contexts. To evaluate cognition, we administered the following tasks: (a) the Attention subtest of CUMANIN Battery (Portellano et al., 2000), (b) the inhibitory control Day and Night like-stroop test (Tardif et al., 2007), (c) the working memory Forward digit span subtest of Wechsler Intelligence Scale for Children III (Wechsler, 1991), (d) the flexibility test Dimensional Change Card Sort (Zelazo, 2006), (e) the Test of Non-verbal Intelligence 4 (Brown et al., 2010), and (f) the processing speed subtest Coding of Intelligence Scale Wechsler Preschool and Primary III (Wechsler, 2004). Sociodemographic information was collected through parent interviews. To measure SES, we count the number of poverty indicators (including housing, education, health, public infrastructure). The results of this study are shown in Fig. 1.

As the main general result across most cognitive measures, rural context and higher number of poverty indicators were associated with lower cognitive performance. However, the standardized beta scores indicated that context is more important for inhibitory control, working memory, and intelligence (more weight in the regression equation) than poverty indicators. Specifically, to live in an urban context is associated with 0.299 more points on inhibitory control score, 0.359 points on working memory score, and 0.538 points on intelligence score, when poverty indicators are constant. The interaction between context and poverty indicators was not significant. The analyses described above indicate that context is indicative of risks in cognitive development, even more than poverty indicators, or, in other words, suggest that poverty measures are not sensitive enough to describe the scarcity of rural contexts.

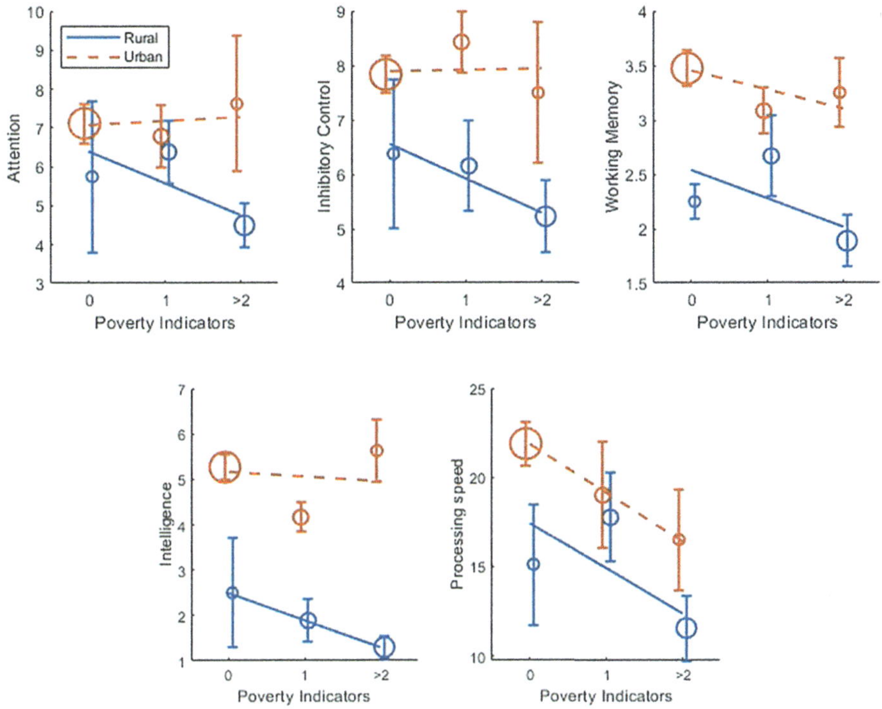

Fig. 1 Cognitive achievement as a function of context (urban or rural) and SES (measured as the number of poverty indicators)

Note. Lines correspond to linear fits of each context separately; points' sizes correspond to subsample sizes. Effects of poverty on flexibility are not shown in the figure because it is an ordinal variable. (This figure has been adapted from Hermida et al. (2019a), under the terms of https://creativecommons.org/licenses/by/4.0/)

Why Do Rural Contexts Present More Risks for Child Cognitive Development?

In the next step of the same study (Hermida et al., 2019a), we analyzed which of the variables differ by context and actually predicted child cognition. Results showed that three of these variables had a significant effect on children's cognitive performance:

(a) Months of past preschool attendance was positively correlated with Attention ($R^2 = 0.194$, $F = 3.119$, $\beta = 0.285$, $p = 0.027$). The time spent in preschool had a very strong effect on attentional performance (ranging from 5 points to 10 points) in a 30-second cancellation task. This finding resonates with a long tradition of investigation that has shown that the first years of education have an enormous impact on future cognitive development (Barnett, 2008; Campbell et al., 2002; Magnuson et al., 2004; Temple & Reynolds, 2007).

(b) Fathers' completed level of education was correlated positively with Flexibility ($R^2 = 0.109$, $Z = 3.23$, $\beta = 0.264$, $p < 0.001$). There is evidence that father interactions provide children with unique, enriching experiences that do not occur in mother–child interactions (Grossmann et al., 2008; Pancsofar et al., 2010); thus, our result is consistent with that of studies that associate parental education with cognitive development (Grantham-McGregor et al., 2007). Also, the fathers' education level is strongly correlated with family income, which in turn is associated with children's cognition (Khanam & Nghiem, 2016).

(c) Having more government subsidies was negatively associated with working memory ($R^2 = 0.459$, $F = 3.318$, $\beta = -0.239$, $p = 0.044$). Government subsidies are a proxy measure of the precariousness of living conditions (it is given only to families with unemployed parents living in vulnerable living conditions). Hence, homes that receive more government subsidies are also homes that have the most vulnerable living conditions (Roca, 2011). This indicates that while government subsidies are effective (Roca, 2011), their effect measured in variability in cognitive performance is insufficient to compensate—at least in Argentina—the original differences in income and social resources.

In essence, this factorial study of dimensions of poverty confirms that living in a rural area does not limit cognitive opportunities per se. Instead, certain forms of low SES that are typical of rural areas have a strong impact on cognitive development. This indicates that attempting to collapse poverty in a single numerical indicator might be misleading on the grounds that different contexts pose distinct and singular risks to cognitive development (Duncan et al., 2017; Lipina, 2016). Poverty in rural and urban contexts stands for qualitatively and quantifiable different forms of scarceness, with rural populations suffering more deprivation. Following the model of risk accumulation proposed by Evans (2013), to live in a poor place that, also, is rural would imply even greater risk. The sum of effects would produce a worse scenario for child development, given a situation of less available resources for an adequate cognitive development. In consequence, the literature on urban poverty could not necessarily explain what happens in the context of rural poverty.

The Homogeneity of Rural Poverty

In another study (Hermida et al., 2019b), we investigated, in a poor rural sample, which individual and socio-environmental variables make the difference in child cognitive performance in a processing speed task. In that study, 44 five-year-old children were evaluated with a processing speed task (Coding subtest of Intelligence Scale Wechsler Preschool and Primary III, Wechsler, 2004). Also, individual and socio-environmental information was obtained from parents' interviews (including SES information). Results indicated that higher scores in the effortful control dimension of temperament were associated with higher performance in the processing speed task. Interestingly, we did not find a relation between processing speed and the other individual nor socio-environmental variables. This result disagrees with previous studies that showed a relation between these factors and cognitive performance: parental education (Foulkes & Mori, 2009; Mykerezi et al., 2014; Tine, 2017), father's occupation and overcrowding (Ngure et al., 2014), and past preschool attendance (Castro & Rolleston, 2015; Gouin et al., 2015) have been associated with rural child cognitive achievement. In fact, our results suggested an effect of SES (linked with socio-environmental variables) since these children obtained a mean processing speed performance lower than expected for their age. But within this group, there was no association between socio-environmental variables and performance. Why did these socio-environmental variables show no impact on cognition performance inside our rural sample?

Two possible first explanations emerged: (a) finding effects on general cognition does not exactly mean finding effects on processing speed and associations may differ in distinct rural contexts and (b) each rural area is particular, so the absence of impact that we found might be replicable only in similar extreme low SES context. However, we would like to accentuate another explanation based on our data: since all children in this sample have very low performance, it might be possible that in this range the socio-environmental effect on performance has saturated. In other words, if we had included children with a higher SES, we would probably find an effect of the environmental variables on the processing speed score. But this extremely low SES context could not produce enough variability to show the association of SES-processing speed. This alerts us to an important point: in LA rural contexts, socioeconomic variables are likely not to have the same dispersion than in urban contexts. Figure 2 shows the distribution of variables typically associated with child cognitive development across contexts: parental education, father's occupation, and past preschool attendance not only tend to have lower values in rural samples, but also have lower dispersion (i.e., rural data is concentrated in a few values). This homogeneity of LA rural contexts seems logical, considering that the extreme poverty scenario we described depends mostly on the lack of access to public resources, a characteristic that is common to all subjects living the same rural contexts. For example, mother education (crucial to cognition in urban places) is likely to have no effect on rural places due to the fact that all mothers have extremely low educational levels. In consequence, a pool of variables that make a lot of sense to measure in an urban place lose their interest in predicting cognition in LA rural contexts.

Part 2: Recommendations for Researchers Analyzing Child Development in LA Rural Poverty

Considering Part 1 data, in this section we will draw some recommendations for researchers, educators, and policy makers interested in studying rural poverty in LA.

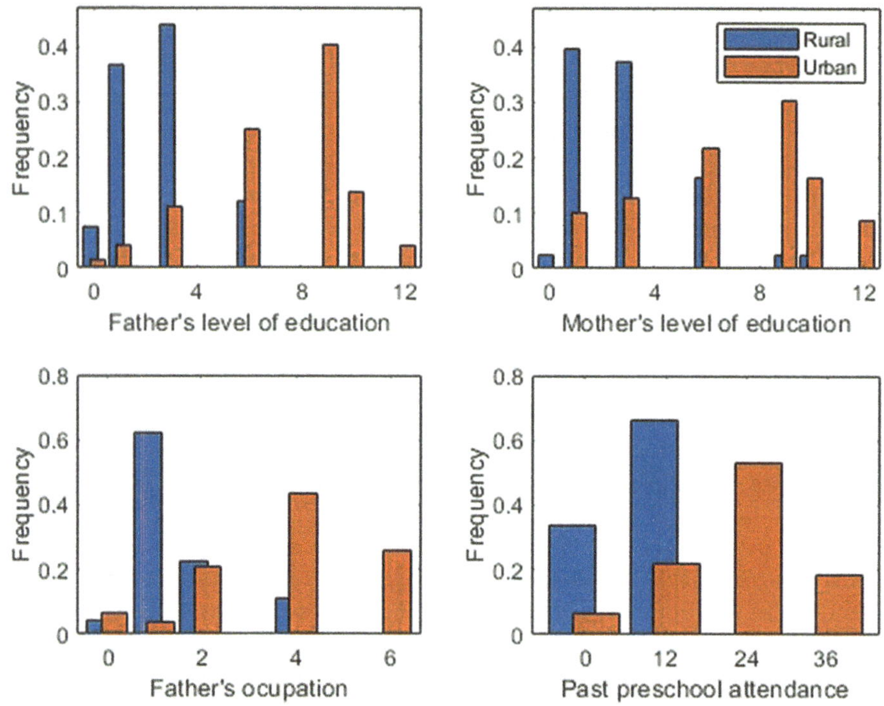

Fig. 2 Percentages for variables typically affecting child cognitive development, across rural and urban population

Note. Fathers' and Mothers' level of education are a code used by the INDEC (2010) where 0 = not studied; 1 = primary school uncompleted; 3 = primary school completed; 6 = high school uncompleted; 9 = high school completed/college uncompleted; 10 = college completed/graduate school uncompleted; and 12 = graduate school completed. Fathers' occupation is a code used by the INDEC (2010) to measure occupation in function of salary where 0 = unemployed; 1 is, e.g., peddler; 2 is, e.g., a street sweeper; 4 is, e.g., taxi driver; etc. Past preschool attendance is measured in months

Preparing a Study in LA Rural Contexts

As different child cognitive development researchers have claimed since various decades ago (Ogbu, 1981), tasks should be culturally appropriate (i.e., tasks should be familiar enough for the novelty to not interfere with cognition evaluation). The conventional tools to evaluate child-rearing practices and children's competencies as standards are not always useful in understanding rural child-rearing and competencies. In the case of rurality, besides the cultural appropriateness, we need to have in mind the disproportionate poverty rates of rural context we have explained before. The extreme scarcity of LA rural contexts became central not only when planning and conducting the studies, but also when interpreting results. The whole design of the study will be signed by poverty rates at the rural place selected.

Considering this, we can give some general recommendations to be applied if they are possible:

- Conducting pilot studies in order to test cultural appropriateness of tasks is extremely useful in rural contexts. There are multiple social, cultural, economic, and practical barriers that you will detect better when piloting tasks. However, as sometimes it is not possible, let's see how we could reduce problems in the following points.
- Although today, most LA rural children have access to primary school, which makes tasks including pencil and paper, language, or math calculation, more familiar for them, the quality of education is still much poorer than in urban places (as we previously showed based on our and other's data). Thus, if you can choose between tasks including academic or schooling processes (such as reading, writing, and math calculations) and tasks with the lower influence of schooling to evaluate the same processes, choose the last ones.
- Most standardized tests have norms based only on urban population; therefore, we recommend avoiding comparisons with standardized measures when it is not the objective of the research.
- Another important issue for testing in LA rural contexts is to consider the lack of availability of materials needed for an evaluation (e.g., no electricity, no internet, no copy machine, no printer). The use of technology is significantly lower in LA rural contexts (which means a cultural barrier, but also a practical one); availability of internet is not guaranteed in various rural communities; moreover, sometimes electricity is not available. Consequently, for cultural appropriateness and practical issues, we discourage the use of computer testing and recommend planning all materials without taking for granted that you will find them in the rural place itself.
- Linked with the previous point, it is extremely difficult to conduct studies at the brain level (e.g., EEG), since all tools to explore the brain are extremely unfamiliar for those children, but mostly because of the absence of the minimum conditions that equipment needs.
- Consider the lack of local professionals available to assist researchers during the study. For example, it is very unlikely to find a psychologist in a LA rural place. If you can train people from the same contexts to test children, which is the best option, you will diminish the influence of the cultural distance between the tester and the children. But if you need professionals, probably they should come from other places only for your study.
- When calculating the minimum sample size, take into account the low population concentration and that, as a consequence, one school can have children of very different ages. Maybe you will need to collect information in various rural schools to get the minimum sample size.
- When writing the informed consent, take into account that many rural parents are illiterate, so you can try non-writing ways of giving consent (i.e., verbal consent recorded in audio). Also, take into account the average educational level of the context to write an information sheet culturally appropriated. It is useful to check how to write the context with a rural teacher.

- Access to rural populations is not easy, especially if you have no previous contact with people living in the rural place. A general recommendation is to identify a community leader and invite him to participate as part of the study. The leader is the person that could mediate between you and rural people, helping you to understand which is the best way of organizing your research activities and helping participants to understand the objective and activities of the study. Another option of the way to contact potential participants is a public institution (e.g., school, health primary care center).

Interpreting Results of Studies in LA Rural Contexts

- Ideally, include an anthropologist analyzing researchers' and community's prejudices. This point is critical for not to interpret results ethnocentrically. The anthropologist could help you to adapt tasks and activities in order to follow research objectives in a culturally appropriate way.
- Regarding statistical analyses, consider that models and predictors that are crucial in an urban context cannot be important in rural ones, for instance, for mere mathematical conditions, e.g., floor effects. In other words, and as data of the study we mentioned (Hermida et al., 2019b) suggested, rural contexts tend to have lower variability in socioeconomic factors important for child cognitive development. Thus, what makes a lot of sense to measure in urban places might not make it in rural places.
- Factors that affect cognition in rural places are mostly linked with economic and social variables. Based on our experience, it is probable that cognitive training (at the individual level) will not reach big effects in the absence of changes in economic and social conditions (i.e., creating schools, increasing employment, etc.). Instead, interventions at the economic or social level, such as cash transfer programs, have much more probability of reaching the objectives because those programs could improve factors that are present in rural contexts and affect cognition (parental education, occupation, and general SES level). Ending rural poverty requires commitment and action by a broad set of local, national, and international actors. Punctual interventions could help, but in this scenario of high scarcity, the more comprehensive the intervention is, the higher impact is likely to have.

Conclusion

This chapter addressed an important and quite unexplored issue: rural child cognitive development. The chapter started with a description of the rural context in LA, and its differences with rurality in the WEIRD world: LA rural context is significantly less developed (in socioeconomic terms) than urban places. As a consequence, living in a LA rural context (in comparison with an urban one) could be riskier for child cognitive development.

The Argentinian case showed how children of the same age and SES differ according to context, with rural children obtaining lower scores at all cognitive tasks. Later, also in Argentina, we saw that past preschool attendance, parental education, and the number of subsidies (a proxy for vulnerability) all accounted for that cross-contextual difference. Finally, we observed that all the variables that are typically associated with cognition showed no effect if we take only rural low SES children's cognition: a possible explanation for these is the socioeconomic characteristic or rural contexts. They are very homogeneous. The lack of opportunities resulted in a lack of variability in factors such as maternal education, SES, parental occupation, etc.

Based on those examples, we draw the chapter to a close by sharing some recommendations for researchers interested in conducting studies in rural LA contexts. Such suggestions are based on the data presented here and are mainly focused on taking into consideration the high level of scarcity characteristic of these settings. It is our hope this chapter will encourage more research in rural child development, enhancing what is known about an important but invisible issue.

Acknowledgments The authors thank Fundación Mundo Sano for funding and Nahuel José Nicolás Weimer for his contributions.

References

Barnett, W. S. (2008). Preschool education and its lasting effects: Research and policy implications. In Boulder & Tempe (Eds.), *Education and the public interest center & education policy*. Retrieved [09/01/21] from http://epicpolicy.org/publication/preschooleducation

Bradley, R. H., & Corwyn, R. F. (2002). Socioeconomic status and child development. *Annual Review of Psychology, 53*(1), 371–399. https://doi.org/10.1146/annurev.psych.53.100901.135233

Brown, L., Sherbenou, R. J., & Johnsen, S. K. (2010). *Toni-4, test of nonverbal intelligence*. Pro-Ed.

Campbell, F. A., Ramey, C. T., Pungello, E., Sparling, J., & Miller-Johnson, S. (2002). Early childhood education: Young adult outcomes from the abecedarian project. *Applied Developmental Science, 6*(1), 42–57.

Carrasco Quintero, M. D. R., Ortiz Hernández, L., Roldán Amaro, J. A., & Chávez Villasana, A. (2016). Desnutrición y desarrollo cognitivo en infantes de zonas rurales marginadas de México. *Gaceta Sanitaria, 30*(4), 304–307. https://doi.org/10.1016/j.gaceta.2016.01.009

Castaneda, A., Doan, D., Newhouse, D., Nguyen, M. C., Uematsu, H., Azevedo, J. P. (2016). Who are the poor in the developing world? *Policy research working paper; no. 7844*. Washington, DC: World Bank. Retrieved [09/01/2021] from https://openknowledge.worldbank.org/handle/10986/25161. License: CC BY 3.0 IGO.

Castro, J. F., & Rolleston, C. (2015). Explaining the urban-rural gap in cognitive achievement in Peru: The role of early childhood environments and school influences. *Young lives working paper 139*. Oxford: Department of International Development at the University of Oxford. Retrieved [09/01/21] from https://assets.publishing.service.gov.uk/media/57a08986ed915d3cfd0002be/YL-WP139_Explaining-the-urban-rural-gap-in-cognitive-achievement-in-peru.pdf

Duncan, G. J., Magnuson, K., & Votruba-Drzal, E. (2017). Moving beyond correlations in assessing the consequences of poverty. *Annual Review of Psychology, 68*, 413–434. https://doi.org/10.1146/annurev-psych-010416-044224

Economic Commission for Latin America and the Caribbean, ECLAC. (2018). *Panorama Social de América Latina 2017*. Santiago de Chile. Retrieved [09/01/21] from: https://repositorio.cepal.org/bitstream/handle/11362/42716/7/S1800002_es.pdf

Evans, G. W., Li, D., & Whipple, S. S. (2013). Cumulative risk and child development. *Psychological Bulletin, 139*(6), 1342.

Food and Agriculture Organization, FAO. (2019). *Overview of rural poverty in Latin America and the Caribbean, solutions for eliminating rural poverty in the 21st century* (114 pp). FAO.

Farah, M. J., Shera, D. M., Savage, J. H., Betancourt, L., Giannetta, J. M., Brodsky, N. L., … Hurt, H. (2006). Childhood poverty: Specific associations with neurocognitive development. *Brain Research, 1110*(1), 166–174.

Foulkes, D. M., & Mori, E. S. (2009). Formación de la capacidad cognitiva en México: Impactos económicos. *Estudios Económicos, número extraordinario*, (1), 83–122. Retrieved [09/01/2021] from http://www.redalyc.org/pdf/597/59724201004.pdf

Gouin, M., Flamant, C., Gascoin, G., Rouger, V., Florin, A., Guimard, P., … Hanf, M. (2015). The association of urbanicity with cognitive development at five years of age in preterm children. *PLoS One, 10*(7), e0131749. https://doi.org/10.1371/journal.pone.0131749

Grantham-McGregor, S., Cheung, Y. B., Cueto, S., Glewwe, P., Richter, L., … International Child Development Steering Group. (2007). Developmental potential in the first 5 years for children in developing countries. *The Lancet, 369*(9555), 60–70. https://doi.org/10.1016/S0140-6736(07)60032-4

Grossmann, K., Grossmann, K. E., Kindler, H., & Zimmermann, P. (2008). *A wider view of attachment and exploration: The influence of mothers and fathers on the development of psychological security from infancy to young adulthood*. Guilford Press.

Hermida, M. J., Shalom, D. E., Segretin, M. S., Goldin, A. P., Abril, M. C., Lipina, S. J., & Sigman, M. (2019a). Risks for child cognitive development in rural contexts. *Frontiers in Psychology, 9*, 2735.

Hermida, M. J., Segretin, M. S., Shalom, D. E., Lopez-Rosenfeld, M., Abril, M. C., Lipina, S. J., & Sigman, M. (2019b). Temperament predicts processing speed in low socioeconomic status rural preschoolers. *Mind, Brain, and Education, 14*(1), 61–70.

Hermida, M. J., Shalom, D. E., Segretin, M. S., Goldin, A. P., Abril, M. C., Lipina, S. J., & Sigman, M. (2021). Corrigendum: Risks for child cognitive development in rural contexts. *Frontiers in Psychology, 12*, 647039. https://doi.org/10.3389/fpsyg.2021.647039

International Labour Organization, ILO. (2016). Trabajar en el campo en el siglo XXI. Realidad y perspectivas del empleo rural en América Latina y el Caribe. *Panorama Laboral Temático, 3*. Retrieved [09/01/2021] from https://www.ilo.org/global/about-the-ilo/newsroom/news/WCMS_530327/lang%2D%2Den/index.htm

Joint Research Centre, JRC. (2008). *Urbanization: 95% of the world's population lives on 10% of the land*. ScienceDaily. ScienceDaily, 19 December 2008. Retrieved [09/01/2021] from: www.sciencedaily.com/releases/2008/12/081217192745.htm

Khanam, R., & Nghiem, S. (2016). Family income and child cognitive and noncognitive development in Australia: Does money matter? *Demography, 53*(3), 597–621.

Lichter, D. T., & Johnson, K. M. (2007). The changing spatial concentration of America's rural poor population. *Rural Sociology, 72*(3), 331–358.

Lipina, S. J. (2016). Critical considerations about the use of poverty measures in the study of cognitive development. *International Journal of Psychology, 52*(3), 241–250. https://doi.org/10.1002/ijop.12282

Lyu, H., Dong, Z., Roobavannan, M., Kandasamy, J., & Pande, S. (2019). Rural unemployment pushes migrants to urban areas in Jiangsu Province, China. *Palgrave Communications, 5*(1), 1–12.

Macias, P. G. G., & Cueto-Calderon, J. S. (2020). *Movilidad humana en Latinoamérica: Nuevos patrones, nuevos retos*. Transnational Press.

Macours, K., Schady, N., & Vakis, R. (2012). Cash transfers, behavioral changes, and cognitive development in early childhood: Evidence from a randomized experiment. *American Economic Journal: Applied Economics, 4*(2), 247–273.

Magnuson, K. A., Meyers, M. K., Ruhm, C. J., & Waldfogel, J. (2004). Inequality in preschool education and school readiness. *American Educational Research Journal, 41*(1), 115–157. https://doi.org/10.3102/00028312041001115

Miller, P., Votruba-Drzal, E., & Setodji, C. M. (2013). Family income and early achievement across the urban–rural continuum. *Developmental Psychology, 49*(8), 1452.

Mykerezi, E., Kostandini, G., Jordan, J. L., & Melo, I. (2014). On rural-urban differences in human capital formation: Finding the 'bottlenecks'. *Journal of Rural Social Sciences, 29*(1), 2.

Ngure, F. M., Reid, B. M., Humphrey, J. H., Mbuya, M. N., Pelto, G., & Stoltzfus, R. J. (2014). Water, sanitation, and hygiene (WASH), environmental enteropathy, nutrition, and early child development: Making the links. *Annals of the New York Academy of Sciences, 1308*(1), 118–128.

Noble, K. G., Houston, S. M., Brito, N. H., Bartsch, H., Kan, E., Kuperman, J. M., ... Sowell, E. R. (2015). Family income, parental education and brain structure in children and adolescents. *Nature Neuroscience, 18*(5), 773–778.

Ogbu, J. U. (1981). Origins of human competence: A cultural-ecological perspective. *Child Development*, 413–429.

Olinto, P., Beegle, K., Sobrado, C., Uematsu, H., et al. (2013). The state of the poor: Where are the poor, where is extreme poverty harder to end, and what is the current profile of the world's poor? *Economic Premise, 125*(2), 1–8.

Pancsofar, N., Vernon-Feagans, L., & Family Life Project Investigators. (2010). Fathers' early contributions to children's language development in families from low-income rural communities. *Early Childhood Research Quarterly, 25*(4), 450–463.

Portellano, J., Mateos, R., Martínez, R., Tapia, A., & Granados, M. (2000). *Manual CUMANIN, Cuestionario de Madurez Neuropsicológica Infantil*. Tea Ediciones.

Roca, E. (2011). Asignación universal por hijo (AUH): Extensión de las asignaciones familiares. *Revista Debate Público*. Retrieved [09/01/2021] from http://www.edumargen.org/docs/curso46-/unid03/apunte03_03.pdf

Rossel, C., & Food and Agriculture Organization, FAO. (2012). *Protección social y pobreza rural en América Latina*. FAO. Retrieved [09/01/2021] from http://www.fao.org/3/a-au333s.pdf

Schreuder, R. (2010). Unstacking the deck: rural poverty and the effects on childhood development. *Michigan Journal of Social Work and Social Welfare 1*, 45–55.

Schady, N. (2011). Parents' education, mothers' vocabulary, and cognitive development in early childhood: Longitudinal evidence from Ecuador. *American Journal of Public Health, 101*(12), 2299–2307.

Schady, N., Behrman, J., Araujo, M. C., Azuero, R., Bernal, R., Bravo, D., ... Vakis, R. (2014). *Wealth gradients in early childhood cognitive development in five Latin American countries*. The World Bank.

Sianes, A., Simpson, I. C., & Ibáñez-Alfonso, J. A. (2021). Cognitive interventions in children and adolescents from low socioeconomic status backgrounds: A systematic review protocol of randomized controlled trials. *Systematic Reviews, 10*(1), 1–10.

Tardif, T., So, C. W.-C., & Kaciroti, N. (2007). Language and false belief: Evidence for general, not specific, effects in cantonese-speaking preschoolers. *Developmental Psychology, 43*(2), 318. https://doi.org/10.1037/0012-1649.43.2.318

Temple, J. A., & Reynolds, A. J. (2007). Benefits and costs of investments in preschool education: Evidence from the child–parent centers and related programs. *Economics of Education Review, 26*(1), 126–144. https://doi.org/10.1016/j.econedurev.2005.11.004

Tine, M. (2014). Working memory differences between children living in rural and urban poverty. *Journal of Cognition and Development, 15*(4), 599–613. https://doi.org/10.1080/15248372.2013.797906

Tine, M. (2017). Growing up in rural vs. urban poverty: Contextual, academic, and cognitive differences. In G. Staicu (Ed.), *Poverty, inequality and policy* (1st ed). InTechOpen, Retrieved [9/01/2021] from https://www.intechopen.com/chapters/55118

United Nations International Children's Emergency Fund, UNICEF, & World Bank Group. (2016). *Ending extreme poverty: A focus on children*. UNICEF & World Bank Group.

United Nations International Children's Emergency Fund, UNICEF. (2012). *The state of the world's children 2012: Children in an urban world*. UNICEF.

Vargascruz, S. L., Rodriguez Escobar, G., Ibañez Pinilla, E. A., Matiz Salazar, M. I., & Overgaard, H. J. (2016). Relación entre el estado nutricional y el ausentismo escolar en estudiantes de escuelas rurales. *Revista de Salud Pública, 17(6)*, 861–873. https://doi.org/10.15446/rsap.v17n6.48709

Wechsler, D. (1991). *Wisc-iii: Wechsler intelligence scale for children: Manual*. Psychological Corporation.

Wechsler, D. (2004). *Wppsi-iii. Wechsler Preeschool and primary scale of intelligence-third edition manual*. The Psychological Corporation.

World Bank. (2018). *Poverty and shared prosperity: Piecing together the poverty puzzle*. World Bank.

Zelazo, P. D. (2006). The dimensional change card sort (dccs): A method of assessing executive function in children. *Nature Protocols -Electronic Edition, 1*(1), 297. https://doi.org/10.1038/nprot.2006.46

Concepts, Language, and Early Socialization in the Indigenous Wichi Perspective: Toward a Relational–Ecological Paradigm

Andrea Taverna, Migdalia Padilla, Matías Fernández Ruiz, and María Celeste Baiocchi

Introduction

One of the crucial problems in most current approaches in developmental psychology has been the identification of the factors that explain development. This is an intellectual strategy originated in modern thought—the Cartesian-Split-Mechanistic worldview (Lakatos, 1978)—which sharply separates processes considered internal to the mind (innate dispositions, representations, etc.) from the external world (e.g., social determinations). As a result, "Cartesian psychology" is still facing unsolvable dichotomies such as nature versus nurture, development separated from evolution, child-centrism separated from culture, etc.

At the same time, there has been a rising movement toward an integrative developmental science based on relational thinking (Lerner, 2006, 2011; Lerner & Overton, 2008; Overton, 2006, 2010, 2012; Overton & Lerner, 2012). Generally speaking, this relational–ecological paradigm understands development in terms of the organism–environment econiche, encouraging attention to system-level dynamics rather than focusing on components in isolation. Centrally, the different theoretical versions within Relationism attribute the source of the change to the interactions of the system in question at those different dynamic levels, confirming its

We would like to especially thank our colleagues and native speakers from Wichi Lawet community (Formosa, Argentina), Aurelia Pérez, Élida María Pérez, María Segundo, Modesto Palma, and Luisa Pérez, for their valuable commitment to the project. We are also grateful to children and their families for their willingness and for sharing their native language and cultural knowledge. This work received funding from Argentina, PICT-2018-02516, and Coop International (CONICET-NIH), which was granted to the first author.

A. Taverna (✉) · M. Padilla · M. Fernández Ruiz · M. C. Baiocchi
Consejo Nacional de Investigaciones Científicas y Técnicas, Buenos Aires, Argentina
e-mail: taverna@irice-conicet.gov.ar

M. V. Alves et al. (eds.), *Cognitive Sciences and Education in Non-WEIRD Populations*, https://doi.org/10.1007/978-3-031-06908-6_6

self-movement (Castorina, 2002). By focusing on development of interdependent elements as part of a system, this movement leads to the healing of the classic fundamental antinomies (e.g., subject–object, mind–body, nature–nurture, culture–individual) providing concepts that are inclusive and represented, not as pure forms, but as forms that flow across fuzzy boundaries (Overton, 2013a, b). Within the framework of this relationshipism, an alternative paradigm to Cartesian psychology—relational–ecological psychology—is opening up, in which a coalition of different perspectives and theoretical initiatives coexist, such as Developmental Ecological Psychology (Szokolszky & Read, 2018), Developmental Systems Theory (Oyama et al., 2001), Dynamic Systems Approaches (e.g., Lewis, 2010; Thelen & Smith, 2006), Sociocultural and Ecobehavioral Perspectives (e.g., Cole, 1996; Gauvain et al., 2011; Heft, 2001; Nelson, 1996; Rogoff, 2014; Valsiner, 1998), and Approaches on Embodied Intersubjectivity (e.g., Di Paolo & De Jaegher, 2016; Gallagher, 2005) (see Szokolszky & Read, 2018, for a detailed description).

We agree that ecological–relational psychology reveals strong possibilities for addressing the problems that the splitting Cartesian psychology has presented to the field of developmental science, in particular, the excessive focus on the causal factors that "explain or cause" development resulting in an individual child split from their context/culture. However, despite this important contribution, Cartesian rationality tracks are still found in developmental perspectives. The first is the over-representation of only one cultural group: children of middle-class European–American descendants (e.g., Gauvain et al., 2011). Second, even among the cross-cultural approaches, a bulk of studies often treat culture as a variable that influences, but is not constitutive of, the individual and development (see Mistry et al., 2013; Overton, 2013a, b).

In this chapter, we argue that these limitations, both methodological and epistemological, can be addressed bidirectionally: under the umbrella of the relational paradigm (Overton, 2013a, b) on the one hand and from evidence other than dominant populations—such as indigenous communities—on the other, populations that often happen to exhibit epistemological orientations aligned with the foundations of relational thinking. Different contributions from both Psychology and Anthropology have described indigenous epistemological orientations, that is, their way of knowing, as "relational epistemologies" (Bird-David, 1999; Medin et al., 1997, 2002, 2006, 2013, 2015; Pierotti, 2011). Briefly, these epistemologies are about knowing the world by focusing primarily on relatedness, developing the skills of being-in-the-world with other things (Bird-David, 1999).

In this work, we focus on two crucial developmental processes—concepts and language—from the Wichi perspective, an indigenous group living in the Chaco region in South America. First, we will describe their ways of knowing, attention to, and interaction with the world, identifying what aspects of their orientations make them relational epistemologies. Second, we will reconsider the Wichi relationality in light of our psychological evidence from two perspectives: (a) how Wichi children and adults conceptualize and reason about their environment, particularly about the *hunhat lheley* (inhabitants of the earth), and (b) how Wichi infants become native speakers and competent social participants of their culture. Finally, based on this evidence, which, as we shall see, is well aligned with relational thinking, we will describe the emerging ecological–relational paradigm, which brings the relationshipism front and center.

The Wichi People: A Relational Epistemology on the *husek*

The Wichi are an indigenous community living in the Chaco Forest located in Northern Argentina (Fig. 1). We have focused on this population because they are an indigenous community with a strong native language, a constellation of experiences, and cultural orientations that differ considerably from those of Westerners (see Taverna et al., 2012 for a review). Interestingly, the Wichi language (*Wichi Lhamtes*, or "the words of the people") is the first language in the family and the primary language in the community (Taverna & Waxman, 2020; Vidal & Kuchenbrandt, 2015). Classified as a member of the Mataco-Mataguayan language

Fig. 1 Geographical setting of Wichi. Areas highlighted in gray identify Wichi communities as well as other indigenous groups of the Chaco Region

family (Tovar, 1964), this language has approximately 40,000 native speakers of Wichi in Argentina and Bolivia. In our research, we focus on the Wichi population living in Laguna Yema, Formosa, and Argentina, where the dialect known as del Teuco o Bermejo is spoken (Gerzenstein, 2003; Nercesian, 2014).

Anthropological documentations show that relationality among a great deal of species, environments (such as forests, rivers and lagoons), and "spiritual" beings are central to the Wichi (Palmer, 2005). This relationality is well captured by an overarching category *hunhat lheley* (inhabitants of the earth), which is composed of different inhabitants such as humans (the Wichi itself, the *siwele* or not Wichi people, and other ethnic groups) (Fig. 2), distinct animal categories (*tshotoy, tshotoy inot lheley, tshotoy fwiy'ohen, laloy*), several categories of plants (*hal'o, tokos,* etc.), and spiritual beings (*wuk, ahot*), all living in ecological habitats such as the "monte" (*tayhi*), domestic environments, and water habitats (*inot*) (Palmer, 2005; Suárez & Montani, 2010; Taverna et al., 2012). All *hunhat lheley* are perceived by the Wichi in the frame of a relational epistemology that is organized around the notion of *husek* (goodwill) as an agent of vitality and socialization. Several distinct kinds of *husek* have been noted (Palmer, 2005). First, *husek* invokes a notion of vitality (or "vital will"); thus, all *hunhat lheley*—humans, plants, animals, spirits, etc.—are related by means of this vital will (the blood, the greenness), which is absent in other entities (metal, stones, soil, etc.). Vital will is central to life processes as fundamental as growth, decomposition, and death. In addition, Palmer (2005) notes

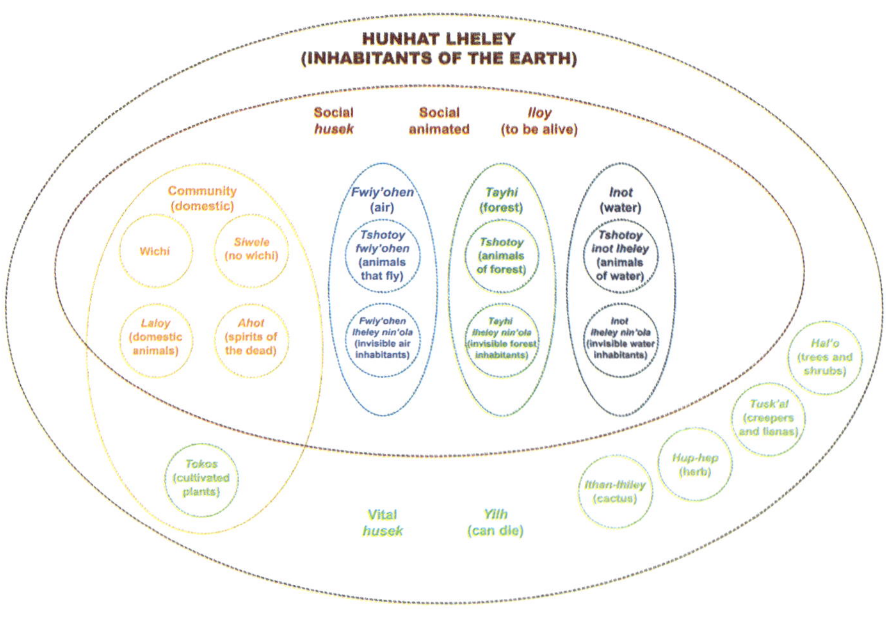

Fig. 2 Schematic representation of *hunhat lheley*. The colors indicate the different categories and environments. The dotted nodes indicate the categories and environments that are the subject of our current investigations

that *husek* also invokes a notion of socialization ("social will" or goodwill), which relates humans, non-human animals, and spiritual entities, but not to the plant kingdom (but see Suárez and Montani (2010) for a distinct interpretation for certain plants with animistic properties) and is deemed central to the process of socialization in the Wichi community. According to Wichi people, this agent defines socialization within the ecosystem. Socialization is a process in which "social *husek*" affects a transition from a natural, pre-social aggressive state to a more mature prosocial, peaceful one (Palmer, 2005). According to Palmer (2005) and our own native consultants, after a few months of life, human infants acquire social *husek*, reducing their natural pre-social aggressive tendencies and providing the social cooperation and pro-social behaviors required for membership within the Wichi community (Palmer, 2005). Thus, the *hunhat lheley* are perceived in their relational perspective, either from their vital will (vital *husek*) or from their social will (social *husek*). In what follows, we will show how this relationality shapes both, the representational means that the Wichi invoke when they reason about *hunhat lheley* and the language socialization and acquisition process that make the Wichi competent participants in their culture.

Concepts and Reasoning About *Hunhat lheley*

Concepts have commanded the agenda of cognitive and developmental sciences aligned with Cartesian psychology, occupying a prime position in research on the nature and representation of knowledge. Along with more recent approaches (Medin et al., 2013, 2015), in this chapter we argue that concepts should be considered in light of the cultural orientations in which they emerge. This position represents a shift from studies that have focused on concepts and categories such as whether they are acceptable or lousy examples of the category (e.g., Rosch & Mervis, 1975) or as part of universal theoretical frameworks such as folkbiology, folkpsychology, and folkphysics (for a more extended discussion, see Medin et al., 2013, 2015; Ojalehto & Medin, 2015; Ojalehto et al., 2017).

For over a decade, in our research program, we have adopted an ecological as well as an emergentist–constructionist approach with respect to how our cultural orientations permeate the conceptual systems which underlie knowledge across development (Medin et al., 2013, 2015; Taverna et al., 2020). Particularly, we have focused on how the Wichi, both children and adults, conceptualize and reason about *hunhat lheley* (inhabitants of the earth) (Baiocchi, 2019; Baiocchi et al., 2019; Fernández Ruiz, 2021; Fernández Ruiz et al., in press; Taverna et al., 2012, 2014, 2016, 2018). Here, we present this evidence organized on two cognitive processes: (a) categorization—how children and adults decide which entities are of the same type (Solomon et al., 1999); and (b) causal reasoning—how people perceive, represent, and reason about causality (Bender et al., 2017; Waldmann, 2017).

Categorization

One of the key cognitive processes in conceptual representation is categorization, the process by which people decide which entities are of the same type (Murphy, 2002; Smith & Medin, 1981; Solomon et al., 1999). Particularly, concepts and categories about the natural world have been pervasive in the developmental sciences, with decades of work dedicated to identifying the conceptual frameworks (taxonomic, thematic, functional ecological) and core concepts (e.g., animate, living things) that children use to organize and reason about nature (Margolis & Laurence, 2015; Murphy, 2002; Smith & Medin, 1981). In our pioneering studies in this community (Taverna et al., 2012, 2014, 2016, 2018), we focus on this process, studying through classification and reasoning tasks, how Wichi children (5–6-year-olds, 10–11-year-olds) and adults decide which inhabitants and/or entities "are alive" (*iloy)*, "can die" (*yilh*) and share *husek* (social, vital will) and how these attributions permeate the way they reason about the relationships among inhabitants and entities (people, animals, plants, artifacts, etc.) (see Table 1 for the proportion of *husek* attribution as a function of each inhabitant or entity; Taverna et al., 2012). We found clear findings: (a) only the Wichi and animals—but not plants—possess *iloy* (to be alive) reflecting an animistic perspective on life; (b) nevertheless, the three inhabitants—Wichi, animals, and plants—can *yilh* (to die); (c) both living categories—humans and animals—are equally central in their epistemology, rejecting any anthropocentric perspective in reasoning (Carey, 1985, 2009); (d) the social *husek* is attributed to humans and animals—but not plants—emphasizing pro-social intentions and animistic behaviors between them; (e) these patterns were found rather early in childhood, showing that the influence of the Wichi cultural orientations are sufficiently strong to maintain access to the animate interpretation of alive (*iloy*) across development.

Based on those findings, more recently (Baiocchi, 2019; Baiocchi et al., 2019), we ask how Wichi children (5-year-olds, 10-year-olds) and adults conceptually

Table 1 Proportion of *husek* attribution for each inhabitant/entity

Test item	*M*
Human	0.94
Jaguar	0.82
Dog	0.82
Snake	0.82
Vinchuca (*Triatoma infestans*)	0.76
Mosquito	0.82
Bottle tree	0.29
*Chágua*r (*Bromelia* sp.)	0.17
Sun	0.05
Chair	0

Reproduced from Taverna et al. (2012)

organize the animals of the Chaco Forest, *tshotoy*.[1] All of them were asked to do a sorting task used in previous studies (Bailenson et al., 2002; López et al., 1997; Medin et al., 2006), adapted to the native perspective and the Wichi language. The participants, after identifying each animal, had to group the 41 photographs of *tshotoy* (see Fig. 3) in numerous rounds of classification, according to how they thought and argued "the animals were in nature." On every round of sorting, participants were free to form as many groups as they wished (see Baiocchi et al., 2019 for a detailed description of the task).

Three important findings were obtained about how the Wichi represent and conceptually organize *tshotoy*. First, we identified an ecological organization—not taxonomic (e.g., mammals, reptiles)—among *tshotoy*, based primarily on interactions among the animals of the Chaco forest. The distance among *tshotoy*, that is, how close and how far they are in the Wichi ecosystem, was defined primarily by specific ecological relationships and rarely by taxonomic ones (e.g., the morphological

Fig. 3 Complete set of photographs used in the study, in alphabetical order with each animal labeled with its scientific name

---

[1] *Tshotoy* is made up of various species of mammals and reptiles (among others). As such, this animal category, which is imbued with powerful cultural significance because the Wichi identify themselves as descendants of *tshotoy*, provided an opportunity to examine the salience of taxonomic similarity (e.g., morphological or other perceptual commonalities among *tshotoy* like species, kind, size) and Wichi ecological frameworks about the organization of the animals of their ecosystem (e.g., food chain, habitat, social relations, utility to humans).

similarity between animals). Second, in this ecological organization, we observe that the Wichi classify the animals as aggressive (e.g., snakes, cats, "they are dangerous, they can attack each other and the rest of the animals, can attack people") away from the peaceful ones (e.g., armadillos, cows, rats, pigs, "they are defenseless animals, they live together without problem, you find them anywhere"). In addition, aligned with this organization, the Wichi formed subgroups based on other ecological principles (e.g., utility of animals, habits, food chain, habitat).

We co-interpret these classifications based on a socioecological principle that emphasizes the relationships of sociability vs. (pre-social) aggressiveness on the basis of which peaceful, defenseless friendly animals would be conceptually different from aggressive, dangerous hostile ones (Baiocchi, 2019; Baiocchi et al., 2019). These findings challenge universal postulates (e.g., Berlin et al., 1973, 1974; Osherson et al., 1990) since the Wichi adults did not show taxonomic relations as a priority to organize *tshotoy;* instead, they showed culture-specific principles in the conceptual organization of the animals of their native forest. In addition, when analyzing the sorting through development, we observed that this socioecological principle (peaceful vs. hostile animals) is already present in 5-year-old children and progresses with greater specialization as age advances. For 5-year-old children, all *tshotoy* are aggressive, for 10-year-olds, *tshotoy* are simultaneously aggressive and peaceful, while in adulthood the aggressive animals are far away from the peaceful ones.

Causal Reasoning

Causal reasoning is a central cognitive process of human cognition through which people perceive, represent, and reason about causal events and their interactions (Bender et al., 2017; Waldmann, 2017). Recent research explored causal representations and framework theories of Wichi people when they reason about causal relationships between agents, entities, and environments of their native ecosystem (Fernández Ruiz, 2021; Fernández Ruiz & Taverna, in prep.). Based on open interviews with key consultants and articulating the previous anthropological and ethnobiological evidence (Arenas, 2003; Montani, 2018; Palmer, 2005; Suárez, 2020), we focused on ecosystem events because they are rich in relationships and involve different types of entities characterized by complexity, emergent processes, and circular causality. As a result, we first identified *hunhat lheley*, entities, properties, and native features that make up the most significant events of the Wichi ecosystem from the perspective of their native speakers. On the basis of these insights, in a following stage, we explore causal reasoning among the Wichi, focusing on two dimensions: (a) the *type of causes* (*natural*, *supernatural*) that the Wichi invoked when they reason about the *hunhat lheley*[2] and (b) the relationships between *native*

[2] Since ethnographic background documented that Wichi people do not categorically differentiate between "natural" and "supernatural" beings, we tested empirically the validity of these assumptions, showing an intimate relationship between "spiritual" and biological worlds (Palmer, 2005; Suárez, 2014).

features of *hunhat lheley* (*husek*—will; *inot*—water; *nin'ola*—invisible) and *ecological levels* (*inhabitant, environment, annual climate cycle*) from which causes come. To study this, we employed a causal attribution task adapted to the Wichi epistemology and their language. In this task, participants were asked to infer the causes of 38 events that represented different common phenomena among *hunhat lheley* (humans, animals, plants, and non-human beings), environments (forest, lagoon, etc.), and other entities of their ecosystem (e.g., inert matter) (e.g., "master spirit of forest is angry," why do you think this happens?) (Table 2). The events were accompanied with images of native animals and plants extracted from the books "*Hunhat Lheley*" (inhabitants of the earth), prepared by Wichi speakers and members of our team (Pérez et al., 2017a, b, c, d, e, 2021).

Results confirmed our hypothesis that the Wichi people attributed "natural" causes (e.g., food) to all *hunhat lheley* and entities, even those considered "supernatural" by the Western ontology (e.g., master spirit of forest), evidencing that these latter are ecosystem's inhabitant like any other. In addition, Wichi people were more likely to attribute causes of the inhabitant level (e.g., mood) to *hunhat lheley* with social *husek*, but causes of the annual climatic cycle level (e.g., rains) to *hunhat lheley* and entities without social *husek*. They were also more likely to attribute causes of the environment level (e.g., lagoon pollution) to aquatic *hunhat lheley* and entities—*inot*—, rather than to *hunhat lheley* and entities that live on land or in the air.

Table 2 List of events of Wichi ecosystem used to study causal reasoning

Events	
Carob tree grows	*Master spirit of forest dies*
Carob tree bears fruits	*Master spirit of water grows*
Carob tree is sick	*Master spirit of water is angry*
Carob tree is dry	*Master spirit of water is sick*
Chaco chachalaca grows	*Master spirit of water dies*
Chaco chachalaca flies	*Wichi cuts a carob tree and then grows*
Chaco chachalaca sings	*Wichi hunts*
Chaco chachalaca is sick	*Wichi drowns in lagoon*
Chaco chachalaca dies	*Wichi fishing*
Fish grows	*Wichi grows*
Fish swims	*Wichi is angry*
Fish is sick	*Wichi is sick*
Fish dies	*Wichi dies*
Fruit is delicious	*Wood floats*
Lagoon is dry	*Wood falls*
Lagoon has fish	*Yacare caiman grows*
Master spirit of forest grows	*Yacare caiman swims*
Master spirit of forest is angry	*Yacare caiman is sick*
Master spirit of forest is sick	*Yacare caiman dies*

Reproduced from Fernández Ruiz (2021)

From results emerged a native framework for causal reasoning with three interactive causal principles: *relational*—all inhabitants, "natural" and "supernatural," are connected; *socioecological*—the distinction between animate–social *hunhat lheley*—social *husek*—and inanimate–pre-social entities; *environmental*—the distinction between aquatic *hunhat lheley*/entities—*inot*—and nonaquatic. These causal principles could delimit different domains to the supposedly universal domains of Cartesian psychology (folkphysics, folkbiology, and folkpsychology). For example, aquatic environments and their inhabitants could constitute a potentially different domain, made up of physical entities (e.g., lagoon, river), biological beings (e.g., fish), and "supernatural" beings (e.g., master spirit of water), establishing specific domain limits of this ecological human group.

In sum, the findings on concepts and reasoning among the Wichi strengthen the relational perspective on conceptual development (Medin et al., 2013, 2015; Taverna et al., 2020), showing how cultural knowledge permeates the conceptual system—conceptual representations and processes—throughout development. Within the relational epistemology of the *hunhat lheley*, these inhabitants are perceived in their relational perspective, either from their vital will (vital *husek*) or social animistic will (social *husek*). These two relational perspectives, evident in their belief system, shape the Wichi's conceptual representations about living things (*iloy*), animate (social–animist *husek*), and animals (pre-social–aggressive *tshotoy* vs. social and peaceful ones) and the type of causes the Wichi invoke in explaining the behavior of the inhabitants and entities of their ecosystem ("all inhabitants, even "supernatural ones" are affected by natural causes"). In addition, certain relevant environmental features of the *hunhat lheley* (e.g., belonging or not to an aquatic environment) also affect their causal reasoning. Many of these culturally specific conceptual patterns emerge at early stages of childhood (e.g., the ecological character of *thsotoy*), changing with greater specialization as age advances. Precisely, the interactions of the learning system at different levels—epistemological orientations, representational resources, cognitive processes—are those that drive changes within the system in question and its conceptual representations. That is why the Wichi perspective, strongly aligned with the relational–ecological paradigm (Overton, 2013a, b; Read & Szokolszky, 2018), challenges the Cartesian-divided-mechanistic worldview that sees child's mind separated from the context/culture and development as a result of causal factors, providing interpretations that explain representational emergence and cultural knowledge synergically.

Language Acquisition and Early Socialization in the Wichi Language

Language acquisition, and more specifically that of grammar, is a pampered subject within developmental science as it is a formal development produced by a pre-logical cognitive system that is observable in the child's production behavior (López

Ornat, 1999; for a review of this discussion, see MacWhinney, 2004). However, most of the studies on language acquisition on which the current psycholinguistic theories are based have been carried out with Indo-European languages and in speaking communities that usually belong to the Euro-American middle class urban cultural model (see De León Pasquel, 2005, 2012). Regarding the Wichi, there is an absence of the documentation on the acquisition of the great variety of indigenous languages during childhood that exist in this area of our country (except for recent studies on Qom/Toba language (Audisio et al., 2021)).

In this context, the study of the acquisition of grammatical knowledge in the Wichi, a polysynthetic and agglutinating language, is central. According to the constructivist–emergentist perspective (López Ornat, 1999), the grammatical knowledge is a logical system that emerges and is formed in and by the learning task and is acquired during and due to the acquisition process. In this frame, since the acquisition of grammar occurs in the human environment, the child is socially and simultaneously exposed to the linguistic forms and the referents of their meanings within the language learning task (López Ornat, 1999). Consequently, the problem of language acquisition in Wichi is addressed along with the properties of the environmental language being acquired in conjunction with the early socio-cognitive competencies that emerged in social interaction.

Our methodological approach consists mainly of corpus data. It consists of 101 hours of video recordings of 16 children from two cultural groups (Spanish-speaking from urban context and indigenous Wichi from Chaco Forest) in their natural environment at different points of development during their first 4 years. This corpus was gathered in the framework of an intense fieldwork carried out in the Wichi Lawet community, Laguna Yema, Formosa, and with Spanish-speaking families from the city of Rosario, Santa Fe, between the years 2012 and 2017 (Taverna et al., 2020; Taverna, 2021).

Coordinating Attention and Mental State Attributions in Caregiver–Infant Interaction

The study of the mind has been conceived under biology and sociology as divided realms; also, the mind has been assumed as imperceivable and opaque to others in the early stages of development (Szokoloszky & Read, 2018). It is not until around the first year that a crucial developmental change occurs: infants understand others as intentional agents who can direct their (infants') attention and whose attention can, in turn, be directed by an adult, and thus share communication about objects or events in the world. Semiotic resources for drawing an interlocutor's attention everywhere include speech, gaze, body touching and postures, pointing gestures, and other actions. In turn, adults deploy a series of resources as well, ones that allow them to interpret and attribute intentions to the significant actions that infants perform. The issue is whether adults deploy semiotic resources in comparable ways with infants in different settings.

In an ongoing set of studies, we address this question by seeking evidence of the process through which mother–infants from the two cultural communities under study—Wichi and Spanish-speaking infants and their families—come to be able to coordinate attention and mental states attribution within social interaction. We focus on the infant's bodily manifestations, movements, and gaze patterns and those directed toward the infant. For its analysis, based on the Constant Comparative Method (Strauss & Corbin, 1990), we encode semiotic body forms directed toward and from the infant, encoding the gaze pattern and identifying different categories that describe an attentional gradient of the interaction participants. This procedure allows us to know how attention is organized in these interactions cross-culturally (Taverna & Padilla, 2020).

Preliminary results show that unlike mothers from other indigenous communities studied to date, who tended to develop a less child-centered profile (Schieffelin & Ochs, 1986), Wichi mothers tend to focus their attention on the child. However, in contrast to Spanish-speaking communities, joint attention episodes seem to be scarce. Interestingly, Wichi caregivers—but not their Spanish-speaking counterparts—tend to display a sustained attentional ability when infants are interacting with objects or events in their environment, intervening only if necessary (Taverna & Padilla, in prep.). Based on these first results, we hypothesize that this sustained generalized and outstanding observational "talent" of the Wichi mothers when the child is interacting with objects could configure a cultural-specific triadic attentional modality—child–object–mother or caregiver—one in which the caregiver coordinates her attention by offering it and enabling it to the child with a more "lateral" (Clark, 1996; De León Pasquel, 2005) and less interventionist participation. We propose to call this culturally specific engagement pattern "affordance attention," in which shared attention and speech would not be compulsory, although the attentive presence of the caregiver would be.

In the frame of these engagement formats, we are also exploring how caregivers are Mind Minded (Meins et al., 2001) to infants cross-culturally. To accomplish this, we first interactively analyze the speech and actions directed at the infant, seeking to detect actions, verbalizations, and gestures of the mother as indicators that she is interpreting motivations, intentions, wishes, and propositional attitudes of the infant (Pérez & Español, 2014), and not only through mentalistic comments (Mind Mindedness) (Meins et al., 2001), which is the common perspective in the studies conducted in the field. So far, we have obtained two important results: (1) the Mind Mindedness (MM) construct is being revisited; (2) a multimodal and culturally sensible category system was built for the analysis of the Maternal Mental States Attributions (MMSA) with subcategories aimed to identify how the mother (or caregiver) attributes motivations, intentions, wishes, and propositional attitudes to the infant and how it occurs (Table 3).

Results are a novel set of categories that integrate smart social perception, context, an expansion from the dyad, and shared bodily awareness and understand mentalistic interactions embedded in system-level dynamics rather than its components

Table 3 Set of observational categories for the analysis of the Maternal Mental States Attributions

Category	Definition	Examples
Mental state interpretation Mentalistic Interaction (I) No Mentalistic Interaction (NI)	An adult or peer attributes wishes, beliefs, or intentions to a target infant	The caregiver sees the infant reaching for an object and hands it to him/her and says "*is this what you want?*"
Modality Verbal (VI) Corporeal (CI) Multimodal (MI)	The semiotic modality of a mentalistic interpretation	Verbal: the caregiver says "*don't touch that*" while the infant is reaching for an object Corporeal: the caregiver physically removes the infant before he or she touches the object Multimodal: is a combination of the verbal and corporeal modalities
Direction Child directed (CDI) Overheared (OI)	That is to whom the mentalistic interpretation is directed	Directed: the caregiver asks the target infant what he/she wants Overheard: the caregiver asks another person in the room what he/she wants
Source Mother (Ma) Other adult (Oa) Other Child (Ch)	Who is attributing the mentalistic behavior	The caregiver says "*don't touch that*" (source) and the infant retreats
Executor Mother (Ma) Other adult (Oa) Other Child (Ch)	Who enforces the mentalistic interaction	From the previous example: another person removes the infant before it touches the prohibited object
Type of Discursive Strategy	The type of pragmatic statement that is used in the mentalistic interpretation (Taverna, 2021)	
Prescriptive	*Warn*: negative, harmful consequences//kelh	*Suwanas tajlhy* (Come, there are ants)
	Order: directives that tend to regulate the child's action	*Yajlhek* (Don't go)
Referential-denotative	*Establish reference:* draws attention to entities, environmental situations	*Wepa neche* (Listen to the seriema)
	Name: entities, people, or situations	*Catita* (Parrot)
Conversational	*Affirm-deny:* verbal and gestural	Yes/No expressions or nods
	Comment: situations, events, objects, actions	*Am ka huyey escuela* (You don't go to school)
	Encourage: proposes and encourages the performance of actions	*Tsiteyej hulu tolothila* (Play with the earth, there you have the cup)
	Question: about states, emotions, actions	*¿Lawenhe Tito?* (Did you sew Tito?)//*¿Tenlo inot?* (Do you want water?)

in isolation (Rietveld & Kiverstein, 2014). Precisely, within relationism, analysis is about creating categories, not about cutting nature at its joints (Overton, 2013a). Moreover, by developing an observational coding system based on the data corpus itself, instead of applying an external coding system, we could not only identify the specific semiotic resources—speech, gaze, touch, body postures, gestures—used by caregivers from populations under study but also how these resources are combined in novel interactive configurations that serve to coordinate attention and mental state interpretations in mother–infant interaction at the different target cultural settings.

The Acquisition of the Wichi Language: First Outcomes

Previous studies in morphologically complex languages are not very numerous (for a review, see Kelly et al., 2014) but within the available evidence the focus is fundamentally on the learning mechanisms that are put into play in the process of acquisition of these languages, which are presented to the infant as a real puzzle (Peters, 1981). Our ongoing longitudinal research seeks precisely to generate the first psycholinguistic evidence in this native language. Particularly, we focus on the grammaticalization process, that is, the transition between the first pre-grammatical verbalizations to the simple but completely grammatical productions.

Our preliminary results show changes in children's Wichi speech productions and in the properties of the linguistic information to which infants are exposed. Both changes delineate a clear evolutionary trajectory, one that goes from a pre-grammatical stage toward the emergence of grammatical or morphosyntactic knowledge. At the monomorphemic period (MLU (Mean Length Utterance) =1), infants seem to produce mostly nominal and verbal stems stripped of affixes, and nominal stems outstrip verbal ones in children's vocabulary, even though the number of verbs far outstrips the number of nouns in the adult input. More precisely, when the MLU barely exceeds 1 (1.05–1.17), nominal and verbal affixes (type and token) represent only 4% of the total number of nouns and verbs produced in this period.

At the same time, at this pre-grammatical stage, a "motherese" or a stable speech register employed by mothers (and other caregivers) when addressing young children in this culture was identified (Taverna, 2021). The Wichi "motherese" was characterized by a constellation of prosodic, lexical, and pragmatic-discursive features showing differences and similarities with the ones observed in other linguistic communities. First, in contrast to baby talk from dominant linguistic populations, prosodically the Wichi "motherese" shows the same normal tone as in adult–adult interactions. Second, a set of nine Wichi baby talk lexical items were discovered that thus far had not been captured in the literature on this language. Semantically, these are terms equivalent to those of adult speech that refer to aspects of interest to the child (see Table 4). In addition, at a discursive-pragmatic level, the Wichi "motherese" at the pre-grammatical stage focused on discursive strategies with directive

Table 4 Wichi baby talk, semantically organized with its equivalent adult term in Wichi and English

Baby talk Adult term		
	Wichi	English
Entities and people		
Chuku	*Asinoj*	Dog
Lulu	*Tulu*	Cow, meat
Tata	*Lafwcha*	Dad
Titit		Little car
Toto		Toy
Emotional states		
Kuku	*Nuway*	To feel scared
Nana	*Oytaj*	To feel pain
Basic needs		
Chichi		Maternal breast
Mimi	*Inot*	Water

Reproduced from Taverna (2021)

functions (prescriptions and/or denotations). For example, a group of prescriptive strategies in the here and now are characterized by orders referring to concrete actions (1050: *yajnencho* [don't come down]), prescriptions in the near future in the form of warnings (1006: *che suwele hin'am* [the non-Wichi person is watching you]), and denotations that label objects or events in the child's surrounding world. This "labeling" function ranges from names of objects of interest to the child (1058: *titit* [little car]) through the use of Spanish loans (2037: *jutu* [foto—photo]) to names that denote people (5249: *siwele* [non-Wichi person]) and animals (5018: *cheche* [parrot]//5232: *neche* [seriema].

When infants reach an MLU of 1.5, the number of verbal stems (both type and token) increases, so children's early noun-advantage characteristic of the previous period decreases and comes into closer alignment with the patterns in the linguistic input. At this stage, as the number of verbal stems increases, so do the number of verbal and nominal affixes that encode the grammatical categories that characterize the morphology of both nouns and verbs in Wichi. In fact, at this point of development, both nominal and verbal affixes represent 20% of the total of the nouns and verbs, five times more than the previous period. So, as soon as children approach an MLU of 1.50, there is an increasing productive command over the grammatical categories often expressed in Wichi morpho-syntactically within the word level for both nouns (e.g., possession, deixis, augmentative, etc.) and verbs (pronominal pronouns, mode, etc.), emerging the first combinatory morphosyntactic forms (see Taverna & Waxman, 2020, for a more detailed description of the emergent grammatical categories at this phase).

Regarding the "motherese" at this stage, during the beginning of the combination of morphemes (M = 1.5 and beyond), maternal speech shows discursive strategies that compose more elaborated functions than those from the previous stage, so conversational routines or proto-conversations emerged. Thus, among the most frequent discursive strategies, we observed comments, which tend to denote "showing the world" that surrounds the child (Tsep) (5260: *hin we suwanas* [look at the ants]//5218: *neche wepan we* [listen to the seriema]), interrogatives (Dqst) (5070: *¿apihi Norberto?* [where is Norberto?]), responses to questions from the young child (Drsp) (5209: *chayhu* [it's hot]), and assertions and/or negations (Dan) (1130: *is* [Good]//1005: *kha* [no]) in the context of events situated in the present and related to the child's surrounding world (Tsep).

Overall, infants from all the world's languages are expected to acquire their environmental language at any point between their 18 months and three and a half years. Predicting how grammatical knowledge of the language under acquisition will develop means to show how the language learning system uses what information (social, cognitive, linguistic, etc.) and, also, how both (information and system) change as a result of the acquisition process until they become a grammatical knowledge system. The first responses to the Wichi acquisition problem show qualitative changes both in the child language—from a pre-grammatical period toward first morphosyntactic combinations—and in certain aspects of the socialization process, the maternal speech, "motherese" Wichi. This linguistic transition occurs in a social environment, one that would be characterized by a child-centered context, in which mothers, instead of engaging children exclusively in joint attention formats, seem to coordinate attention in a culturally specific fashion creating "attentional affordances" episodes, where speech and joint gazes would not be compulsory, and mothers will intervene only if it is necessary (Taverna & Padilla, in prep.). When and how Wichi infants' speech becomes grammatical is still a matter of our current research efforts. Both knowing (process) and known (products) are considered as equal and indissociable complementary processes in the construction, acquisition, and growth of knowledge, in this case, the Wichi grammatical knowledge.

The Wichi Perspective and the Relational Turn on Developmental Science

Our work on thought and language across development among the Wichi reveals the power of relationshipism in two cross-fertilized native-scientific epistemological directions.

From the native toward the scientific path, the discovery of the mutual and synergistic relationship between the emerging Wichi representational resources and their native epistemology (commonly viewed as mental-cognitive vs. external and contextual polarities in Cartesian psychology) took us toward the adoption of relationism as a paradigm with three main characteristics. First, the different polarities (e.g., mind/culture, learning-system/context, etc.) are not considered exclusive contradictions as in the split epistemology but as differentiated polarities (e.g., co-iguals) of a unified inclusive matrix of relations, each pole defines and is defined recursively by its opposite (Overton, 2013a). Second, in order to set a positive agenda for empirical inquiry, opposites (e.g., conceptual and language-learning systems vs. cultural orientations, lingüistic input) are considered levels of analysis, the focus of attention, in recognition that they do not neglect absolute foundations (Latour, 1993, 2004) but perspectives in a multiperspective world (Rogoff, 1992). In the third movement, a new "middle space" (Overton, 2013a), which Latour (1993, 2004) called "the middle kingdom position", away from the extremes of the Cartesian splits is created to then discover a novel system—a synthesis—that will coordinate the two-only conflicting polarities. For example, in this middle position, the language learning system and the different interacting levels (social, cognitive, linguistic, etc.) change as a result of the acquisition process.

Under the umbrella of this relationism, the constructionism–emergentist perspective on language acquisition (López Ornat, 1999) as well as recent cognitive-ecological accounts on concepts (Medin et al., 2015) constitutes theoretical approaches under which we framed our research among the Wichi. Within these perspectives, key notions such as econiche, affordances, and representational emergence work as crucial entry points.

The term "ecological," increasingly popular in recent years used in several different contexts (Jensen & Greve, 2019), emphasizes the profoundly relational character of entities or elements in the world. Within the cognitive-ecological approach on concepts and conceptual representations (Medin et al., 2015), it is believed that, like some species in an ecosystem, certain ideas may grow better in certain ecologies than in others, persisting and achieving a widespread distribution. In addition, these stabilized "ideas-habitats" might work as "cultural affordances" (Ramstead et al., 2016). Typically understood as possibilities for actions the environment offers to a creature (such as reaching, grasping, sitting, walking, and so on) (Gibson, 1979/1986), the affordance construct is fundamentally interactional. In a recent sociocultural understanding of the concept (Ramstead et al., 2016; Rietveld & Kiverstein, 2014; and see Chemero, 2003; Costall, 1995; Costall & Richards, 2013; Heft, 2017; Michaels, 2003; Withagen & Chemero, 2012) taking Gibson's (1979/1986) ideas and applying it to the human econiche, cultural affordances are precisely those stabilized cultural cues (cultural knowledge, values, socialization and linguistic practices, etc.), which constitute different behavioral patterns (or forms of life) that characterized particular groups relatively robust on sociocultural time scales, and acquired through sociocultural processes such as joint-intentionality or shared-expectations, cultural conventionality, and perspective-taking abilities (see Ramstead et al., 2016). The ecological niche is, then, a network of interrelated cultural affordances. At the learning-system (conceptual and language system) level, the recursive interaction between the system in question, the learning mechanisms, and the cultural affordances in each human econiche (practices, values, cultural knowledge, etc.) are those that drive changes at the representational resources within the learning system synergically.

Finally, from the scientific toward the native direction, relationism and its three key concepts—econiche, affordances, and representational emergence—have illuminated the Wichi ecosystem as an ecological niche made up of interconnected cultural affordances, relatively widespread and robust within the population. The *hunhat lheley*, *tshotoy*, spiritual inhabitants, and the Wichi itself coexist fully integrated with the Chaco forest, and it is precisely the Wichi cognitive (e.g., categorization, causal reasoning), linguistic and social competencies as stabilized behavioral patterns that contribute to building this native organism–environment econiche as it is. Likewise, it is his human organism–environment econiche that offers an open system of stabilized cultural cues or "cultural affordances" (the *hunhat lheley* notion and its epistemological orientations as well as the environmental language and its referents/meanings/forms, socialization, cultural practices) based on the specific competencies of this forms of life, what shapes, in conjunction with general learning mechanisms, the acquisition, and developmental processes.

Certainly, these insights are far from new. First of all, they are in line with the developmental ecological psychology and a "coalition" of relational–ecological developmental theories, which from a different epistemological and theoretical basis, all focus on an anti-dualistic agenda for the field of developmental science (see Szokolszky & Read, 2018, for a detailed description). The cross-fertilization among these disciplines and perspectives is an indispensable effort for the full realization of the relational–ecological turn in development. In this enterprise, addressing one remaining challenge is urgent: overcoming the over-representation of the Anglo-speaking children of middle-class European American descent and including a broader range of cultural childhoods. In our case, both conjugated relationshipisms, the native and the scientific, pushed us toward the anti-dualist epistemic framework (Baiocchi et al., 2019; Fernández Ruiz, 2021, in prep.; Taverna, et al., 2020). The active synergy between ecological psychology (econiche, affordances), constructionism–emergentist perspectives (learning-system, representational emergence), cultural–cognitive approaches (ecosystem, epistemological orientations, cultural practices), and the relational epistemology of the hunhat lheley (Wichi perspective) collaborates in an open and evolving movement that seeks and finds interactive conjunction of new insights. We are on that way.

References

Arenas, P. (2003). *Etnografía y alimentación entre los Toba-Ñachilamole#ek y Wichí-Lhuku'tas del Chaco Central (Argentina)*. Pastor Arenas.

Audisio, C., Cúneo, P., Ojea, G., & Rosemberg, C. (2021). Indicadores sintácticos y morfológicos de la transitividad en el input toba/qom a niños pequeños y niñas pequeñas de Argentina. *Forma y Función, 34*(2). https://doi.org/10.15446/fyf.v34n2.88628

Bailenson, J. N., Shum, M. S., Atran, S., Medin, D. L., & Coley, J. D. (2002). A bird's eye view: Biological categorization and reasoning within and across cultures. *Cognition, 84*(1), 1–53. https://doi.org/10.1016/S0010-0277(02)00011-2

Baiocchi, M. C. (2019). *Categorizar en la cultura: la sociabilidad intuitiva como marco conceptual para representar a los animales del monte chaqueño (tshotoy) en niños y adultos wichí* (Unpublished doctoral thesis). Universidad Nacional de Córdoba.

Baiocchi, M. C., Waxman, S., Pérez, E. M., Pérez, A., & Taverna, A. (2019). Social-ecological relations among animals serve as a conceptual framework among the Wichi. *Cognitive Development, 52*, 100807. https://doi.org/10.1016/j.cogdev.2019.100807

Bender, A., Beller, S., & Medin, D. L. (2017). Causal cognition and culture. In M. R. Waldmann (Ed.), *The Oxford handbook of causal reasoning* (pp. 717–738). Oxford University Press. https://doi.org/10.1093/oxfordhb/9780199399550.013.34

Berlin, B., Breedlove, D. E., & Raven, P. H. (1973). General principles of classification and nomenclature in folk biology. *American Anthropologist, 75*(1), 214–242. https://doi.org/10.1525/aa.1973.75.1.02a00140

Berlin, B., Breedlove, D. E., & Raven, P. H. (1974). *Principles of Tzeltal plant classification: An introduction to the botanical ethnography of a Mayan-speaking people of Highland Chiapas*. Academic Press.

Bird-David, N. (1999). "Animism" revisited: Personhood, environment, and relational epistemology. *Current Anthropology, 40*(1), 67–91. https://doi.org/10.1086/200061

Carey, S. (1985). *Conceptual change in childhood*. MIT Press.

Carey, S. (2009). *The origin of concepts*. Oxford University Press.

Castorina, J. A. (2002). El Impacto de la Filosofía de la Escisión en la Psicología del Desarrollo Cognoscitivo. *Psykhe, 11*(1), 25–57. https://doi.org/10.7764/PSYKHE.11.1.453

Chemero, A. (2003). Information for perception and information processing. *Minds and Machines, 13*, 577–588. https://doi.org/10.1023/A:1026209002908

Clark, H. (1996). *Using language*. Cambridge University Press.

Cole, M. (1996). *Cultural psychology: A once and future discipline*. Harvard University Press.

Costall, A. (1995). Socializing affordances. *Theory & Psychology, 5*(4), 467–481. https://doi.org/10.1177/0959354395054001

Costall, A., & Richards, A. (2013). Canonical affordances: The psychology of everyday things. In P. Graves-Brown, R. Harrison, & A. Piccini (Eds.), *The Oxford handbook of the archaeology of the contemporary world* (pp. 59–80). Oxford University Press. https://doi.org/10.1093/oxfordhb/9780199602001.013.047

De León Pasquel, L. (2005). *La llegada del alma: lenguaje, infancia y socialización entre los mayas de Zinacantán*. CIESAS.

De León Pasquel, L. (2012). Language socialization and multiparty participation frameworks. In A. Duranti, E. Ochs, & B. Schieffelin (Eds.), *The handbook of language socialization* (pp. 81–111). Wiley-Blackwell. https://doi.org/10.1002/9781444342901.ch4

Di Paolo, E., & De Jaegher, H. (2016). Neither individualistic, nor interactionist. In C. Durt, T. Fuchs, & C. Tewes (Eds.), *Embodiment, enaction, and culture* (pp. 87–105). MIT Press. https://doi.org/10.7551/mitpress/9780262035552.003.0005

Fernández Ruiz, M. (2021). *Cognición causal wichí. Un estudio acerca de las representaciones de la causalidad entre los wichí del Gran Chaco*. Unpublished MS thesis. Facultad Latinoamericana de Ciencias Sociales-Universidad Autónoma de Madrid.

Fernández Ruiz, M., Baiocchi, M. C., & Taverna, A. (in press). *Socioecología como teoría marco distintiva: aportes cognitivos al entendimiento de la relación naturaleza/cultura entre los wichí*. Actas del VI Congreso de la Asociación Latinoamericana de Antropología (ALA).

Fernández Ruiz, M., & Taverna, A. (in prep.). Causal cognition among Wichi people.

Gallagher, S. (2005). *How the body shapes the mind*. Clarendon Press.

Gauvain, M., Beebe, H., & Zhao, S. (2011). Applying the cultural approach to cognitive development. *Journal of Cognition and Development, 12*(2), 121–133. https://doi.org/10.1080/15248372.2011.563481

Gerzenstein, A. (2003). Variaciones dialectales de algunas unidades del sistema consonántico wichí. In A. Tisera & J. Zigarán (Eds.), *Lenguas y culturas en contacto* (pp. 69–81). CEPIHA-Universidad Nacional de Salta.

Gibson, J. J. (1979/1986). *The ecological approach to visual perception*. Lawrence Erlbaum Associates Publishers.

Heft, H. (2001). *Ecological psychology in context: James Gibson, Roger Barker, and the legacy of William James's radical empiricism*. Erlbaum.

Heft, H. (2017). Perceptual information of "an entirely different order": The "cultural environment" in the senses considered as perceptual systems. *Ecological Psychology, 29*(2), 122–145. https://doi.org/10.1080/10407413.2017.1297187

Jensen, T. W., & Greve, L. (2019). Ecological cognition and metaphor. *Metaphor and Symbol, 34*(1), 1–16. https://doi.org/10.1080/10926488.2019.1591720

Kelly, B., Wigglesworth, G., Nordlinger, R., & Blythe, J. (2014). The acquisition of polysynthetic languages. *Language and Linguistics Compass, 8*(2), 51–64. https://doi.org/10.1111/lnc3.12062

Lakatos, I. (1978). *The methodology of scientific research programmes. Philosophical papers, Volume 1*. Press.

Latour, B. (1993). *We have never been modern*. Harvard University Press.

Latour, B. (2004). *The politics of nature*. Harvard University Press.

Lerner, R. M. (2006). Developmental science, developmental systems, and contemporary theories of human development. In R. M. Lerner & W. Damon (Eds.), *Handbook of child psychology: Theoretical models of human development* (pp. 1–17). John Wiley & Sons. https://doi.org/10.1002/9780470147658.chpsy0101

Lerner, R. M. (2011). Structure and process in relational, developmental systems theories: A commentary on contemporary changes in the understanding of developmental change across the life span. *Human Development, 54*(1), 34–43. https://doi.org/10.1159/000324866

Lerner, R. M., & Overton, W. F. (2008). Exemplifying the integrations of the relational developmental system: Synthesizing theory, research, and application to promote positive development and social justice. *Journal of Adolescent Research, 23*(3), 245–255. https://doi.org/10.1177/0743558408314385

Lewis, M. (2010). The emergence of consciousness and its role in human development. In W. F. Overton & R. M. Lerner (Eds.), *Handbook of life-span development: Cognition, biology, and methods across the lifespan* (Vol. 1, pp. 628–670). Wiley. https://doi.org/10.1196/annals.1279.007

López, A., Atran, S., Coley, J. D., Medin, D. L., & Smith, E. E. (1997). The tree of life: Universal and cultural features of folkbiological taxonomies and inductions. *Cognitive Psychology, 32*(3), 251–295. https://doi.org/10.1006/cogp.1997.0651

López Ornat, S. (1999). La adquisición del lenguaje: nuevas perspectivas. In F. Cuetos Vega & M. de Vega Rodríguez (Eds.), *Psicolingüística del Español* (pp. 469–534). Trotta.

MacWhinney, B. (2004). A multiple process solution to the logical problem of language acquisition. *Journal of Child Language, 31*(4), 883–914. https://doi.org/10.1017/S0305000904006336

Margolis, E., & Laurence, S. (2015). *The conceptual mind. New directions in the study of concepts*. MIT Press.

Medin, D. L., Lynch, E. B., Coley, J. D., & Atran, S. (1997). Categorization and reasoning among tree experts: Do all roads lead to Rome? *Cognitive Psychology, 32*(1), 49–96. https://doi.org/10.1006/cogp.1997.0645

Medin, D. L., Ross, N., Atran, S., Burnett, R., & Blok, S. (2002). Categorization and reasoning in relation to culture and expertise. *Psychology of Learning and Motivation, 41*, 1–41. https://doi.org/10.1016/S0079-7421(02)80003-2

Medin, D. L., Ross, N. O., Atran, S., Cox, D., Coley, J., Proffitt, J. B., & Blok, S. (2006). Folkbiology of freshwater fish. *Cognition, 99*(3), 237–273. https://doi.org/10.1016/j.cognition.2003.12.005

Medin, D. L., Ojalehto, B., Marin, A., & Bang, M. (2013). Culture and epistemologies: Putting culture back into the ecosystem. In Y. Hong, M. J. Gelfand, & C. Chiu (Eds.), *Advances in culture and psychology* (Vol. 4, pp. 177–217). Oxford University Press. https://doi.org/10.1093/acprof:osobl/9780199336715.003.0004

Medin, D. L., Ojalehto, B., Waxman, S. R., & Bang, M. (2015). Relations: Language, epistemologies, categories, and concepts. In E. Margolis & S. Laurence (Eds.), *The conceptual mind. New directions in the study of concepts* (pp. 349–378). MIT Press.

Meins, E., Fernyhough, C., Fradley, E., & Tuckey, M. (2001). Rethinking maternal sensitivity: Mothers' comments on infants' mental processes predict security of attachment at 12 months. *The Journal of Child Psychology and Psychiatry and Allied Disciplines, 42*(5), 637–648. https://doi.org/10.1111/1469-7610.00759

Michaels, C. F. (2003). Affordances: Four points of debate. *Ecological Psychology, 15*(2), 135–148. https://doi.org/10.1207/S15326969ECO1502_3

Mistry, J., Contreras, M., & Dutta, R. (2013). Culture and child development. In R. M. Lerner, M. A. Easterbrooks, J. Mistry, & I. B. Weiner (Eds.), *Handbook of psychology: Developmental psychology* (pp. 265–285). John Wiley & Sons, Inc. https://doi.org/10.1002/9781118133880.HOP206011

Montani, R. (2018). Imágenes indígenas del bosque chaqueño: animales y plantas en el universo visual wichí. *Caravelle. Cahiers du monde hispanique et luso-brésilien, 110*, 65–86. https://doi.org/10.4000/caravelle.2897

Murphy, G. L. (2002). *The big book of concepts*. MIT Press.

Nelson, K. (1996). *Language in cognitive development: Emergence of the mediated mind*. Cambridge University Press. https://doi.org/10.1017/CBO9781139174619

Nercesian, V. (2014). *Wichi lhomtes. Estudio de la gramática y la interacción fonología-morfología-sintaxis-semántica*. LINCOM.

Ojalehto, B. L., & Medin, D. L. (2015). Perspectives on culture and concepts. *Annual Review of Psychology, 66*, 249–275. https://doi.org/10.1146/annurev-psych-010814-015120

Ojalehto, B. L., Medin, D. L., & García, S. G. (2017). Conceptualizing agency: Folkpsychological and folkcommunicative perspectives on plants. *Cognition, 162*, 103–123. https://doi.org/10.1016/j.cognition.2017.01.023

Osherson, D. N., Smith, E. E., Wilkie, O., López, A., & Shafir, E. (1990). Category-based induction. *Psychological Review, 97*(2), 185–200. https://doi.org/10.1037/0033-295X.97.2.185

Overton, W. F. (2006). Developmental psychology: Philosophy, concepts, methodology. In R. M. Lerner & W. Damon (Eds.), *Theoretical models of human development. Handbook of child psychology* (Vol. 1, 6th ed., pp. 18–88). Wiley.

Overton, W. F. (2010). Life-span development: Concepts and issues. In W. F. Overton & R. M. Lerner (Eds.), *Cognition, biology, and methods across the lifespan. The handbook of life-span development* (Vol. 1, 6th ed., pp. 1–29). Wiley. https://doi.org/10.1002/9780470880166.hlsd001001

Overton, W. F. (2012). Evolving scientific paradigms: Retrospective and prospective. In L. L'Abate (Ed.), *Paradigms in theory construction* (pp. 31–65). Springer. https://doi.org/10.1007/978-1-4614-0914-4_3

Overton, W. F. (2013a). A new paradigm for developmental science: Relationism and relational-developmental-systems. *Applied Developmental Science, 17*(2), 94–107. https://doi.org/10.1080/10888691.2013.778717

Overton, W. F. (2013b). Chapter two - Relationism and relational-developmental systems: A paradigm for developmental science in the post-Cartesian era. *Advances in Child Development and Behavior, 44*, 21–64. https://doi.org/10.1016/b978-0-12-397947-6.00002-7

Overton, W. F., & Lerner, R. M. (2012). Relational developmental systems: A paradigm for developmental science in the postgenomic era. *Behavioral and Brain Sciences, 35*(5), 375–376. https://doi.org/10.1017/S0140525X12001082

Oyama, S., Griffiths, P. E., & Gray, R. D. (2001). *Cycles of contingency: Developmental systems and evolution*. MIT Press.

Palmer, J. (2005). *La buena voluntad wichí. Una espiritualidad indígena*. APCD.

Pérez, D., & Español, S. (2014). Intersubjetividad y atribución psicológica. In P. Quintanilla, C. Mantilla, & P. Cépeda (Eds.), *Cognición social y lenguaje. La intersubjetividad en la evolución de la especie y en el desarrollo del niño* (pp. 371–392). Fondo Editorial de la Pontificia Universidad Católica del Perú.

Pérez, A., Pérez, E. M., Taverna, A., & Baiocchi, M. C. (2017a). *Hal'o*. EDUNaF.

Pérez, A., Pérez, E. M., Taverna, A., & Baiocchi, M. C. (2017b). *Laloy*. EDUNaF.

Pérez, A., Pérez, E. M., Taverna, A., & Baiocchi, M. C. (2017c). *Tshotoy*. EDUNaF.
Pérez, A., Pérez, E. M., Taverna, A., & Baiocchi, M. C. (2017d). *Tshotoy fwiy'ohen*. EDUNaF.
Pérez, A., Pérez, E. M., Taverna, A., & Baiocchi, M. C. (2017e). *Tshotoy inot lheley*. EDUNaF.
Pérez, A., Pérez, E. M., Taverna, A., & Baiocchi, M. C. (2021). *Hunhat lheley-Habitantes de la tierra*. EDUVIM-EDUNaF.
Peters, A. M. (1981). Language typology and the segmentation problem in early child language acquisition. *Proceedings of the seventh annual meeting of the Berkeley linguistics society* (pp. 236–248). https://doi.org/10.3765/bls.v7i0.2078
Pierotti, R. (2011). *Indigenous knowledge, ecology and evolutionary biology*. Routledge Press.
Ramstead, M. J., Veissière, S. P., & Kirmayer, L. J. (2016). Cultural affordances: Scaffolding local worlds through shared intentionality and regimes of attention. *Frontiers in Psychology, 7*, 1090. https://doi.org/10.3389/fpsyg.2016.01090
Read, C., & Szokolszky, A. (2018). Developmental ecological psychology: Changes in organism-environment systems over time, part II. *Ecological Psychology, 30*(2), 125–128. https://doi.org/10.1080/10407413.2018.1439084
Rietveld, E., & Kiverstein, J. (2014). A rich landscape of affordances. *Ecological Psychology, 26*(4), 325–352. https://doi.org/10.1080/10407413.2014.958035
Rogoff, B. (1992). Three ways to relate person and culture: Thoughts sparked by Valsiner's review of apprenticeship in thinking. *Human Development, 35*(5), 316–320. https://doi.org/10.1159/000277225
Rogoff, B. (2014). Learning by observing and pitching in to family and community endeavors: An orientation. *Human Development, 57*(2–3), 69–81. https://doi.org/10.1159/000356757
Rosch, E., & Mervis, C. B. (1975). Family resemblances: Studies in the internal structure of categories. *Cognitive Psychology, 7*(4), 573–605. https://doi.org/10.1016/0010-0285(75)90024-9
Schieffelin, B. B., & Ochs, E. (1986). Language socialization. *Annual Review of Anthropology, 15*(1), 163–191. https://doi.org/10.1146/annurev.an.15.100186.001115
Smith, E. E., & Medin, D. L. (1981). *Categories and concepts*. Harvard University Press.
Solomon, K. O., Medin, D. L., & Lynch, E. (1999). Concepts do more than categorize. *Trends in Cognitive Sciences, 3*(3), 99–105. https://doi.org/10.1016/s1364-6613(99)01288-7
Strauss, A., & Corbin, J. (1990). *Basics of qualitative research*. Sage Publications.
Suárez, M. E. (2014). *Etnobotánica wichí del bosque xerófito en el Chaco semiárido salteño*. Autores de Argentina.
Suárez, M. E. (2020). Morfología botánica wichí: un estudio etnobiológico. *Revista del Museo de Antropología, 13*(3), 443–443. https://doi.org/10.31048/1852.4826.v13.n3.27844
Suárez, M. E., & Montani, R. M. (2010). Vernacular knowledge of Bromeliaceae species among the Wichí people of the Gran Chaco, Argentina. *Journal of Ethnobiology, 30*(2), 265–289. https://doi.org/10.2993/0278-0771-30.2.265
Szokolszky, A., & Read, C. (2018). Developmental ecological psychology and a coalition of ecological–relational developmental approaches. *Ecological Psychology, 30*(1), 6–38. https://doi.org/10.1080/10407413.2018.1410409
Taverna, A. S. (2021). Motherese in the Wichi language (El maternés en la lengua wichí). *Journal for the Study of Education and Development/Infancia y Aprendizaje, 44*(2), 303–335. https://doi.org/10.1080/02103702.2021.1889290
Taverna, A. S., Waxman, S. R., Medin, D. L., & Peralta, O. A. (2012). Core-folkbiological concepts: New evidence from Wichí children and adults. *Journal of Cognition and Culture, 12*, 339–358. https://doi.org/10.1163/15685373-12342079
Taverna, A. S., Waxman, S. R., Medin, D. L., Moscoloni, N., & Peralta, O. A. (2014). Naming the living things: Linguistic, experiential and cultural factors in Wichí and Spanish speaking children. *Journal of Cognition and Culture, 14*, 213–233. https://doi.org/10.1163/15685373-12342122
Taverna, A. S., Medin, D. L., & Waxman, S. R. (2016). "Inhabitants of the earth": Reasoning about folkbiological concepts in Wichi children and adults. *Journal of Early Education and Development, 27*(8), 1109–1129. https://doi.org/10.1080/10409289.2016.1168228

Taverna, A. S., Medin, D. L., & Waxman, S. R. (2018). "Inhabitants of the earth": Reasoning about folkbiological concepts in Wichí children and adults. In P. J. Marshall & K. Brenneman (Eds.), *Young children's developing understanding of the biological world* (pp. 7–27). Routledge. https://doi.org/10.1080/10409289.2016.1168228

Taverna, A. S., Medin, D. L., & Waxman, S. R. (2020). Tracing culture in children's thinking: a socioecological framework in understanding nature (Rastreando la cultura en el pensamiento infantil: una socioecología para comprender la naturaleza). *Journal for the Study of Education and Development/Infancia y Aprendizaje, 43*(2), 247–270. https://doi.org/10.1080/02103702.2020.1723277

Taverna, A. S., & Padilla, M. I. (2020). Adquisición del wichí: hacia una metodología para el estudio del lenguaje infantil en una lengua nativa argentina. In C. P. Tramallino (Ed.), *Homenaje a Zulema Solana. Estudios sobre lingüística computacional, adquisición y enseñanza de lenguas* (pp. 94–191). Universidad Nacional de Rosario.

Taverna, A., & Waxman, S. (2020). Early lexical acquisition in the Wichi language. *Journal of Child Language, 47*(5), 1052–1072. https://doi.org/10.1017/S0305000919000898

Taverna, A. S., & Padilla M. I. (in preparation). The cultural organization of attention: evidence from Wichí and Spanish-speaking mother-infant interactions.

Thelen, E., & Smith, L. B. (2006). Dynamic systems theories. In R. M. Lerner & W. Damon (Eds.), *Handbook of child psychology: Theoretical models of human development* (pp. 258–312). John Wiley & Sons. https://doi.org/10.1002/9780470147658.chpsy0106

Tovar, A. (1964). El grupo mataco y su relación con otras lenguas de América del Sur. *Actas del 35° Congreso Internacional de Americanistas, II*, 439–452.

Valsiner, J. (1998). *The guided mind: A sociogenetic approach to personality*. Harvard University Press.

Vidal, A., & Kuchenbrandt, I. (2015). Challenges of linguistic diversity in Formosa. In C. Stolz (Ed.), *Language empires in comparative perspective (pp. 89–112).* De Gruyter. https://doi.org/10.1515/9783110408362.89

Waldmann, M. R. (2017). Causal reasoning: An introduction. In M. R. Waldmann (Ed.), *The Oxford handbook of causal reasoning*. Oxford University Press. https://doi.org/10.1093/oxfordhb/9780199399550.013.1

Withagen, R., & Chemero, A. (2012). Affordances and classification: On the significance of a sidebar in James Gibson's last book. *Philosophical Psychology, 25*(4), 521–537. https://doi.org/10.1080/09515089.2011.579424

The Role of Local Violence on Children's Affective Decision-Making

Hernán Delgado and Alejandra Carboni

Introduction

Self-regulation is defined, in a basic sense, as a set of abilities to control attention, emotion, and behavior in pursuit of long-term goals (Posner & Rothbart, 2000). This concept is closely linked to executive functions (EFs) and comprises cognitive, behavioral, and affective skills that are important to manage sustained actions toward environment adaptation (Inzlicht et al., 2021; McClelland et al., 2015). Self-regulation develops through early experiences and social interactions in which caregivers, educators, and other significant individuals structure and shape children's trajectories (Grolnick & Farkas, 2002). Both the cognitive and emotional aspects related to self-regulation are good predictors of academic competence, especially in the transition to formal education, because an adaptive development in the school context depends on children's ability to manage their reactions and their task-related behaviors (Trentacosta & Izard, 2007; Blair & Razza, 2007). Remembering and using information, attending to and understanding what others are saying, directing motor actions, and persisting toward goals are all indicators of adaptive behavioral regulation (McClelland et al., 2007).

Self-Regulation and Executive Functions and Decision-Making

Traditionally, research on self-regulation and EFs have followed separate paths (Baddeley, 2007). However, in the last decade, several studies attempted to link both areas, proposing that "hot" EFs integrate cognition and emotion (Zelazo et al., 2010;

H. Delgado · A. Carboni (✉)
Centro de Investigación Básica en Psicología, Universidad de la República, Montevideo, Uruguay

M. V. Alves et al. (eds.), *Cognitive Sciences and Education in Non-WEIRD Populations*, https://doi.org/10.1007/978-3-031-06908-6_7

Blair & Razza, 2007; Bull & Scerif, 2001). The authors have also argued for the importance of bottom-up regulatory modulation in the regulation of top/down EF processes (Cunningham & Zelazo, 2007; Blair, 2016; Lewis & Todd, 2007). Specifically, some authors focus on decision-making about situations that have emotionally significant consequences (i.e., meaningful rewards and/or losses), or what might be called "affective decision-making" (Kerr & Zelazo, 2004; Garon & Moore, 2007).

Decision-making is a complex skill that depends on reasoning capacity, but also on emotional control (Byrnes, 2002). Some authors posit that it is helpful to think of decision-making as a least four-step process: first, it is needed to set a goal, second to consider the possible options to meet that goal, third to rank the options, and finally to select the better one (Byrnes, 2002; Hsu et al., 2012). However, in daily situations, decision-making is mostly more dynamic and recursive than step-like (Byrnes, 2013). This adaptive ability is not solely dependent on logical reasoning but on emotional skills, and it is also associated with the developmental context (Mata et al., 2013). Children at an early age must face decisions that involve the ability to evaluate the future consequences of their actions (Garon & Moore, 2007), and the context in which they make decisions is relevant.

One of the most widely used paradigms to assess the development of decision-making is the Children Gambling Task (CGT) designed by Kerr and Zelazo (2004). In this task, children must learn the immediate and long-term payoffs of their choices by exploring different options (Bechara et al., 2000), for instance, when a child is asked to decide between a long-term "better" option and an immediate but risky reward, is the same in terms of adaptation, if the decision is taken by a child who is developing in a safe, predictable, and resource-rich home environment or by a child who is living in a harsh, unpredictable, and resource-scarce context?

Classical Developmental Perspective vs Evolutionary Perspective

The classical developmental psychopathology perspective (see Ellis et al., 2012) posits that supportive environments tend to favor "optimal" or "adaptive" outcomes across the lifespan, whereas stressful environments tend to undermine the probability of healthy growth, fostering "dysfunctional" or "maladaptive" developmental outcomes. The prevailing conceptual framework from which these tags emerge has been structured based on the evidence accumulated in studies nurtured by middle-class, European American samples (Lerner, 2006). Along these lines, the adaptive nature of developmental outcomes has tended to be assessed taking as a point of comparison the constellation of results that are socially desirable according to the dominant beliefs and values. Thus, the term "adaptive" alludes to traits or behaviors that promote health and well-being. In contrast, those traits considered harmful to health are referred to as "maladaptive".

From an evolutionary life history perspective, organisms must face trade-offs in the allocation of time and energy across a set of fitness-relevant activities, including growth, maintenance, and reproduction (Ellis et al., 2009). At the broadest level of analysis, these resource-allocation decisions that are made throughout ontogenetic development are the ones that shape the individual's life history strategy (i.e., a group of correlated life history traits), which are proposed to exist along a fast-slow continuum. Typically, and in a narrow sense, individuals on the fast end exhibit short early reproduction, small body size, and high mortality rates, whereas those who adopt slow strategies have the opposite characteristics (Del Giudice et al., 2011).

For instance, according to the traditional deficit-based model, risky and impulsive decision-making is considered a deviant behavior. On the other hand, despite the long-term consequences of growing up in resource-poor or stressful environments, the evolutionary perspective considers that risky behaviors could be adaptive under certain circumstances (Fenneman & Frankenhuis, 2020; DelGiudice, 2009; Ellis et al., 2009). In other words, the "get what you can while you can" attitude could reflect an adaptive strategy in response to difficult or unpredictable early social environments. This is reasonable considering that, compared with people living in resource-plentiful environments, people living in resource-scarce ones more often face threats and challenges to their well-being and have few opportunities to overcome poverty. These opportunities, moreover, are unstable over time, so that what is presented today as a benefit may not be present in the future.

Adaptation strategies develop early in life when children are particularly open to incorporating information from their local environments into their biology. It has been proposed that individual differences in life history strategies develop in response to two basic dimensions of environmental stress: unpredictability and harshness (Ellis et al., 2009). Unpredictability is the difficulty to foretell fluctuations in the harshness of environmental conditions across time (Belsky et al., 2012). On the other hand, the harshness of the context refers to morbidity and mortality caused by external factors that escape individual control (for instance, the possibility of death from a stray bullet in the context of a dispute between neighborhood gangs).

Regarding environmental harshness, the presence of extrinsic morbidity-mortality cues tends to be linearly related to lower levels of socioeconomic status (SES) in Western societies (McLoyd, 1998). It has been proposed that exposure to violence constitutes an important cue of extrinsic morbidity-mortality. For instance, experiencing violent acts, such as being robbed, or witnessing a murder case, are relevant cues of the level of violence in the environment. Furthermore, direct exposure to violence would not be essential for it to act as a stressor. In violent environments, the daily circulation of information about violent events is more common (Chen et al., 2002).

The study of the potential role of extrinsic morbidity-mortality cues in the early modulating of life history strategies is of particular interest for Latin America and the Caribbean (LAC) countries, given that it is the most violent region in the world. With a homicide rate of 24 per 100,000 inhabitants in 2015, there is no other region that exhibits a higher homicide rate than LAC. Homicides in LAC represent 33% of homicides in the world, even though this region is inhabited by 9% of the world's population (Jaitman, 2017). The annual homicide rate is around four times higher in LAC than the world average. The outcome of the comparison is more dramatic if it is done with Western, Educated, Industrialized, Rich, and Democratic (WEIRD) countries. For instance, the murder rate in LAC is six times greater than that of the United States, and 20 times higher than in the United Kingdom. Even Uruguay, which has historically been considered, along with Chile and Argentina, a country with relatively low levels of violence (World Health Organization, 2002), has experienced an alarming rise in the murder rate in the last 10 years. To put it in perspective, homicide rates per 100,000 inhabitants in WEIRD societies like Western European countries (e.g., Portugal, Spain, or Germany) are currently more than 10 times lower than in Uruguay. Thus, the high number of homicide crimes—relatively well documented by LAC governments—constitutes a proxy of the violence to which people in this region are exposed and has the potential to be an important factor in the modulation of the early development of cognition.

As mentioned earlier, it is well documented that people living in poverty face a higher risk of violent crime (Imbusch et al., 2011). From an evolutionary-developmental perspective, it has been suggested that this ecological circumstance would be associated with the development of "fast" strategies characterized by impulsivity, short-term opportunism, and immediate reward-oriented strategies (Del Giudice, 2014; Fawcett et al., 2012; Griskevicius et al., 2013; Frankenhuis et al., 2016; Fenneman & Frankenhuis, 2020).

To the best of our knowledge, only two studies have addressed the relation between SES and affective decision-making using the CGT: Mata et al. (2013) and Delgado et al. (2022). Both studies found an effect of SES in the last blocks of the task. More specifically, the high-SES group made more advantageous choices than their low-SES peers in the final blocks of the task. Added to this, in the last decade, several studies have evaluated the association between mortality cues in the environment and different psychological developmental outcomes (Griskevicius et al., 2011a, b), but none that explores the decision-making development and environmental harshness association in the light of life history theory.

The present work explores the association between self-regulation development (by using the CGT), SES, and environmental harshness in the light of life history theory. Taking into account that evolutionary theory has emphasized the importance of the first 5–7 years of life as a sensitive window in which ecological circumstances modulate development (Belsky et al., 1991), we assessed children between the ages of 5 and 7 years and consider as a measure of environmental harshness the cumulative homicide rate in the 3 years before the completion of the cognitive assessment. It has been suggested that the Iowa Gambling Task can be subdivided into two stages that assess two types of decision-making. While the first half of the task corresponds to ambiguous choices, where the risk associated with the different decks is still unknown, the second half of the task corresponds to risky choices, since the probability of outcome has been learned (Brand et al., 2007). Furthermore, performance in this second stage of the task would be mostly linked to top-down control mechanisms demanded in affectively charged situations (i.e., hot executive functions) (Brand et al., 2006). According to life history theory, childhood harshness promotes the development of present-oriented and risk-taking behaviors. Therefore, we hypothesized that the performance in the last trials of the CGT would be negatively correlated with local mortality levels.

Methods

Participants

A total of 224 children (109 girls; overall mean age = 75.32 months, SD = 6.96) attending kindergarten (level 5) and first grade at public schools in Montevideo-Uruguay participated in the study. Informed consent was obtained from parents/caregivers, and the study was approved by the Research Ethics Committee of Psychology School (Udelar). All methods were performed in accordance with the Declaration of Helsinki (General Assembly of the World Medical Association, 2014).

Procedure

All children were tested individually in a quiet room specially arranged for assessment at their schools. CGT application lasted between 20 and 35 minutes. Parents or caregivers were required to answer the Socioeconomic Level Index (SLI) questionnaire through telephone interviews.

Instruments

Socioeconomic Status

SES was determined using the SLI, developed by the Uruguayan Economic Research Center (Perera & Cazulo, 2016). The SLI consists of 12 items: place of residence, familiar composition, type of health coverage (public or private) of the family members, number of income earners in the households and educational level of the household head, presence of elements of comfort, and frequency of maid help in the home. Based on this scale, an index is constructed that allows the ordering of households in three economic classes: high (51–100), middle (32–50), and low (0–31).

Neighborhood Homicide Rates

The rate of neighborhood violent crime was calculated with data from the Department of Justice. The National Observatory of Violence and Crime has an open-access statistical memory that contains the total annual number of reported homicides distributed according to the limits of the jurisdiction of each Montevideo police station (Ministry of Interior, 2019). Specifically, we got the total number of homicides in the last six semesters before the date of the cognitive assessment. Each child's neighborhood was assigned the number of cases calculated for their local police station. For cases in which the neighborhood was located in the jurisdiction of more than one police station, an average was made.

Rapines were not considered in the analysis because it is a type of crime that presents a high rate of underreporting, especially in neighborhoods from low-SES backgrounds.

Children's Gambling Task

A computerized two-deck version of CGT was adapted from Kerr and Zelazo's version (2004). The task was applied using a computer (monitor and keyboard). The CGT included two decks of 60 cards each (Fig. 1). The back of one deck was covered with black and white vertical stripes, while the back of the other deck was covered with horizontal ones. In the front of the cards, gains were reported through the use of happy yellow faces, while angry red faces represented losses. Depending on the earnings/losses ratio, one deck was advantageous over trials while the other was disadvantageous (the pattern of lines on the back of the cards and the position of the decks on the screen were randomized). The number of earnings remained

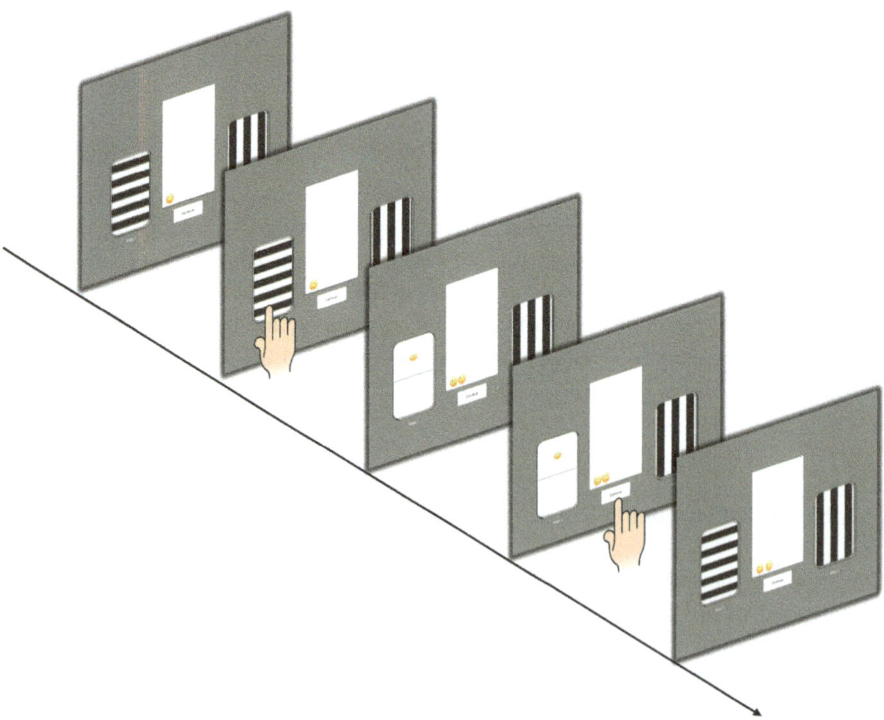

Fig. 1 Sequence of screens of the Children's Gambling Task

constant across cards in both decks, but the number of losses was variable between cards and decks. The advantageous deck always provided gains of one point, accompanied by either no loss or the loss of one point. The disadvantageous deck always provided an earning of two points together with losses of 0, 4, 5, or 6 points. Therefore, consistently selecting from the advantageous deck yielded a net gain in the long run, whereas selecting from the disadvantageous deck yielded a net loss. The order of the cards in each deck was fixed, as described in Kerr and Zelazo (2004).

The CGT version used in this study involved the child trying to win as many points as possible (represented by happy faces). The child was told that if he or she accumulated enough points (an amount never specified), the points could be exchanged for a prize from a table containing a variety of prizes varying in size and desirability (e.g., a bubble blower, a magnetic drawing board or a fidget spinner). However, if they got a small number of points, they could only exchange the points for an unattractive prize (e.g., a pencil or an eraser). This method was developed based on previous research (Wilson et al., 2009) to ensure that children could identify at least one award that was attractive enough to motivate them to earn the most points in the task. Thus, before starting the task, children were led to the prize table and encouraged to select the prize they were going to try to achieve in the game.

The task was administered according to the instructions outlined in Kerr and Zelazo (2004) and Mata (2013). Thus, once the desired prize was selected by the child, the task began with six demonstration trials. This stage implied the experimenter selecting three cards consecutively from one of the decks, followed by three consecutive cards from the other one. Each time a card was selected, the experimenter announced the number of points earned and then deposited the points in the plastic container. Besides, the points were added to the cylinder presented between the decks on the computer screen. Similarly, when points were lost, they were removed from the plastic container and returned to the experimenter box (a process also observed in the virtual container). After the demonstration trials, children were given 15 happy faces to start with. The test trials then started and selections were made across 60 test trials, which were divided into five blocks of twelve card choices. The test trials were administered exactly like the demonstration trials. A bonus stage was programmed at the end of the task for those children who did not get too many points. This stage ensured that all the children achieved enough points to select the prize of their choice.

Statistical Analyses

All analyses were performed on the R platform (Team, 2013). We conducted two-level multilevel models using the lme function in package "nlme" (Pinheiro et al., 2012). Based on previous research (Preston et al., 2007; Brand et al., 2006), we divided the task into the learning phase (Blocks 1, 2, 3—i.e., Trials 1–36) and the performance phase (Blocks 4 and 5—i.e., Trials 37–60) to discriminate the influence of local mortality on learning the gain/loss schedule versus on risk-taking behavior after the frequencies and magnitudes of wins and losses have been learned. Thus, Phase was included as the independent variable at the first level of the model (a categorical variable with 2 levels (Learning; Performance). The independent variables at the second level of the model consisted of Homicides and SES (low-SES vs. middle/high-SES). In order to control for the potential effect of age and sex on task performance, both variables were entered as fixed effects into models, but were only kept if they significantly improved the model fit. The dependent variable entailed proportion scores across the two phases. The proportion score of each phase was calculated by subtracting the proportion of disadvantageous choices from the proportion of advantageous choices, which yielded difference scores ranging from −1 to 1. A net score above zero implied that children were selecting cards advantageously, and a net score below zero implied a disadvantageous selection. Regression parameters were estimated using maximum likelihood. A random intercept (for Participant) and random slope (for Phase) were also included in the model; that is, children were allowed to have different intercepts and rates of learning (slopes) through the two phases.

Results

Extrinsic Mortality and Performance on the Children's Gambling Task

The multilevel analysis consisted of multiple steps of model comparisons. At each step, the different models were compared by means of −2log likelihood tests to aid decisions about including specific terms. We also compared the fit of the model using AIC and BIC indices. Results showed a significant main effect of Phase since a significant reduction of the deviance was observed when comparing the unconditional growth model (model containing repeated measures) (Model B) with the unconditional means model (Model A), $\chi^2(1) = 15.04$, $p < 0.001$. This suggests that in the overall sample there was a significant increase in the proportion of advantageous choices as the experimental task progressed. Adding age to the model (Model C) revealed that being older was associated with better performance, $\chi^2(1) = 12.41$, $p < 0.001$. No significant improvement in the model was detected for the inclusion of Sex (Model D), $\chi^2(1) = 0.12$, $p = 0.72$; therefore, it was omitted from the model. When the random slope was considered (Model E), it resulted in a better fitting model, $\chi^2(2) = 53.39$, $p < 0.0001$, suggesting that the effect of Phase significantly varied across Participants. Next, when the variable Homicides was added to extend the model (Model F), the likelihood ratio test did not show a significant improvement of the model $\chi^2(1) = 2.45$, $p = 0.12$. However, adding the interaction between Homicides and Phase (Model G) resulted in measurable improvements in the goodness of fit, $\chi^2(1) = 5.89$, $p < 0.05$, indicating that Homicides is associated with performance as the task unfolded (Table 1 provides a summary and ANOVA results of

Table 1 Model summary and ANOVA results for Model G, Model I, and Model J

Model	Model G			Model I			Model J		
Fixed effects	Phase + Homicides + Age + Phase × Homicides			Phase + SES + Age + Phase × SES			Phase + SES + Homicides + Age + Phase × SES + Phase × Homicides + SES × Homicides + Phase × SES × Homicides		
Random effects[a]	Phase/Participant			Phase/Participant			Phase/Participant		
	df	*F*	*p*	df	*F*	*p*	df	*F*	*p*
Main effects									
Phase	894	12.65	0.00***	894	12.58	0.00***	892	12.72	0.00***
SES	–	–	–	221	9.20	0.00**	219	6.24	0.01*
Homicides	221	3.52	0.06	–	–	–	219	3.73	0.05
Age	221	12.64	0.00**	221	12.00	0.00***	219	11.61	0.00***
Interaction effects									
Phase × Homicides	894	5.88	0.02*	–	–	–	892	5.91	0.01*
Phase × SES	–	–	–	894	3.82	0.05	892	1.24	0.26
SES × Homicides	–	–	–	–	–	–	219	0.09	0.76
Phase × SES × Homicides	–	–	–	–	–	–	892	1.70	0.19

Note. The organization of parameters in the models (as either fixed or random effects) is indicated in the top rows, and the effects (either alone or in an interaction) of each parameter with a fixed effect are evaluated in the bottom rows. Significant effects are shown in bold

SES socioeconomic status

$^{*}p < 0.05$; $^{**}p < 0.01$; $^{***}p < 0.001$

[a]The specification Phase/Participant means that the effect of Phase (i.e., the slope) varies across children

model G). We then assessed the statistical significance of simple slopes of Homicides by each level of the categorical moderator Phase. This analysis revealed that the average slope estimated for the performance phase ($b = -0.0044$, CI = −0.0075, −0.0014) was negative and significantly different from zero. This means that for each increase in the number of homicides the estimated proportion score decreases by 0.0044 for the performance phase. A contrast was conducted to test the difference of the Homicides slope between the two phases of the task. This analysis showed that the slope of the performance phase was significantly different and more negative than the slope of the learning phase ($b = 0.0030$, $p < 0.05$).

SES and Performance on the Children's Gambling Task

Since SES is also considered a measure of environmental harshness, to determine whether it is related to performance, we first added SES (as a variable with two levels) (Model H) to Model E (the model prior to the inclusion of Homicides).

Adding these terms significantly improved the fit of the model, $\chi^2(1) = 7.00$, $p < 0.01$, suggesting that SES accounts for a fraction of the variance in the score that the children obtained on the whole task. Then, with the aim of assessing whether the effect of SES differs between the two phases of the task, we added the SES × Phase interaction (Model I). The comparison between models I and H barely missed reaching the significance level $\chi^2(1) = 3.83$, $p = 0.051$ (see Table 1).

Extrinsic Mortality, SES, and Performance on the Children's Gambling Task

In the first place, we examined zero-order correlation between homicides and SES scores. As expected, the number of homicides and the SES score were correlated ($r = -0.45$, $p < 0.0001$), indicating that the number of homicides increases while SES scores decrease in our sample.

We then built a final model including the interaction term of Homicides × Phase × SES (Model J) to determine whether the effect of Homicides on task performance differed across phases between the SES groups. The summary and ANOVA results for Model J and previous ones can be seen in Table 1. We did not observe a significant three-way interaction $F(1,892) = 0.19$. This indicates that the Phase × Homicides interaction described previously did not differ across SES groups. A key result of our study is that the Phase × Homicides effect seen in model G persisted in model J despite the inclusion of SES. The effect of Homicides in the whole task did not differ between SES groups. The test of simple slopes of Homicides indicated that among the middle/high-SES group, only the simple slope of the performance phase differed from zero ($b = -0.0061$, CI = −0.0118, −0.0004), while among the low-SES group, no differences were found across phases (Fig. 2). This indicates that the influence of environmental harshness on affective decision-making was specific to children from middle/high-SES background. Specifically, for the middle/high-SES group, as the cases of homicides increase (per unit), the proportion score decreases by 0.0061 in the performance phase (Fig. 2).

Discussion

Although the evidence from the traditional models of early life socioeconomic adversity and cognitive development suggests that individuals from low-SES backgrounds tend to show a propensity to choose smaller, immediate rewards over larger, delayed rewards, compared with people from high-SES backgrounds, studies that have assessed decision-making in the context of the life history theory are still scarce. The life history theory proposes that life strategies are arrayed on a fast-slow continuum. While the cluster of fast strategies is proposed to promote the focus on immediate or short-term goals, the slow strategies are more associated with

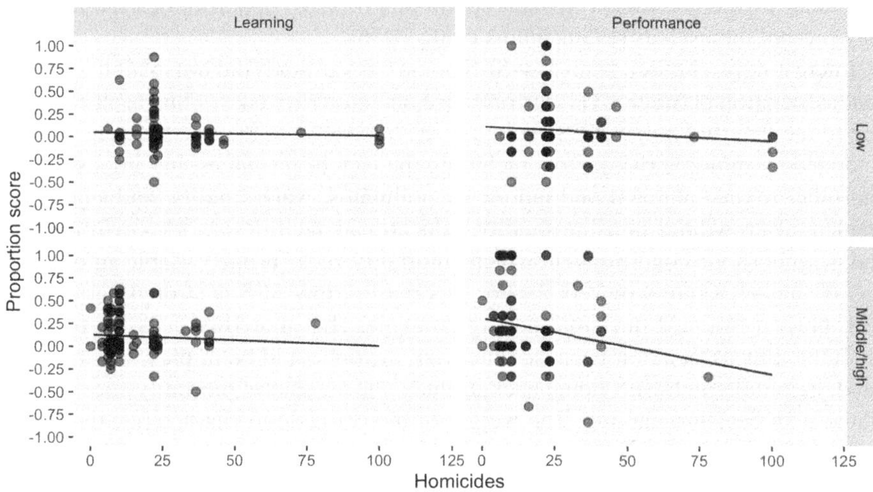

Fig. 2 Plot of slopes of the influence of phase and socioeconomic status (SES) on associations between the rate of violent crime and affective decision-making

future-oriented behaviors and longer-term goals. In the present study, we hypothesized that the presence of high mortality cues in the neighborhood during the early stages of development would be associated with a present-oriented and risky behavior pattern of decision-making in children (a fast life history strategy). In support of our hypothesis, we found an association between the rate of neighborhood violent crime and affective decision-making assessed using the CGT, although different depending on children's family SES background. Specifically, while we found a significant negative association of violent crime with performance in the performance phase of the task for the middle/high-SES group, violent crime had no significant effect on CGT performance in the low-SES group. We also observed a main effect of SES on CGT performance. Specifically, we observed that middle/high-SES children performed higher in the task than their low-SES peers.

As expected, the negative correlation between the levels of extrinsic morbidity-mortality at the neighborhood level and performance on the CGT was observed in the performance phase of the task. Considering the fact that the latter trials of the gambling task are supposed to assess decisions under risk, this finding suggests that once children have progressively learned the outcome probabilities (Brand et al., 2007), higher violent crime rates are associated with a risk-taking propensity in the middle/high-SES group. This finding is consistent with the notion that high levels of environmental harshness promote the development of fast life history strategies. Such strategy entails the preference for risk and for smaller and immediate rewards over larger future rewards. Although the deficit perspective describes this preference as maladaptive or dysfunctional, this tendency would be adaptive in harsh environments (Fenneman & Frankenhuis, 2020; Ellis et al., 2009). Because rewards in harsh environments are unstable and uncertain over time, it can be adaptive and beneficial for people raised in such conditions to seize opportunities the moment they present themselves (Frankenhuis et al., 2016).

Interestingly, children from low-SES home backgrounds did not exhibit a significant association between CGT performance and mortality cues. This result could be due to the fact that in low-SES ecologies the harshness of the environment is characterized by multiple risk factors including the lack of resources, hazardous wastes and other toxins, ambient and indoor air pollutants, water quality, ambient noise, residential crowding, housing quality, educational facilities, work environments, and neighborhood conditions (in which "mortality cues" is one variable among others) (Evans & Kantrowitz, 2002). For this reason, low SES is strongly associated with present-oriented decision-making strategies, but mortality cues seem not to modulate this association because other risk factors could be in play. On the other hand, in middle/high-SES home backgrounds, the number of mortality cues is associated with decision-making. The weight of this variable to characterize the harshness of the environment appears to be more relevant in middle/high-SES contexts because the other factors are not always in play and this neighborhood characteristic seems to impact more strongly in children from middle/high-SES homes. Added to this, violent crime could be more unpredictable in middle/high SES ecologies, adding another dimension that might modulate cognitive development (Griskevicius et al., 2011a).

The findings of our study add to a growing body of evidence indicating that different forms of environmental harshness are associated with life history outcomes. Indeed, although little is known about the role of violent crime in affective decision-making in children, there is some evidence regarding other related risk-taking behaviors drawn on life history theory. For instance, our study extends previous evidence showing that higher levels of local mortality are associated with faster reproductive strategies, like reproducing at an early age (Griskevicius et al., 2011b; Wilson & Daly, 1997). Although it has not been interpreted from the evolutionary lens, previous evidence has also shown that children exposed to the acute effects of local homicides exhibited lower levels of attention and impulse control (Sharkey et al., 2012), and vocabulary and reading scores (Sharkey, 2010). Interestingly, Sharkey et al. (2012) suggested that parental distress could be one important mechanism mediating the effects of local violence on cognitive development. Furthermore, the effect of violent crime has also been assessed at the biological level. Theall et al. (2017) found that violent crime is associated with increased biological markers of stress (e.g., shorter telomere length). Despite the evidence from this work, identifying the effect of community violence on cognitive development is an arduous task. Associations between violent crime rates and children's outcomes could be related to other home environment variables that are also in play (i.e., parenting style, cognitive stimulation, home SES) and might influence children's likelihood of school success (Sharkey et al., 2012).

Finally, although our findings in non-WEIRD population are in line with those observed in studies composed of WEIRD samples, a critical aspect should be taken into account when assessing the role of local violence in child development in Latin American societies. The homicide rate is only a proxy for the total violence to which people are exposed and, as such, it does not allow us to get a completely clear picture of the frequency and heterogeneity of urban violent crimes. One of the most important problems in LAC, either for the design of public policies toward preventing violent crime or for the assessing of its social consequences, is the high level of crimes not officially recorded. Recent estimates suggest that the dark figure of crime (the gap between crimes reported in victimization surveys and crimes registered in administrative police records) in LAC non-WEIRD countries is approximately 40% higher than in developed WEIRD countries (Jaitman & Anauati, 2019). This point is important, as it suggests that the official data used to assess the impact of extrinsic morbidity-mortality cues on cognitive development in LAC societies may not necessarily capture the true reality that children face daily since official statistics tend to underestimate the number of violent crimes. That is why caution should be exercised when comparing the impact of urban violence on child development between WEIRD and non-WEIRD societies when using statistics from official reports.

Our results add to a still scarce set of reports that highlight the need to deepen our understanding of the impact that neighborhood violence can have on child development. We are still far from understanding how the development of cognition can be the result of the integration of environmental cues during early childhood. It is necessary to continue accumulating evidence in order to discover the evolutionary logics that underlie psychological development.

References

Baddeley, A. D. (2007). *Working memory, thought, and action*. Oxford University Press.

Bechara, A., Damasio, H., & Damasio, A. R. (2000). Emotion, decision making and the orbitofrontal cortex. *Cerebral Cortex, 10*(3), 295–307.

Belsky, J., Steinberg, L., & Draper, P. (1991). Childhood experience, interpersonal development, and reproductive strategy: An evolutionary theory of socialization. *Child Development, 62*(4), 647–670.

Belsky, J., Schlomer, G. L., & Ellis, B. J. (2012). Beyond cumulative risk: Distinguishing harshness and unpredictability as determinants of parenting and early life history strategy. *Developmental Psychology, 48*(3), 662.

Blair, C. (2016). Executive function and early childhood education. *Current Opinion in Behavioral Sciences, 10*, 102–107.

Blair, C., & Razza, R. P. (2007). Relating effortful control, executive function, and false belief understanding to emerging math and literacy ability in kindergarten. *Child Development, 78*(2), 647–663.

Brand, M., Labudda, K., & Markowitsch, H. J. (2006). Neuropsychological correlates of decision-making in ambiguous and risky situations. *Neural Networks, 19*(8), 1266–1276.

Brand, M., Recknor, E. C., Grabenhorst, F., & Bechara, A. (2007). Decisions under ambiguity and decisions under risk: Correlations with executive functions and comparisons of two different gambling tasks with implicit and explicit rules. *Journal of Clinical and Experimental Neuropsychology, 29*(1), 86–99.

Bull, R., & Scerif, G. (2001). Executive functioning as a predictor of children's mathematics ability: Inhibition, switching, and working memory. *Developmental Neuropsychology, 19*(3), 273–293.

Byrnes, J. P. (2002). The development of decision-making. *Journal of Adolescent Health, 31*(6), 208–215.

Byrnes, J. P. (2013). *The nature and development of decision-making: A self-regulation model.* Psychology Press.

Chen, E., Matthews, K. A., & Boyce, W. T. (2002). Socioeconomic differences in children's health: How and why do these relationships change with age? *Psychological Bulletin, 128*(2), 295.

CINVE. M. I. (2019). Statistical Report of the National Observatory on Violence and Crime. Uruguay. Ministry of Interior. https://www.minterior.gub.uy/observatorio/index.php/estadisticas

Cunningham, W. A., & Zelazo, P. D. (2007). Attitudes and evaluations: A social cognitive neuroscience perspective. *Trends in Cognitive Sciences, 11*(3), 97–104.

Del Giudice, M. (2009). Sex, attachment, and the development of reproductive strategies. *Behavioral and Brain Sciences, 32*(1), 1–21.

Del Giudice, M. (2014). An evolutionary life history framework for psychopathology. *Psychological Inquiry, 25*(3–4), 261–300.

Del Giudice, M., Ellis, B. J., & Shirtcliff, E. A. (2011). The adaptive calibration model of stress responsivity. *Neuroscience & Biobehavioral Reviews, 35*(7), 1562–1592.

Delgado, H., Aldecosea, C., Menéndez, Ñ., Rodríguez, R., Nin, V., Lipina, S., & Carboni, A. (2022). Socioeconomic status differences in children's affective decision-making: The role of awareness in the Children's Gambling Task. Developmental Psychology.

Ellis, B. J., Figueredo, A. J., Brumbach, B. H., & Schlomer, G. L. (2009). Fundamental dimensions of environmental risk. *Human Nature, 20*(2), 204–268.

Ellis, B. J., Del Giudice, M., Dishion, T. J., Figueredo, A. J., Gray, P., Griskevicius, V., Hawley, P. H., Jacobs, W. J., James, J., Volk, A. A., & Wilson, D. S. (2012). The evolutionary basis of risky adolescent behavior: Implications for science, policy, and practice. *Developmental Psychology, 48*(3), 598–623. https://doi.org/10.1037/a0026220

Evans, G. W., & Kantrowitz, E. (2002). Socioeconomic status and health: The potential role of environmental risk exposure. *Annual Review of Public Health, 23*(1), 303–331.

Fawcett, T. W., McNamara, J. M., & Houston, A. I. (2012). When is it adaptive to be patient? A general framework for evaluating delayed rewards. *Behavioural Processes, 89*(2), 128–136.

Fenneman, J., & Frankenhuis, W. E. (2020). Is impulsive behavior adaptive in harsh and unpredictable environments? A formal model. *Evolution and Human Behavior, 41*(4), 261–273.

Frankenhuis, W. E., Panchanathan, K., & Nettle, D. (2016). Cognition in harsh and unpredictable environments. *Current Opinion in Psychology, 7*, 76–80.

Garon, N., & Moore, C. (2007). Developmental and gender differences in future-oriented decision-making during the preschool period. *Child Neuropsychology, 13*(1), 46–63.

General Assembly of the World Medical Association. (2014). World Medical Association Declaration of Helsinki: Ethical principles for medical research involving human subjects. *The Journal of the American College of Dentists, 81*(3), 14–18.

Griskevicius, V., Tybur, J. M., Delton, A. W., & Robertson, T. E. (2011a). The influence of mortality and socioeconomic status on risk and delayed rewards: A life history theory approach. *Journal of Personality and Social Psychology, 100*(6), 1015–1026. https://doi.org/10.1037/a0022403

Griskevicius, V., Delton, A. W., Robertson, T. E., & Tybur, J. M. (2011b). Environmental contingency in life history strategies: The influence of mortality and socioeconomic status on reproductive timing. *Journal of Personality and Social Psychology, 100*(2), 241.

Griskevicius, V., Ackerman, J. M., Cantú, S. M., Delton, A. W., Robertson, T. E., Simpson, J. A., Thompson, M. E., & Tybur, J. M. (2013). When the economy falters, do people spend or save? Responses to resource scarcity depend on childhood environments. *Psychological Science, 24*(2), 197–205.

Grolnick, W. S., & Farkas, M. (2002). Parenting and the development of children's self-regulation. In M. H. Bornstein (Ed.), *Handbook of parenting: Vol. 5. Practical issues in parenting* (2nd ed., pp. 89–110). Erlbaum.

Hsu, W. K., Tseng, C. P., Chiang, W. L., & Chen, C. W. (2012). Risk and uncertainty analysis in the planning stages of a risk decision-making process. *Natural Hazards, 61*(3), 1355–1365.

Imbusch, P., Misse, M., & Carrión, F. (2011). Violence research in Latin America and the Caribbean: A literature review. *International Journal of Conflict and Violence (IJCV), 5*(1), 87–154.
Inzlicht, M., Werner, K. M., Briskin, J. L., & Roberts, B. W. (2021). Integrating models of self-regulation. *Annual Review of Psychology, 72*, 319–345.
Jaitman, L. (Ed.). (2017). *The costs of crime and violence: New evidence and insights in Latin America and the Caribbean* (Vol. 87). Inter-American Development Bank.
Jaitman, L., & Anauati, V. (2019). The dark figure of crime in Latin America and the Caribbean. *Journal of Economics, Race, and Policy, 3*(1), 76–95.
Kerr, A., & Zelazo, P. D. (2004). Development of "hot" executive function: The children's gambling task. *Brain and Cognition, 55*(1), 148–157.
Lerner, R. M. (2006). Developmental science, developmental systems, and contemporary theories of human development. In W. Damon & R. M. Lerner (Series Eds.) & R. M. Lerner (Vol. Ed.), *Handbook of child psychology: Vol 1. Theoretical models of human development* (6th ed., pp. 1–17). Wiley.
Lewis, M. D., & Todd, R. M. (2007). The self-regulating brain: Cortical-subcortical feedback and the development of intelligent action. *Cognitive Development, 22*(4), 406–430.
Mata, F., Miranda, D. M., Sallum, I., Bechara, A., & Malloy-Diniz, L. F. (2013). Do general intellectual functioning and socioeconomic status account for performance on the Children's gambling task? *Frontiers in Neuroscience, 7*(68), 1–9. https://doi.org/10.3389/fnins.2013.00068
McClelland, M. M., Cameron, C. E., Connor, C. M., Farris, C. L., Jewkes, A. M., & Morrison, F. J. (2007). Links between behavioral regulation and preschoolers' literacy, vocabulary, and math skills. *Developmental Psychology, 43*(4), 947–959. https://doi.org/10.1037/0012-1649.43.4.947
McClelland, M. M., John Geldhof, G., Cameron, C. E., & Wanless, S. B. (2015). Development and self-regulation. In R. M. Lerner (Org.), *Handbook of child psychology and developmental science* (pp. 1–43). Wiley. https://doi.org/10.1002/9781118963418.childpsy114
McLoyd, V. C. (1998). Socioeconomic disadvantage and child development. *American Psychologist, 53*(2), 185.
Perera, M., & Cazulo, P. (2016). Socioeconomic Level Index. Update Proposal.
Pinheiro, J., Bates, D., DebRoy, S., & Sarkar, D. (2012). Nonlinear mixed-effects models.
Posner, M. I., & Rothbart, M. K. (2000). Developing mechanisms of self-regulation. *Development and Psychopathology, 12*(3), 427–441.
Preston, S. D., Buchanan, T. W., Stansfield, R. B., & Bechara, A. (2007). Effects of anticipatory stress on decision making in a gambling task. *Behavioral Neuroscience, 121*(2), 257.
Sharkey, P. (2010). The acute effect of local homicides on children's cognitive performance. *Proceedings of the National Academy of Sciences, 107*(26), 11733–11738.
Sharkey, P. T., Tirado-Strayer, N., Papachristos, A. V., & Raver, C. C. (2012). The effect of local violence on children's attention and impulse control. *American Journal of Public Health, 102*(12), 2287–2293.
Team, R. C. (2013). *R: A language and environment for statistical computing*. Austria.
Theall, K. P., Shirtcliff, E. A., Dismukes, A. R., Wallace, M., & Drury, S. S. (2017). Association between neighborhood violence and biological stress in children. *JAMA Pediatrics, 171*(1), 53–60.
Trentacosta, C. J., & Izard, C. E. (2007). Kindergarten children's emotion competence as a predictor of their academic competence in first grade. *Emotion, 7*(1), 77.
Wilson, M., & Daly, M. (1997). Life expectancy, economic inequality, homicide, and reproductive timing in Chicago neighbourhoods. *BMJ, 314*(7089), 1271.
Wilson, A. C., Lengua, L. J., Tininenko, J., Taylor, A., & Trancik, A. (2009). Physiological profiles during delay of gratification: Associations with emotionality, self-regulation, and adjustment problems. *Journal of Applied Developmental Psychology, 30*(6), 780–790.
World Health Organization. (2002). World report on violence and health.
Zelazo, P. D., Qu, L., & Kesek, A. C. (2010). Hot executive function: Emotion and the development of cognitive control. In S. Calkins & M. A. Bell (Eds.), *Child development at the intersection of emotion and cognition* (pp. 97–111). American Psychological Association. https://doi.org/10.1037/12059-006

The Role of Peer Pressure in Adolescents' Risky Behaviors

Carlos Andrés Libisch, Flavio Marsiglia, Stephen Kulis, Olalla Cutrín, José Antonio Gómez-Fraguela, and Paul Ruiz

Risky Behaviors in Adolescence

Adolescence is a developmental period associated with concentrated change, resulting in an expansion of social networks and exposure to new risks and protective influences for youth. Involvement in some type of minor antisocial behavior in adolescence, such as substance use, is quite common (Le Blanc, 2015). Adolescence is the stage of development in which risky behaviors are most likely to be initiated, and such behavior commonly occurs within peer networks (Moffitt, 2018). Research has consistently found that risky behaviors, including those considered antisocial behaviors, are most prevalent around 15–17 years old (DeLisi, 2015). Cognitive and social development during preadolescence leads adolescents to exhibit less aggressive behavior than in later adolescence, and more covert behavior, which is difficult for parents and other adults to detect (Burt, 2012). Some of the most common risky behaviors at this phase are substance use, shoplifting, or running away from home (Le Blanc, 2015; Patterson & Yoerger, 1999). This chapter focuses on substance use and related factors, guided by this main research question: What role does peer pressure play in adolescents' substance use initiation?

C. A. Libisch (✉) · O. Cutrín · J. A. Gómez-Fraguela
Universidade de Santiago de Compostela, Santiago de Compostela, Spain
e-mail: andres.libisch@rai.usc.es

F. Marsiglia · S. Kulis
Arizona State University, Tempe, FL, USA

P. Ruiz
Universidad de la República, Facultad de Psicología, Montevideo, Uruguay

M. V. Alves et al. (eds.), *Cognitive Sciences and Education in Non-WEIRD Populations*, https://doi.org/10.1007/978-3-031-06908-6_8

Influence of Peer Affiliations in Adolescent Behavior

Over the last few decades, several theoretical models have been proposed in order to explain antisocial behavior in adolescence and understand its origins and consequences. The developmental perspective on the study of antisocial behavior is one of the most integrative approaches. A main characteristic of the developmental perspective is the holistic-integrative theoretical approach (Le Blanc, 2015; Lerner, 2006). This approach considers antisocial behavior as resulting from a process, which links different risk factors to one another, within different systems and contexts. Some of the key contexts during adolescence include family, school, and peers (Bronfenbrenner, 2005; Magnusson & Stattin, 2006). The very interaction of such systems may constitute a risk factor when it comes to developing maladjusted outcomes (Granic & Patterson, 2006). For example, school personnel might be prejudicial about the academic potential of a student of a racial/ethnic minority group if they show negative attitudes toward that group. Such attitudes may negatively affect the overall school climate, the school staff's relationship with the student's family and with the individual adolescent. A student who feels unwelcomed at school may look for a sense of belonging and acceptance from peer groups and be more likely to engage in misconduct at school (Demanet & Van Houtte, 2012).

Social interactions and relationships with peers provide psychosocial experiences from an early age (Bagwell, 2004). At the start of adolescence, friendships become more stable and intimate than in previous stages and peers become the main social reference group (Rubin et al., 2008). Adolescents usually belong to small groups freely formed among similar individuals (Brown, 2004; Collins & Steinberg, 2008). Homophily is an important feature of peer groups in adolescence, that is, the tendency to be affiliated with like-minded peers who share similar attitudes, aspirations, and behaviors (Brown & Larson, 2009; Dodge et al., 2008). As a result, adolescents engaging in antisocial or aggressive behaviors most likely belong to antisocial or aggressive peer groups (Brown, 2004; Steinberg, 2017).

The routine activity approach explains deviant behavior as the result of sharing unstructured free time and unsupervised activities with deviant peers (Cohen & Felson, 1979; Osgood et al., 1996). The social learning approach considers the role of the antisocial peer group as one of modeling and reinforcing the antisocial behaviors among its members (Akers & Jennings, 2009; Dishion et al., 1996; Sutherland, 1972). For example, the peer group can label and reinforce antisocial verbal statements, endorse favorable attitudes toward antisocial behavior, or teach skills and techniques to engage in risky behaviors, such as learning from each other how to access or use drugs. These two perspectives support the hypothesis that adolescents who interact and are involved in antisocial peer groups have higher probabilities of exhibiting risky behaviors (Brown & Klute, 2003).

Peer Relationships and Risky Behaviors in Adolescents

Affiliation with peers who engage in antisocial behaviors is one of the strongest risk factors related to engaging undesirable behaviors in adolescence (Bonta & Andrews, 2017; Brauer & De Coster, 2015). Several factors related to friendships within peer groups appear to influence adolescent antisocial behavior. The literature identifies the intensity and closeness of friendships, the stability of the relationships, the frequency and amount of time spent with peers, and the proportion of at-risk peers integrated into the peer group (Eassey & Buchanan, 2015; Kreager et al., 2011; Laursen et al., 2012). Peer influence also appears to influence educational achievement and expectations, a key predictor of substance use. Peer influence has a mediating effect on high school graduation and college enrollment of students in low socioeconomic status (SES) schools (Palardy, 2013). All these factors in different ways measure the intensity and frequency of exposure to peer influence, which is the ultimate variable intended to be measured (McGloin, 2009).

Antisocial behavior within peer groups is directly and strongly associated with higher levels of a broad range of adolescent antisocial behaviors, from minor, non-violent risky behaviors to more severe, violent behaviors (Herrenkohl et al., 2012; Snyder et al., 2012). More specifically, peers' substance use appears to be more associated with the individual adolescent's own substance use behavior than with other types of risky behaviors (Cutrín et al., 2019; Monahan et al., 2014; Tompsett et al., 2013).

Regional research in Latino America has shown that the peer group gives the greatest influence on adolescent decision-making and such influence significantly affects substance use (Malacas et al., 2019; Pilatti et al., 2013). A Mexican study with adolescents shows that women have a greater resistance to peer pressure than men and found that young people who have a deficit in the ability to resist peer pressure reported higher rates of alcohol use (Sanchez et al., 2013). Research specifically conducted in the Uruguayan context has described ritual processes that emerge within youth subcultures in relation to substance use, where group discourses and reinforcement lead to an excessive use of alcohol and other drugs (Filardo, 2002). Other studies in Uruguay also found significant peer effects on antisocial behavior (shoplifting and being involved in fights) (Balsa, Gandelman, & González, 2015a) and risky attitudes (Balsa, Gandelman, & Roldán, 2015b). Although studies tend to focus on peers as a risk factor, peers can also put forth a protective influence against substance use and abuse (Thurow et al., 2020).

Online Peer Influence in Risky Behaviors Like Substance Use

Peer pressure takes place in person and in online social interactions. Social media is a unique social context that influences individuals' thoughts, behaviors, and relationships (McFarland & Ployhart, 2015; Peter & Valkenburg, 2013; Subrahmanyam

& Šmahel, 2011). Research has shown that peer influence processes on social media contribute to adolescents' risky behaviors, especially alcohol use. Adolescents who report a greater number of friends mentioning alcohol on social media are more likely to post such references themselves. Perceptions of peers' posts about alcohol are also a stronger predictor of adolescents' alcohol-related posting than self-reports of their own drinking behavior (Geusens & Beullens, 2017). Adolescents and young adults exposed to peers' alcohol-related content on social media are more likely to initiate and escalate drinking behaviors (Geusens & Beullens, 2016; Nesi et al., 2017).

Descriptive (exaggerated perceptions that most peers engage in risky behaviors) and injunctive norms (expectations that peers would approve one's engagement in risky behaviors) can partially explain the process of peer influence. For example, adolescents exposed to alcohol content on social media are more likely to endorse alcohol-favorable descriptive and injunctive norms (Beullens & Vandenbosch, 2016; Fournier et al., 2013). These norms appear to act as mediators between exposure and alcohol intention to use and alcohol use (Geusens & Beullens, 2016; Nesi et al., 2017). Many of these studies are from the USA or Europe. There are a few studies that also analyze Facebook influence and peer pressure attitudes among Latin Americans young adults, which conclude that the higher the pressure of peers and the greater the exposure to illicit drug content on Facebook, the greater the attitudes that favor the consumption of illicit drugs (Guzmán Facundo et al., 2019).

Other works studying young adult users of social networking sites (SNSs)—such as Facebook—found that interactions between friends are creating new types of friendships and influences (Niland et al., 2015). In the same way, other investigations about Facebook users show that users often experience negative emotions because they feel pressured to access the site frequently due to the fear of missing out and to keep up with relationship maintenance demands (Fox & Moreland, 2015). This is also connecting with the fear of missing out (FoMO) phenomenon. Some studies realize that FoMO also affects the deliberative process of the purchasing behavior of Generation Z consumers, affecting the decision-making process, and has a mediating role in the relationship between social self-efficacy and satisfaction with life (Deniz, 2021).

There is a need to better conceptualize and understand the role of adolescents' online life and specifically the role of peer pressure in the online context. Currently, it seems that the role of peer influence on the emergence of antisocial behaviors such as substance use in adolescence is undeniable. It is less clear how to incorporate these influences in prevention interventions from a neurological and behavioral perspective.

Most children do not use or abuse drugs; this chapter aims at discussing also how peers can model healthy and prosocial behaviors. How can prevention programs capitalize on those influences and can weaken negative peer pressure at the same time that they work on strengthening positive, prosocial relationships?

The following sections further explore these issues from neurobiological and behavioral perspectives and highlight a specific intervention to prevent peer-related risk factors influencing adolescents' behavior.

Contributions from Neurobiology and Comparative Psychology

Peer pressure is a form of social influence that can be studied from a neurobiological and comparative approach by analyzing how different species share information through the socialization process and how this information affects behavior (e.g., imitative learning), cognition (e.g., conceptual elaborations), affectiveness (e.g., emotional contagion), and neurobiology.

For decades, animal-model studies have been showing how individuals exchange information. Peers' dynamics or social facilitation studies the effects of a stimulant on the observer's behavior (e.g., when social pressure contributes to increasing levels of food and fluid consumption). When these processes manipulate behavior, they are known as observational learning or demonstrating effect. This process becomes increasingly important when the exchange of information occurs between individuals of different ages and hierarchies (Choleris et al., 1997; Gonzalez-Pirelli & Ruiz, 2014; Heyes et al., 1994; Zentall & Levine, 1972). As for the demonstrator effect, it is a form of social learning by imitation where the observer learns about behaviors, actions, and emotional responses by imitating a cohabitating demonstrator (Heyes et al., 1994).

In this context, we can consider the different mechanisms by which information is transmitted (cognitive and emotionally), among individuals who interact based on techniques such as observation and imitation (Nicol, 1995). The same techniques are common among numerous animal species and human beings.

In the case of animals when induced into an emotional state, their cohabitation peers become affected as well. For instance, if a rat on its way to being fed sees another rat receive an electrical discharge, it will subsequently stop eating as a way to stop its fellow rat from being electrocuted (De Wall, 2007). This is a sympathetic type of response (Church, 1959; Decety & Chaminade, 2003). Empathy is built upon sympathy, which can be divided into cognitive and emotional. In the first case, it involves a cognitive elaboration, and in the second one an emotional recognition regarding the emotional states that can be observed in the cohabitation individual. On the other hand, emotional contagion is an affective mechanism by which an organism automatically synchronizes its physiological state and behavioral one with others to promote altruistic behavior (Hatfield et al., 1993; Ruiz, 2015). This mechanism is present in almost all phylogeny, human babies, birds, rodents, and non-human primates, among others (Nakahashi & Ohtsuki, 2017). These types of reactions promote prosocial behavior and are related to peer pressure.

Nummenmaa, Hirvonen, Prarkkola, and Hietanen (Nummenmaa et al., 2008) compare emotional contagion with emotional empathy merging both later on in a process called "mirroring system" tied to a system of mirroring neurons emphasizing in the insula as a primary structure. This process takes place by processing and internalizing the emotional states of those around us. Other studies have shown the insula's central role such as the reception of cardiac peacemaker's information in regard to the processing of social cognition. For instance, Couto, Sedeña, Peradejordi, and Battfeld (Couto et al., 2014) observed flaws in tasks that require interoception (emotional regulation) and social cognition (including empathy), but not in other cognitive areas, in the case of patients with an internal as well as external heart (a mechanic device that pumps blood). It is reported that the additional interoceptive stimulation of the device interfered in the normal interceptive sensed in which the insula is a key piece. Moreover, the insula is a structure associated with addiction's development and its alteration is linked to alcohol consumption as seen in preclinical models (Ruiz et al., 2018). Other studies have shown that rats that cohabitate with depressed peers tend to express a depressive phenotype as well as a reduction in dopamine levels in the insula (Ruiz, 2019). These findings also implicate greater alcohol consumption in rats that cohabitate with depressed ones, as opposed to the control group of rats (In-lab unpublished results).

In sum, these findings show how the cited neural behavioral and cognitive elements (particularly the insula) linked to drug consumption are socially influenced, contributing to enlightening discussions around peer influence in substance consumption from comparative psychology and neurobiological perspectives. These neurobiological findings support the need to design and test efficacious prevention interventions for youth.

Prevention Intervention Research, *Mantenete REAL* in Uruguay

Mantenete REAL-Uruguay is a cultural adaptation of *keepin' it REAL* (*kiREAL*), an efficacious school-based drug prevention curriculum originally developed and tested in Arizona, USA (Marsiglia & Hecht, 2005). *kiREAL* is an efficacious school-based universal prevention program targeting substance use and other youth problem behaviors, designed specifically to be implemented at the beginning of adolescence (Hecht et al., 2003). Ecological risk and resiliency theory, communication competence theory, and narrative theory guide the intervention in teaching youth the practical skills they need to avoid engaging in risky behaviors (Marsiglia et al., 2009).

Through risk assessment and by strengthening their communication skills, students acquire concrete peer pressure management skills to avoid or delay their use of alcohol or other drugs and to avoid other risky behaviors. The intervention provides students with the tools to do what they want to do and not to just follow what others want them to do. The main purpose of the program is to teach students how to manage substance use offers of alcohol, tobacco, and other drugs by using the REAL strategies (Kulis et al., 2007). The acronym REAL—Refuse, Explain, Avoid, and Leave—represents four resistance strategies used most commonly by youth who successfully resist using substances: refuse substance offers with a direct no, explain why you decline, avoid substance offer situations, or leave the situation altogether. These offers often take place in peer and family contexts (Marsiglia et al., 2009).

Like other efficacious school-based universal prevention programs, *kiREAL* teaches how to manage risky situations through life skills (Botvin, 1990). It uses an interactive format that allows youth to apply and rehearse the skills, in contrast to programs that only relay information or employ scare tactics. The manualized *kiREAL* program helps youth understand social factors (e.g., peer expectations) that lead to risky behaviors while fostering a variety of adaptive skills including problem-solving, decision-making, risk assessment, cognitive and behavioral coping strategies, and interpersonal skills for resisting substance use. These personal and social skills play an essential role in the healthy psychological adjustment and psychosocial development of adolescents (Sancassiani et al., 2015). Learning these skills has a special relevance in adolescence since this constitutes a critical stage in psychosocial development, especially in terms of susceptibility to social influence from peers and other social actors.

In order to reinforce learning and facilitate the incorporation of the REAL strategies in the behavioral repertoire, the skills training of *kiREAL* comes from the adolescents' context and reflects the youths' communication and decision-making processes. Videos scripted and acted by students illustrate the REAL strategies and are an integral part of the overall intervention (Holleran et al., 2002; Reeves et al., 2008). The videos include recognizable local settings and reflect the cultural norms and experiences of adolescents. Because *KiREAL* approaches culture of origin as a source of resiliency and protection against risk, the videos illustrate risky situations often involving peers and offer specific strategies to behave in ways consistent with the cultural values of the youth (Nagoshi et al., 2020).

Aligned with its theoretical underpinnings, *kiREAL* has been culturally adapted (Marsiglia et al., 2019) and evaluated for Mexico through a national Randomized Control Trial (RCT) with more than 5000 middle school students and their teachers (Kulis, Marsiglia, et al., 2021b). The same binational team replicated and tested the culturally adapted intervention in smaller studies across different locations in Mexico (e.g., Kulis, García-Pérez, et al., 2021a; Marsiglia et al., 2015). It was also culturally adapted and evaluated through feasibility and smaller efficacy and effectiveness of RCTs in other countries such as Guatemala, Kenya, Spain, and Uruguay (Cutrín et al., 2020; Kulis et al., 2019; Marsiglia et al., 2018).

In all the different international studies, *kiREAL* was adapted and evaluated in partnership with local investigators and communities, following rigorous cultural adaptation and evaluation protocols (Marsiglia & Booth, 2015; Marsiglia et al., 2015). Peer influence emerged as a constant across the different studies, but it manifested itself in unique ways in each cultural context. Uruguay presents an interesting case study, as it was the first Latin American nation to decriminalize the use and commercialization of cannabis (Arocena & Aguiar, 2017). In addition, the law that allows this new framework about marihuana use regulates the development of effective drug prevention programs for young adolescents.

Uruguayan Cultural Adaptation

Cultural adaptation is the systematic modification of a treatment or intervention protocol, considering the language, culture of the participants, and their social context. The main goal of these modifications is to make the intervention compatible with the cultural patterns, meaning and values of the target population (Bernal et al., 2009). The cultural adaptation process addresses differences in teaching styles or forms of delivery, adjusts the contents of the program to the different contexts, and addresses differences in communication style and values.

In prevention programs like Mantente REAL, considering differences in each country's educational system is key. For example, in Uruguay as in many other countries, teaching methods tend to be traditional. On average, teachers in Latin America use active instruction only 36% of class time, with passive instruction like copying from the blackboard and seatwork occupying 25% of class time (Alexandrowicz, 2021). Because Mantente REAL requires active participation of the students, the teacher-training component is key. Teachers receiving the training feel comfortable adopting a more active instruction style and enjoy rehearsing with their colleagues this type of facilitation. Latin American teachers are not very accustomed to deliver manualized interventions (Pérez & Mejía, 2015), and this is an important aspect to take into account because Mantente REAL is a manualized program. In order to achieve the desired levels of fidelity, classroom observations and follow-up meetings supported teachers during the implementation. Teachers and students were actively engaged in the cultural adaptation process and provided feedback about format and content. For example, Uruguayan students reported that they enjoyed acting and role-playing the most. Due to the high rates of alcohol use in Uruguay among younger children, participants used role-playing often as a means to rehearsing the REAL strategies when confronted with an offer of alcohol. Differences between and within countries are identified and integrated while protecting the core elements of the intervention that make it evidence-based.

Results from the Pilot Study of *Mantenete REAL* in Uruguay

After culturally adapting *kiREAL* (*Mantenete REAL)* for Uruguay (Marsiglia et al., 2018), a binational team of researchers from Uruguay and the United States collaborated on a test of the prevention program in Montevideo, the national capital and largest city. With the approval and cooperation of the national educational authority, Administración Nacional de Educación Pública (ANEP), six public primary schools were recruited to participate in a small cluster RCT in the 2019 school year. The schools were first stratified—or sorted—into those serving lower-income versus middle-income families and then block randomized into two groups. Three schools were randomly selected to implement *Mantenete REAL*, and the remaining schools were designated to serve as a comparison or control group. Schools in the two groups had matched socioeconomic profiles.

In implementation schools, sixth-grade regular classroom teachers received a 2-day training in delivering the manualized *Mantenete REAL* intervention, including actively preparing and rehearsing the lessons. They delivered the 10 lessons between May and August of 2019, typically one session per week, in their regular sixth-grade classes. In all six schools, sixth-grade students with parental consent completed two questionnaires, timed immediately before and 2 months after *Mantenete REAL* was delivered in the implementing schools. To ensure confidentiality, members of the research team administered the questionnaires and teachers were not present during the survey administration. There were 223 student respondents, 96 in the *Mantenete REAL* schools, and 127 in the control schools. On average, students were 11 years old, nearly gender balanced (49% female, 51% male), and more than half of them (57%) were living with two parents.

The students' pretest and posttest data tracked changes over time in their substance use behaviors, experiences, and attitudes. Seven types of questions asked students about their interactions with and perceptions of their peers, items that are risk factors for adolescent substance use. These questions can provide insights into whether *Mantenete REAL* affected the way that students recognized and dealt with peer influences on substance use. All the measures were either multi-item scales or single items with Likert-scale responses. The measures included the following:

- Descriptive norms: perceptions, often exaggerated, that most peers use substances.
- Injunctive norms: expecting that friends would approve one's use of substances.
- Susceptibility to negative peer pressure to engage in antisocial behavior.
- Receiving substance offers from peers.
- Lack of refusal efficacy: difficulty rejecting substance offers from peers.
- Positive substance use expectancies, such as making it easier to belong to a group.
- Poor decision-making: following others when faced with an important decision.

Figure 1 illustrates the students' responses to these questions at the pretest after dichotomizing the measures at substantively meaningful thresholds. Nearly a third of the students said they would give in to peer pressure in at least one of six scenarios, most commonly to skip school or a class exam. Almost one-fifth said they at times make important decisions by "doing what others do." But on the remaining items only about one in ten or fewer reported negative peer influences, including that most school peers use substances; their friends would react positively if they themselves used substances; peers had offered them substances in the last 30 days; they would have difficulty turning down a cigarette offer from a peer; they agree that using marijuana makes it easier to be part of a group. In assessing whether the

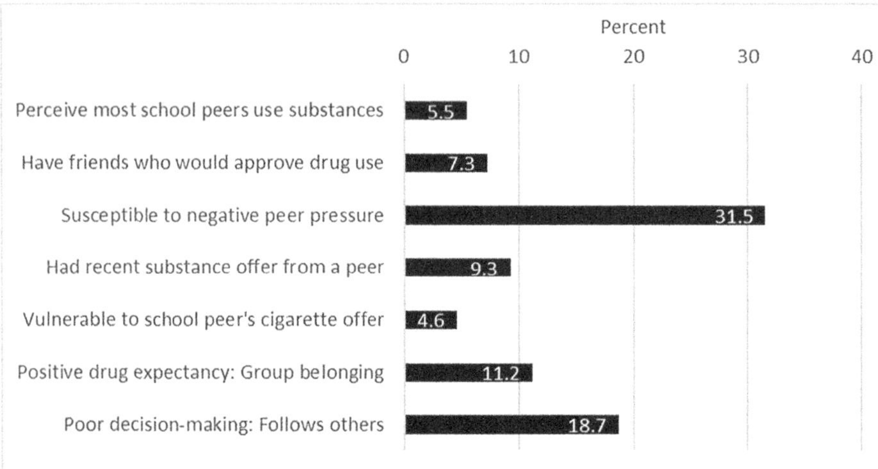

Fig. 1 Negative peer influences at pretest (scales and items dichotomized)

Mantenete REAL program affected these outcomes it is important to remember that most students reported little or no negative peer influence at the start of the study.

Figure 2 summarizes the effects of *Mantenete REAL* on negative peer influences by first contrasting the mean size and direction of changes from the pretest to the posttest in the intervention and control groups. Below these are an estimate of the intervention effect size (Cohen's *d*), which is a standardized comparison of the degree of change in the intervention group relative to a change in the control group. For three outcomes, negative peer influences increased in both groups, but to a larger degree in the control group: perceptions of the extent of peer substance use, susceptibility to peer pressure, and substance offers from peers. In two instances, negative peer influences increased in the control group while decreasing for *Mantenete REAL*: expected friends' approval of substance use and vulnerability to a peer substance offer. For the two final outcomes, reports of negative peer influences declined in both groups, but the decrease was larger in *Mantenete REAL*. These differences between groups were statistically significant (at $p < 0.10$) for four of the outcomes.

The consistent direction of effect sizes in Fig. 1 indicates that on all the outcomes negative peer influences increased less—or decreased more—for *Mantenete REAL* students than for controls. The effect sizes varied from what is considered very small ($<|0.10|$) to medium ($>|0.50|$) in magnitude, but most effects were in the typical range for school-based substance use prevention programs (Tanner-Smith et al., 2018).

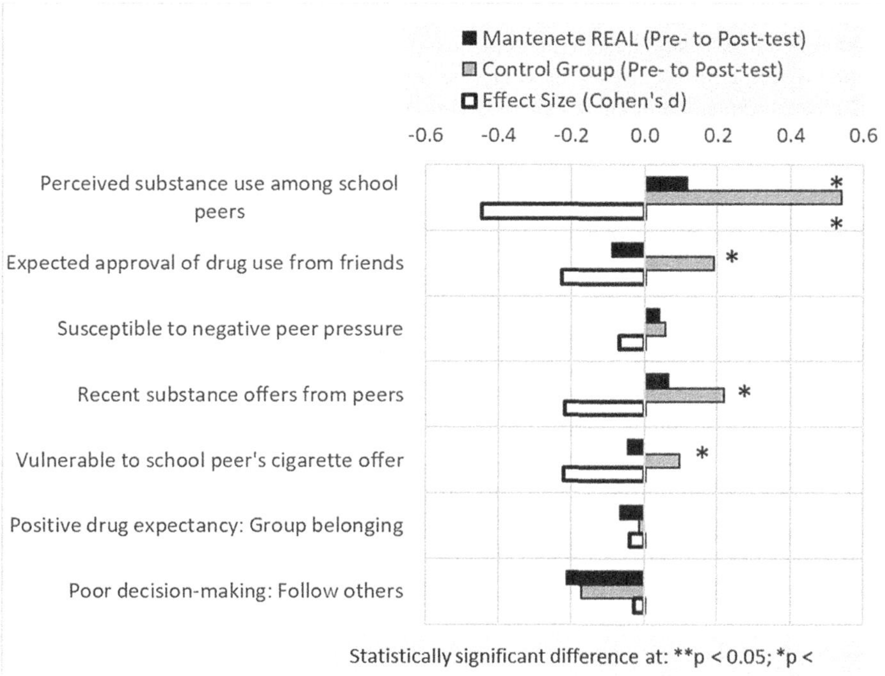

Fig. 2 Changes in negative peer influences from pretest to posttest (means), for *Mantenete REAL* and control groups, and estimated intervention effect size

Discussion and Conclusion

Behavioral and neurobiological studies support the importance of peer pressure as an influence in need of further exploration and integration into efficacious prevention programs (e.g., Malacas et al., 2019; Ruiz, 2019). Research from different disciplines consistently recommends that prevention programs in early adolescence strengthen adolescents' assertiveness and resistance to peer pressure. They also suggest improving problem-solving skills, especially considering the stronger effects that risk taking peers have on adolescents' risky behaviors in late adolescence (Burk et al., 2012; Cutrín et al., 2017). Universal and multimodal school-based prevention programs, such as *Mantenete REAL*, that focus on social skills training and competence enhancement can be useful in strengthening drug refusal skills, problem-solving skills, self-regulation, and self-efficacy (Gardner et al., 2008; Scheier, 2012), all of them essential for enhancing resilience and coping with peer pressure to use substances during adolescence (Leander et al., 2016).

In Uruguay, the pilot test with sixth-grade primary school students showed promising results that the *Mantenete REAL* intervention helps restrain or reverse the accumulation of negative peer influences that are key factors in substance use initiation among preadolescents and early adolescents. Participants in the intervention were less likely to adopt norms and attitudes that increase their risk of using substances, such as exaggerated perceptions that "everyone is doing it," that their friends would approve if they used alcohol and other drugs, and that substance use will make them popular or to "fit in." *Mantenete REAL* participants also reported relatively less exposure to peer offers of alcohol and drugs, less likelihood of accepting them, and less likely to "follow the herd" in decision-making.

Prevention strategies have often focused on managing undesirable affiliations by limiting the amount of unsupervised free time with peers and promoting involvement in structured activities (Mahoney et al., 2004; Meldrum & Barnes, 2017). The Uruguay study indicates that managing attitudes toward drugs and increasing prosocial peer associations, as *Mantenete REAL* does, are effective in decreasing risky and antisocial behaviors (Gottfredson et al., 2004). Promoting prosocial friendships while weakening negative peer pressure is key to modeling healthy and prosocial behaviors. The findings support previous research, which concluded that adolescents affiliated with a stable, cohesive, and extensive prosocial peer group are less likely to be involved in risky behaviors, as positive friends serve as role models for prosocial behavior (Brown et al., 2005; Burt & Klump, 2014). Involvement with prosocial peers also favors the development of prosocial attitudes and beliefs and social skills, such as assertiveness or cooperation (Criss et al., 2009), which in turn can augment the effectiveness of prevention programs such as *Mantenete REAL*.

Limitations and Future Research

One of the limitations of the Uruguay study was its lack of ability to disaggregate online social pressure from in-person peer pressure. In the current digital society, we cannot understand adolescent relationships without taking into account the degree of virtual interaction they have. Future research needs to address the transformative role of social media in adolescent peer relationships (Nesi et al., 2018). Measures of peer pressure need to move beyond traditional interpersonal environments and offer categories to explain how social media transforms peer relation constructs such as peer status, peer influence, friendship, and peer victimization.

Uruguay has a high and very reliable level of access to the internet, and widespread access to devices including among preadolescents. The "Plan Ceibal" is a national program implemented in all public elementary schools in the nation. Created in 2007, Ceibal is a plan for inclusion and equal opportunity with the aim of supporting Uruguayan educational policies with technology (Melgar et al., 2017). Uruguay is an ideal context to conduct in-person and virtual peer pressure research and its relationship to prevention.

If we agree that social media has the power to transform peer relation constructs like peer status, peer influence, friendship, and peer victimization, the strategies that *Mantenete REAL* teaches should be evaluated and adapted to make them more effective in virtual settings and help teens cope with peer online pressure during adolescence (Jules et al., 2021).

Another limitation of the study is that it took place in the largest city of the country without any representation of youth from small towns and rural areas. The current sampling design limits the generalizability of the findings. Future studies need to have a cross-sectional sample with greater geographic representation.

Conclusion

Projects like *Mantente REAL* in Uruguay can make an important prevention contribution for young adolescents in their ability to resist peer pressure and to avoid drug use and other risk behaviors. Improved peer pressure skills can also help especially students from low socioeconomic status advance in their educational attainment and related desired health outcomes. This is important because high rates of exclusion and low rates of completion of secondary education are one of the principal problems in the educational systems in Latin America and the Caribbean. Uruguay, in particular, had a graduation rate of 41% or, in other words, 20 percentage points below the regional average (INEEd, 2020).

The Uruguayan context shows some unique characteristics, such as the decriminalization of cannabis use, a high level of access and use of technologies, and a great need to change the low rates of completion of secondary education, which makes it necessary to develop programs based on evidence like *Mantente REAL*, which focuses on peer pressure skills.

References

Akers, R. L., & Jennings, W. G. (2009). The social learning theory of crime and deviance. In M. D. Krohn, A. J. Lizotte, & G. P. Hall (Eds.), *Handbook on crime and deviance* (pp. 103–120). Springer.

Alexandrowicz, V. (2021). *ESL and cultures resource*. University of San Diego. Downloaded from: https://sites.sandiego.edu/esl/latin-american/

Arocena, F., & Aguiar, S. (2017). Tres leyes innovadoras en Uruguay: Aborto, matrimonio homosexual y regulación de la marihuana/Three innovative laws in Uruguay: abortion, same-sex marriage and marijuana regulation. *Revista de Ciencias Sociales, 30*(40), 43–62.

Bagwell, C. L. (2004). Friendships, peer networks, and antisocial behavior. In J. B. Kupersmidt & K. A. Dodge (Eds.), *Children's peer relations: From development to intervention* (pp. 37–57). American Psychological Association. https://doi.org/10.1037/10653-003

Balsa, A. I., Gandelman, N., & González, N. (2015a). Peer effects in risk aversion. *Risk Analysis, 35*, 27–43. https://doi.org/10.1111/risa.12260

Balsa, A., Gandelman, N., & Roldán, F. (2015b). *Peer Effects in the Development of Capabilities in Adolescence. CAF – Working paper; 2015/09*. CAF. Retrieved from http://scioteca.caf.com/handle/123456789/820

Bernal, G., Jiménez-Chafey, M. I., & Domenech Rodríguez, M. M. (2009). Cultural adaptation of treatments: A resource for considering culture in evidence-based practice. *Professional Psychology: Research and Practice, 40*(4), 361–368. https://doi.org/10.1037/a0016401

Beullens, K., & Vandenbosch, L. (2016). A conditional process analysis on the relationship between the use of social networking sites, attitudes, peer norms, and adolescents' intentions to consume alcohol. *Media Psychology, 19*, 310–333. https://doi.org/10.1080/15213269.2015.1049275

Bonta, J., & Andrews, D. A. (2017). *The psychology of criminal conduct* (6th ed.). Routledge.

Botvin, G. (1990). Substance abuse prevention: Theory, practice, and effectiveness. *Crime and Justice, 13*, 461–519.

Brauer, J. R., & De Coster, S. (2015). Social relationships and delinquency: Revisiting parent and peer influence during adolescence. *Youth & Society, 47*, 374–394. https://doi.org/10.1177/0044118X12467655

Bronfenbrenner, U. (2005). The bioecological theory of human development. In U. Bronfenbrenner (Ed.), *Making human beings human: Bioecological perspectives on human development* (pp. 3–15). Sage Publications.

Brown, B. B., & Klute, C. (2003). Friendships, cliques, and crowds. In G. R. Adams & M. D. Berzonsky (Eds.), *Blackwell handbook of adolescence* (pp. 330–348). Blackwell Publishing. https://doi.org/10.1002/9780470756607.ch16

Brown, B. B. (2004). Adolescents' relationships with peers. In R. M. Lerner & L. Steinberg (Eds.), *Handbook of adolescent psychology* (2nd ed., pp. 363–394). Wiley. https://doi.org/10.1002/9780471726746.ch12

Brown, E. C., Catalano, R. F., Fleming, C. B., Haggerty, K. P., Abbott, R. D., Cortes, R. R., & Park, J. (2005). Mediator effects in the social development model: An examination of constituent theories. *Criminal Behaviour and Mental Health, 15*, 221–235.

Brown, B. B., & Larson, J. (2009). Peer relationships in adolescence. In R. M. Lerner & L. Steinberg (Eds.), *Handbook of adolescent psychology: Vol. 2. Contextual influences on adolescent development* (3rd ed., pp. 74–103). Wiley. https://doi.org/10.1002/9780470479193.adlpsy002004

Burk, W. J., van der Vorst, H., Kerr, M., & Stattin, H. (2012). Alcohol use and friendship dynamics: Selection and socialization in early-, middle-, and late-adolescent peer networks. *Journal of Studies on Alcohol and Drugs, 73*, 89–98. https://doi.org/10.15288/jsad.2012.73.89

Burt, S. A. (2012). How do we optimally conceptualize the heterogeneity within antisocial behavior? An argument for aggressive versus non-aggressive behavioral dimensions. *Clinical Psychology Review, 32*, 263–279. https://doi.org/10.1016/j.cpr.2012.02.006

Burt, S. A., & Klump, K. L. (2014). Prosocial peer affiliation suppresses genetic influences on non-aggressive antisocial behaviors during childhood. *Psychological Medicine, 44*, 821–830.

Choleris, E., Guo, C., Liu, H., Mainardi, M., & Valsecchi, P. (1997). The effect of demonstrator age and number on the duration of socially-induced food preferences in house mice. *Behavioral Processes, 41*, 69–77.

Church, R. (1959). Emotional reactions of rats to the pain of others. *Journal of Comparative and Physiological Psychology, 52*(2), 132–134.

Cohen, L. E., & Felson, M. (1979). Social change and crime rate trends: A routine activity approach. *American Sociological Review, 44*, 588–608. https://doi.org/10.2307/2094589

Collins, W. A., & Steinberg, L. (2008). Adolescent development in interpersonal context. In W. Damon & R. M. Lerner (Eds.), *Child and adolescent development: An advanced course* (pp. 551–590). Wiley.

Couto, B., Salles, A., Sedeño, L., Peradejordi, M., Barttfeld, P., Canales-Johnson, A., Dos Santos, Y. V., Huepe, D., Bekinschtein, T., Sigman, M., Favaloro, R., Manes, F., & Ibanez, A. (2014). The man who feels two hearts: The different pathways of interoception. *Social Cognitive and Affective Neuroscience, 9*(9), 1253–1260.

Criss, M. M., Shaw, D. S., Moilanen, K. L., Hitchings, J. E., & Ingoldsby, E. M. (2009). Family, neighborhood, and peer characteristics as predictors of child adjustment: A longitudinal analysis of additive and mediation models. *Social Development, 18*, 511–535.

Cutrín, O., Gómez-Fraguela, J. A., Maneiro, L., & Sobral, J. (2017). Effects of parenting practices through deviant peers on nonviolent and violent antisocial behaviours in middle- and late-adolescence. *The European Journal of Psychology Applied to Legal Context, 9*, 75–82. https://doi.org/10.1016/j.ejpal.2017.02.001

Cutrín, O., Maneiro, L., Sobral, J., & Gómez-Fraguela, J. A. (2019). Validation of the deviant peers scale in Spanish adolescents: A new measure to assess antisocial behaviour in peers. *Journal of Psychopathology and Behavioral Assessment, 41*, 185–197. https://doi.org/10.1007/s10862-018-9710-6

Cutrín, O., Kulis, S., Maneiro, L., MacFadden, I., Navas, M. P., Alarcón, D., Gómez-Fragüela, J. A., Villalba, C., & Marsiglia, F. F. (2020). Effectiveness of the Mantente REAL program for preventing alcohol use in Spanish adolescents. *Psychosocial Intervention*. https://journals.copmadrid.org/pi/art/pi2020a19

Decety, J., & Chaminade, T. (2003). Neural correlates of feeling sympathy. *Neuropsychologia, 41*, 127–138.

DeLisi, M. (2015). Age-crime curve and criminal career patterns. In J. Morizot & L. Kazemian (Eds.), *The development of criminal and antisocial behavior: Theory, research and practical applications* (pp. 51–63). Springer International Publishing. https://doi.org/10.1007/978-3-319-08720-7_4

Demanet, J., & Van Houtte, M. (2012). School belonging and school misconduct: The differing role of teacher and peer attachment. *Journal of Youth and Adolescence, 41*(4), 499–514.

Deniz, M. (2021). Fear of missing out (FoMO) mediate relations between social self-efficacy and life satisfaction. *Psicologia: Reflexão e Critica, 34*(1), 1–9. https://doi-org.proxy.timbo.org.uy/10.1186/s41155-021-00193-w

De Wall, F. (2007). *Primates y filósofos. La evolución de la moral del simio al hombre*. Barcelona.

Dishion, T. J., Spracklen, K. M., Andrews, D. W., & Patterson, G. R. (1996). Deviancy training in male adolescent friendships. *Behavior Therapy, 27*, 373–390. https://doi.org/10.1016/S0005-7894(96)80023-2

Dodge, K. A., Coie, J. D., & Lynam, D. (2008). Aggression and antisocial behavior in youth. In W. Damon & R. M. Lerner (Eds.), *Child and adolescent development: An advance course* (pp. 437–472). John Wiley & Sons.

Eassey, J. M., & Buchanan, M. (2015). Fleas and feathers: The role of peers in the study of juvenile delinquency. In M. D. Krohn & J. Lane (Eds.), *The handbook of juvenile delinquency and juvenile justice* (pp. 199–216). John Wiley & Sons. https://doi.org/10.1002/9781118513217.ch14

Filardo, V. (Ed.). (2002). *Tribus urbanas en Montevideo: nuevas formas de sociabilidad juvenil* (1ª ed.). Editorial Trilce.

Fournier, A. K., Hall, E., Ricke, P., & Storey, B. (2013). Alcohol and the social network: Online social networking sites and college students' perceived drinking norms. *Psychology of Popular Media Culture, 2*, 86–95. https://doi.org/10.1037/a0032097

Fox, J., & Moreland, J. J. (2015). The dark side of social networking sites: An exploration of the relational and psychological stressors associated with Facebook use and affordances. *Computers in Human Behavior, 45*, 168–176.

Gardner, T. W., Dishion, T. J., & Connell, A. M. (2008). Adolescent self-regulation as resilience: Resistance to antisocial behavior within the deviant peer context. *Journal of Abnormal Child Psychology, 36*, 273–284. https://doi.org/10.1007/s10802-007-9176-6

Geusens, F., & Beullens, K. (2016). The association between social networking sites and alcohol abuse among Belgian adolescents: The role of attitudes and social norms. *Journal of Media Psychology*. https://doi.org/10.1027/1864-1105/a000196

Geusens, F., & Beullens, K. (2017). Strategic self-presentation or authentic communication? Predicting adolescents' alcohol references on social media. *Journal of Studies on Alcohol and Drugs, 78*(1), 124–133. https://doi.org/10.15288/jsad.2017.78.124. [PubMed: 27936372].

Gonzalez-Pirelli, F., Ruiz, P. (2014). Cognición social. En: A Vasquez, Manual de Psicología Cognitiva (pp. 253–276). : Comisión Sectorial de Enseñanza.

Gottfredson, D. C., Gerstenblith, S. A., Soulé, D. A., Womer, S. C., & Lu, S. (2004). Do after school programs reduce delinquency? *Prevention Science, 5*, 253–266. https://doi.org/10.1023/B:PREV.0000045359.41696.02

Granic, I., & Patterson, G. R. (2006). Toward a comprehensive model of antisocial development: A dynamic systems approach. *Psychological Review, 113*, 101–131. https://doi.org/10.1037/0033-295X.113.1.101

Guzmán Facundo, F. R., Vargas Martínez, J. I., Candia Arredondo, J. S., Rodriguez Aguilar, L., & Lopez Garcia, K. S. (2019). Influencia De La Presión De Pares Y Facebook en Actitudes Favorecedoras Al Consumo De Drogas Ilícitas en Jóvenes Universitarios Mexicanos. *Health & Addictions/Salud y Drogas, 19*(1), 22–30. https://doi-org.proxy.timbo.org.uy/10.21134/haaj.v19i1.399

Hatfield, E., Cacioppo, J., & Rapson, R. (1993). Emotional contagion. *Current Directions in Psychological Sciences, 2*, 96–99.

Hecht, M. L., Marsiglia, F. F., Elek-Fisk, E., Wagstaff, D. A., Kulis, S., & Dustman, P. (2003). Culturally grounded substance use prevention: An evaluation of the *keepin't REAL* curriculum. *Prevention Science, 4*, 233–248.

Heyes, C., Jaldow, E., Nokes, T., & Dawson, G. (1994). Imitation in rats (Rattus norvegicus): The role of demonstrator action. *Behavioral Processes, 32*, 173–182.

Herrenkohl, T. I., Lee, J., & Hawkins, J. D. (2012). Risk versus direct protective factors and youth violence: Seattle social development project. *American Journal of Preventive Medicine, 43*, S41–S56. https://doi.org/10.1016/j.amepre.2012.04.030

Holleran, L. K., Reeves, L., Dustman, P., & Marsiglia, F. F. (2002). Creating culturally grounded videos for substance abuse prevention. *Journal of Social Work Practice in the Addictions, 2*, 55–78. https://doi.org/10.1300/J160v02n01_04

INEEd. (2020). *Reporte del Mirador Educativo 6. 40 años de egreso de la educación media en Uruguay*. INEEd.

Jules, M. A., Maynard, D.-M. B., & Coulson, N. (2021). Youth drug use in Barbados and England: Correlates with online peer influences. *Journal of Adolescent Research*, 274--310. https://doi-org.proxy.timbo.org.uy/10.1177/0743558419839226

Kreager, D. A., Rulison, K., & Moody, J. (2011). Delinquency and the structure of adolescent peer groups. *Criminology, 49*, 95–127. https://doi.org/10.1111/j.1745-9125.2010.00219.x

Kulis, S. S., Nieri, T., Yabiku, S., Stromwall, L., & Marsiglia, F. F. (2007). Promoting reduced and discontinued substance use among adolescent substance users: Effectiveness of a universal prevention program. *Prevention Science, 8*, 35–49.

Kulis, S. S., Marsiglia, F. F., Porta, M., Arévalo Avalos, M. R., & Ayers, S. L. (2019). Testing the keepin' it REAL substance use prevention curriculum among early adolescents in Guatemala City. *Prevention Science, 20*, 532–543.

Kulis, S. S., García-Pérez, H. M., Marsiglia, F. F., & Ayers, S. L. (2021a). Testing a culturally adapted youth substance use prevention program in a Mexican border city: Mantente REAL. *Substance Use and Misuse, 56*(2), 245–257. https://doi.org/10.1080/10826084.2020.1858103

Kulis, S. S., Marsiglia, F. F., Medina-Mora, M. E., Nuño-Gutiérrez, B. L., Corona, M. D., & Ayers, S. L. (2021b). Keepin' it REAL – Mantente REAL in Mexico: A cluster randomized controlled trial of a culturally adapted substance use prevention curriculum for early adolescents. *Prevention Science, 22*(5), 645–657. https://doi.org/10.1007/s11121-021-01217-8

Laursen, B., Hafen, C. A., Kerr, M., & Stattin, H. (2012). Friend influence over adolescent problem behaviors as a function of relative peer acceptance: To be liked is to be emulated. *Journal of Abnormal Psychology, 121*, 88–94. https://doi.org/10.1037/a0024707

Leander, N. P., vanDellen, M. R., Rachl-Willberger, J., Shah, J. Y., Fitzsimons, G. J., & Chartrand, T. L. (2016). Is freedom contagious? A self-regulatory model of reactance and sensitivity to deviant peers. *Motivation Science, 2*, 256–267. https://doi.org/10.1037/mot0000042

Le Blanc, M. (2015). Developmental criminology: Thoughts on the past and insights for the future. In J. Morizot & L. Kazemian (Eds.), *The development of criminal and antisocial behavior: Theory, research and practical applications* (pp. 507–535). Springer International Publishing. https://doi.org/10.1007/978-3-319-08720-7_32

Lerner, R. M. (2006). Developmental science, developmental systems, and contemporary theories of human development. In R. M. Lerner & W. Damon (Eds.), *Handbook of child psychology:*

Vol. 1. Theoretical models of human development (6th ed., pp. 1–17). Wiley. https://doi.org/10.1002/9780470147658.chpsy0101

Magnusson, D., & Stattin, H. (2006). The person in context: A holistic-interactionistic approach. In R. M. Lerner & W. Damon (Eds.), *Handbook of child psychology: Vol. 1. Theoretical models of human development* (6th ed., pp. 400–464). Wiley. https://doi.org/10.1002/9780470147658.chpsy0108

Mahoney, J. L., Stattin, H., & Lord, H. (2004). Unstructured youth recreation centre participation and antisocial behaviour development: Selection influences and the moderating role of antisocial peers. *International Journal of Behavioral Development, 28*, 553–560.

Malacas, C., Alfaro, P., & Hernández, R. M. (2019). Factores predictores de la intención de consumo de marihuana en adolescentes de nivel secundaria. *Health and Addictions, 2*, 20–27. https://doi.org/10.21134/haaj.v20i2.481

Marsiglia, F. F., & Hecht, M. L. (2005). *Keepin'it REAL: An evidence-based program.* ETR Associates.

Marsiglia FF, Kulis S, Martinez Rodriguez G, Becerra D, Castillo J. (2009). Culturally specific youth substance abuse resistance skills: Applicability across the US-Mexico.

Marsiglia, F. F., & Booth, J. (2015). Cultural adaptation of interventions in real practice settings. *Research on Social Work Practice, 25*, 423–432.

Marsiglia, F. F., Kulis, S., Booth, J. M., Nuño-Gutiérrez, B. L., & Robbins, D. E. (2015). Long-term effects of the keepin'it REAL model program in Mexico: Substance use trajectories of Guadalajara middle school students. *Journal of Primary Prevention, 36*, 93–104.

Marsiglia, F. F., Kulis, S., Kiehne, E., Ayers, E., Libisch Recalde, C., & Barros Sulca, L. (2018). Adolescent substance use prevention and legalization of Marijuana in Uruguay: A feasibility trial of the keepin' it REAL prevention program. Journal of Substance Use, 23, 457-465. Border. *Research on Social Work Practice, 19*(2), 152–164.

Marsiglia, F. F., Medina-Mora, M. E., Gonzalvez, A., Alderson, G., Harthun, M., Ayers, S., Nuño Gutiérrez, B., Corona, M. D., Mendoza Melendez, M. A., & Kulis, S. (2019). Binational cultural adaptation of the keepin' it REAL substance use prevention program for adolescents in Mexico. *Prevention Science, 20*, 1125–1135.

McFarland, L. A., & Ployhart, R. E. (2015). Social media: A contextual framework to guide research and practice. *Journal of Applied Psychology, 100*, 1653–1677. https://doi.org/10.1037/a0039244

McGloin, J. M. (2009). Delinquency balance: Revisiting peer influence. *Criminology, 47*, 439–477. https://doi.org/10.1111/j.1745-9125.2009.00146.x

Meldrum, R. C., & Barnes, J. C. (2017). Unstructured socializing with peers and delinquent behavior: A genetically informed analysis. *Journal of Youth and Adolescence, 46*, 1968–1981.

Melgar, A. K., Reynés, M. D., & Paredes-Labra, J. (2017). Plan CEIBAL e inclusión social. Un caso paradigmático/Plan CEIBAL and social inclusion. A paradigmatic case./Plano CEIBAL e inclusão social. Um caso paradigmático. *Psicología, Conocimiento y Sociedad, 7*(2), 45–63. https://doi-org.proxy.timbo.org.uy/10.26864/pcs.v7.n2.4

Moffitt, T. E. (2018). Male antisocial behavior in adolescence and beyond. *Nature Human Behavior, 2*, 177–186. https://doi.org/10.1038/s41562-018-0309-4

Monahan, K. C., Rhew, I. C., Hawkins, J. D., & Brown, E. C. (2014). Adolescent pathways to co-occurring problem behavior: The effects of peer delinquency and peer substance use. *Journal of Research on Adolescence, 24*, 630–645. https://doi.org/10.1111/jora.12053

Nagoshi, J. L., Kulis, S., Marsiglia, F. F., & Piña-Watson, B. (2020). Accounting for linguistic acculturation, coping, antisociality and depressive affect in the gender role-alcohol use relationship in Mexican American adolescents: A moderated mediation model for boys and girls. *Journal of Ethnicity in Substance Abuse.* https://doi.org/10.1080/15332640.2020.1781732

Nakahashi, W., & Ohtsuki, H. (2017). Evolution of emotional contagion in group-living animals. *Journal of Theoretical Biology, 15*(440), 12–20.

Nesi, J., Rothenberg, W. A., Hussong, A. M., & Jackson, K. M. (2017). Friends' alcohol-related social networking site activity predicts escalations in adolescent drinking: Mediation by peer norms. *Journal of Adolescent Health, 60*, 641–647. https://doi.org/10.1016/j.jadohealth.2017.01.009

Nesi, J., Choukas-Bradley, S., & Prinstein, M. J. (2018). Transformation of adolescent peer relations in the social media context: Part 2-application to peer group processes and future directions for research. *Clinical Child & Family Psychology Review, 21*(3), 295–319. https://doi-org.proxy.timbo.org.uy/10.1007/s10567-018-0262-9

Nicol, C. (1995). The social transmission of information and behavior. *Applied Animal Behaviour Science, 44*, 79–98.

Niland, P., Lyons, A. C., Goodwin, I., & Hutton, F. (2015). Friendship work on Facebook: Young adults' understandings and practices of friendship. *Journal of Community & Applied Social Psychology, 25*(2), 123–137.

Nummenmaa, L., Hirvonen, J., Parkkola, R., & Hietanen, J. (2008). Is emotional contagion special? An fMRI study on neural systems for affective and cognitive empathy. *NeuroImage, 43*, 571–580.

Osgood, D. W., Wilson, J. K., O'Malley, P. M., Bachman, J. G., & Johnston, L. D. (1996). Routine activities and individual deviant behavior. *American Sociological Review, 61*, 635–655. https://doi.org/10.2307/2096397

Palardy, G. J. (2013). High school socioeconomic segregation and student attainment. *American Educational Research Journal, 50*(4), 714–754. https://doi-org.proxy.timbo.org.uy/10.2307/23526103

Patterson, G. R., & Yoerger, K. (1999). Intraindividual growth in covert antisocial behaviour: A necessary precursor to chronic juvenile and adult arrests? *Criminal Behaviour and Mental Health, 9*, 24–38. https://doi.org/10.1002/cbm.289

Pérez, A., & Mejía, J. (2015). *Evolución de la prevención del consumo de drogas en el mundo y en américa latina*. Corporación Nuevos Rumbos. ISBN: 978-958-57904-21.

Peter, J., & Valkenburg, P. M. (2013). The effects of internet communication on adolescents' psychosocial development: An assessment of risks and opportunities. In E. Scharrer (Ed.), *The international encyclopedia of media studies, volume 5: Media effects/media psychology* (pp. 678–697). Wiley-Blackwell.

Pilatti, A., Brussino, S. A., & Godoy, J. C. (2013). Factores que influyen en el consumo de alcohol de adolescentes argentinos: un path análisis prospectivo. *Revista de Psicología, 22*, 22–36. https://doi.org/10.5354/0719-0581.2013.27716

Reeves, L., Dustman, P., Holleran, L., & Marsiglia, F. F. (2008). Creating culturally grounded prevention videos: Defining moments in the journey to collaboration. *Journal of Social Work Practice in the Addictions, 8*, 65–94.

Rubin, K. H., Bukowski, W. M., Parker, J. G., & Bowker, J. C. (2008). Peer interactions, relationships, and groups. In W. Damon & R. M. Lerner (Eds.), *Child and adolescent development: An advanced course* (pp. 141–180). John Wiley & Sons.

Ruiz, P. (2015). ¿Qué sabemos sobre el contagio emocional?. Definición, evolución, neurobiología y su relación con la psicoterapia. *Cuadernos de Neuropsicología, 9*(3), 15–24.

Ruiz, P., Calliari, A., & Pautassi, R. (2018). Reserpine-induced depression is associated in female, but not in male, adolescent rats with heightened, fluoxetine-sensitive, ethanol consumption. *Behavioral Brain Research, 21*(348), 160–170.

Ruiz, P. (2019). *Estudio de variables cognitivas, neuroendócrinas y consumo de alcohol en modelos animales de trastornos de estados de ánimo, y su contraparte en humanos (Tesis de Doctorado no publicada)*. Universidad Nacional de Córdoba.

Sanchez, C., Andrade, P., Bentancourt, D., & Vital, G. (2013). Escala de Resistencia a la presión de los amigos para el consumo de alcohol. *Actas de Investigación Psicológica, 3*(1), 917–929.

Sancassiani, F., Pintus, E., Holte, A., Paulus, P., Moro, M. F., Cossu, G., & Lindert, J. (2015). Enhancing the emotional and social skills of the youth to promote their wellbeing and positive development: A systematic review of universal school-based randomized controlled trials. *Clinical Practice and Epidemiology in Mental Health, 11*(Suppl 1 M2), 21. https://doi.org/10.2174/1745017901511010021

Scheier, L. M. (2012). Primary prevention models: The essence of drug abuse prevention in schools. In H. Shaffer, D. A. LaPlante, & S. E. Nelson (Eds.), *APA addiction syndrome handbook (Recovery, prevention, and other issues)* (Vol. 2, pp. 197–223). American Psychological Association. https://doi.org/10.1037/13750-009

Snyder, J. J., Schrepferman, L. P., Bullard, L., McEachern, A. D., & Patterson, G. R. (2012). Covert antisocial behavior, peer deviancy training, parenting processes, and sex-differences in the development of antisocial behavior during childhood. *Development and Psychopathology, 24*, 1117–1138. https://doi.org/10.1017/S0954579412000570

Steinberg, L. (2017). *Adolescence* (11th ed.). McGraw-Hill Education.

Subrahmanyam, K., & Šmahel, D. (2011). *Digital youth: The role of media in development.* Springer. https://doi.org/10.1007/978-1-4419-6278-2

Sutherland, E. H. (1972). The theory of differential association. In D. Dressler (Ed.), *Readings in criminology and penology* (pp. 365–371). Columbia University Press. https://doi.org/10.7312/dres92534-039

Tanner-Smith, E. E., Durlak, J. A., & Marx, R. A. (2018). Empirically based mean effect size distributions for universal prevention programs targeting school-aged youth: A review of meta-analyses. *Prevention Science, 19*, 1091–1101. https://doi.org/10.1007/s11121-018-0942-1

Thurow, C. F., Junior, E. L. P., Westphal, R., de São Tiago, F. M. L., & Schneider, D. R. (2020). Risk and protective factors for drug use: A scoping review on the communities that care youth survey. *International Journal of Advanced Engineering Research and Science (IJAERS), 7*. https://doi.org/10.22161/ijaers.711.5

Tompsett, C. J., Domoff, S. E., & Toro, P. A. (2013). Peer substance use and homelessness predicting substance abuse from adolescence through early adulthood. *American Journal of Community Psychology, 51*, 520–529. https://doi.org/10.1007/s10464-013-9569-3

Zentall, T., & Levine, J. (1972). Observational learning and social facilitation in the rat. *Science, 178*, 1220–1221.

The Implications of Exposure to Neurotoxic Metals for Cognitive Development of Children and Adolescents

Cassio Santos-Lima, Chrissie Ferreira de Carvalho, Homegnon Antonin Farréol Bah, and José Antonio Menezes-Filho

Neurodevelopment is subject to a series of interferences, according to the moment of the life cycle, and there is a consensus in the literature that exposure to metals has adverse effects on cognitive functioning, considering its functional use and academic-social insertion throughout life. Children and adolescents, given the moment during life, especially considering the sensitive and critical periods related to the Central Nervous System (CNS), are potentially more vulnerable. In Latin America, several studies have indicated adverse exposure to metals such as mercury, manganese, and lead, among others, which have had negative consequences for developing individuals, especially for those who live in contexts far from urban centers and with increased risk factors for poverty and social marginalization. The purpose of this chapter is to present evidence related to the exposure of some neurotoxic metals in South America (manganese, mercury, and lead) and their effects

C. Santos-Lima (✉)
Curso de Psicologia, Escola Bahiana de Medicina e Saúde Pública (EBMSP), Salvador, Bahia, Brazil

Programa de Pós-Graduação em Psicologia, Instituto de Psicologia, Universidade Federal da Bahia, Salvador, Bahia, Brazil
e-mail: cassiolima@bahiana.edu.br

C. F. de Carvalho
Programa de Pós-Graduação em Psicologia, Departamento de Psicologia, Universidade Federal de Santa Catarina, Florianópolis, Brazil

H. A. Farréol Bah
Programa de Pós-Graduação em Saúde Coletiva, Instituto de Saúde Coletiva, Universidade Federal da Bahia, Salvador, Bahia, Brazil

J. A. Menezes-Filho
Programa de Pós-Graduação em Farmácia, Faculdade de Farmácia, Universidade Federal da Bahia, Salvador, Bahia, Brazil
e-mail: antomen@ufba.br

M. V. Alves et al. (eds.), *Cognitive Sciences and Education in Non-WEIRD Populations*, https://doi.org/10.1007/978-3-031-06908-6_9

on cognition, behavior, and the development of academic competence in children and adolescents.

Mercury and the Its Effects on Human Health and Cognitive Development

Mercury (Hg) is a metal found in three forms—metallic, inorganic, and organometallic compounds—and can be found in air, water, soil, plants, and animals. The elemental mercury is released into the air, primarily by natural sources like volcanic emissions, and by anthropogenic activities, usually from industries and by artisanal gold mining, and is oxidized to the ionic forms, which is converted to methylmercury (MeHg) by anaerobic bacteria in the sediments of lakes, rivers, and sea, entering in the food chain (Davidson et al., 2004; Jedrychowski et al., 2006). Mercury has been recognized as a serious pollutant of aquatic ecosystems (Kehrig et al., 2009), and human exposure to MeHg usually occurs through oral ingestion, with full gastrointestinal absorption, which leads to a rapid distribution to all tissues via the bloodstream, including central nervous system (CNS). The main source of human exposure to MeHg is through consumption of fish and other freshwater and marine animals, such as whale meat and shellfish, which carry variable concentrations of MeHg, associated with carbon radicals (Davidson et al., 2004).

The toxicity of MeHg is related to its association with water-soluble molecules such as proteins or amino acids (Montgomery et al., 2008), and due to its high liposolubility, it targets the myelin sheath where it effectively inhibits neuronal excitability (ATSDR, 1999). MeHg has the ability to cross the blood–brain barrier and accumulate in several areas of the CNS, such as the prefrontal cortex, striatum, hippocampus, and cerebellum (Montgomery et al., 2008; Mourão, 2016). The main reports of Hg poisoning are related to the consumption of fish highly contaminated by MeHg in the 1950s, in Japan, mainly in Minamata and Niigata. From the babies born to pregnant women who consumed the contaminated fish, it became evident that the fetal brain can be highly sensitive to exposure to MeHg (Davidson et al., 2004; Grandjean & Herz, 2015).

MeHg accumulated in the food chain after the biomagnification process poses a significant risk to fish consumers, particularly women of child-bearing age and young children (Hacon et al., 2014). High exposure to MeHg in the prenatal period can cause severe toxic encephalopathy, due to oxidative stress and the possible influence of MeHg in the disarrangement of chromosome distribution during mitosis, affecting the mechanisms of protection and genome integrity (Cano, 2014). Thus, it can be estimated that exposure to MeHg can cause structural changes in the CNS, cognitive impairments, and in social insertion characteristics (Castoldi et al., 2008; Montgomery et al., 2008).

Some factors must be considered to estimate the impact of exposure to MeHg, such as the insertion context, urban or rural area; the age of the group; the source of exposure, and the measure of obtaining accumulation of this metal. Knowing about the type of fish consumed, whether predators or nonpredators, can contribute to the establishment of risk references for the population, since predatory fish, when compared to nonpredatory fish, have a higher concentration of MeHg (Passos et al., 2008). The origin of the population must also be considered, in view of the access to the types of food (Hacon et al., 2014; Vieira et al., 2013), as well as the measures the levels of MeHg in the body, quantification of consumption household fish, and MeHg analysis by hair, blood, and urine collection (Mourão, 2016). For populations with high fish consumption, hair mercury reference values above 10 μg/g are associated with a 5% risk of neurological damage (IPCS, 1990), and in this case, it is worth noting that individual approaches to prevention are limited.

Although there are beneficial effects related to fish consumption, considering its essential role in daily nutrition for more than 2.9 billion people worldwide (Myers et al., 2015), it is highlighted that exposure to MeHg and consumption of fish by women and infants are associated with developmental risks (Sheehan et al., 2014). Grandjean and Landrigan (2014) highlight that MeHg neurotoxicity occurs at exposures much lower than concentrations that affect adult brain function. Ha et al. (2017) carried out a systematic literature review of the literature, which found 815 published articles referring to exposure to Hg. The main results indicate that it is difficult to compare Hg exposure between populations with different biomarkers and estimates; that the fetus is very susceptible to exposure to Hg during organogenesis; and that prenatal exposure to MeHg associated with low birth weight can delay neurological development and growth in children and women of child-bearing age.

Excess metals in the body can cause adverse effects on neuropsychological functioning, with effects on attention, visual memory, and working memory, among others, for example (Carvalho et al., 2014; Grandjean et al., 1999a). Exposure to metals with neurotoxic potential can mainly influence the development of cognitive functions in children and adolescents, as the brain maturation process has not reached its maximum level of development.

Over the last decades, the consequences of contamination by MeHg have been presented, considering different moments of neurodevelopment, such as the prenatal period, childhood, adolescence, and even adulthood. The prospective study of the Faroe Islands is one of the largest and most intensive cohort studies to date to elucidate the possible neurobehavioral effects of exposure to MeHg. Data from this cohort study support the idea that it is important to consider the exposure of a neurotoxic metal as a modifier of neuropsychological aspects related to intelligence, attention, verbal memory, visuospatial memory, and motor function, among others, both during childhood development and along adult life (Debes et al., 2006; Debes et al., 2016). The authors point out that the effects on brain function associated with prenatal exposure to methylmercury appear to be multifocal and permanent, even though fish consumption is associated with beneficial nutrients in human health studies.

In the Amazon region of Brazil, some studies report the adverse effects of exposure to MeHg on neuropsychological functions of riverine populations. Grandjean et al., 1999b) showed that the most exposed communities performed worse on neuropsychological tasks that assessed attention, short-term memory, visuospatial memory, working memory, and motor function. Chevrier et al. (2009), with the same cohort, from a secondary analysis indicate that exposure to mercury was negatively associated with the scores of the drawing task that assessed the episodic visual spatial memory. A score reduction of 1.2 points was observed in the children with hair mercury concentration above 10 μg/g compared to those with hair level below 1 μg/g. From the studies mentioned above, it is possible to see that, considering the research context and the sample characteristics, there may or may not be neuropsychological alterations associated with exposure to MeHg.

Santos-Lima et al. (2020) investigated the relationship between environmental exposure to mercury in riverine communities in the Brazilian Amazon and the neuropsychological functioning of children and adolescents aged 6 to 14 years. The results indicate that high concentrations of total hair Hg were associated with a lower performance in neuropsychological functions, mainly for measures of Verbal Intelligence Quotient (IQ), total IQ, phonological verbal fluency, and visuospatial working memory. Santos-Sacramento et al. (2021), in a review study, found that, for samples from the three main hydrographic basins in the Amazon region, Tapajós, Tocantins, and Madeira, hair mercury levels were two to ten times higher than the recommended reference level, with effects on neurocognitive functioning, in addition to visual, motor, and somatosensory repercussions.

For countries in Latin America and the Caribbean, current levels of exposure to toxicants such as MeHg are dangerous and harmful to children's neurodevelopment, and reducing exposure to neurotoxic metals is essential to protect children's health and development (Dórea, 2020). Brazil as a signatory country to the Minamata Convention must play a central role in the discussion of the effects of exposure to Hg, with a view to minimizing the negative impacts on human health (Crespo-Lopez et al., 2021). It is essential to discuss and adopt policies and actions that prevent and mitigate problems related to the neurodevelopment of children in this region.

The data described above show that even in different contexts, mercury poses a risk to the neuropsychological development of children and adolescents. At the same time, it should be considered that there may be factors that meliorate the effects of exposure on cognitive development, such as the action of nutrients present in fish meat, when this is the route of exposure. Thus, it is important to study and understand how exposure to mercury can affect infants and adolescents, considering the socioenvironmental and economic characteristics, since mercury can affect neuropsychological development in the maturation process.

Children and adolescents with executive difficulties can present learning difficulties, and, in the Brazilian scenario, studies indicate negative outcomes for the educational development of children. The impacts on children's health caused by neurotoxic exposure vary widely across developing countries, and children's rights to a safe environment where they can grow, develop, play, and learn are dangerous when public policy does not consider such related factors (Perlroth & Castelo Branco, 2017). According to Santiago de Vasconcellos (Santiago de Vasconcellos, 2016), based on an estimation study, individuals from the northern region of Brazil are at higher risk of mild intellectual disability associated with exposure to methylmercury. The study data indicate that children can be born with a loss of 2 IQ scores, a loss that can contribute to the maintenance of a low human development scenario, which compromises not only economic growth, but also intellectual growth.

According to the report of the United Nations Children's Fund (Unicef, 2018), in Brazil, there are about seven million children and adolescents with a discrepancy between school grade and age. In the northern region, notably studied regarding exposure to mercury, most states show discrepancies in the early and final years of elementary and high school. For populations environmentally exposed to contaminants such as mercury, unfavorable socioeconomic conditions and relative geographic isolation are added, preventive actions in terms of health and education must be considered to minimize the risks of developmental and social insertion problems.

Manganese Exposure and Children Cognitive and Achievement Performance

Manganese (Mn) is an essential microelement; however, excessive exposures have been associated with neurotoxic effects with cognitive impairment in children. Among the main sources of exposure, there are the digestive routes with the consumption of food and water, through contact and through the respiratory route. Exposure via the respiratory route is considered the fastest and represents the greatest vulnerability since inhaled Mn can be transported directly to the brain through the olfactory nerves (Brenneman et al., 2000; Dorman et al., 2006).

The evidence has shown that children living close to areas with industrial activity or mining residues may be more susceptible to environmental exposure to Mn in Brazil (Carvalho et al., 2014; Menezes-Filho et al., 2011) and Mexico (Catalán-Vázquez et al., 2010; Torres-Agustín et al., 2012). Furthermore, exposure has been reported from drinking water collected from taps in Canada (Bouchard et al., 2018) and from wells in Bangladesh (Khan et al., 2012; Wasserman et al., 2011). An imaging study using fMRI with adolescents in Italy with history of environmental Mn exposure identified reduced response from limbic system, suggesting an alteration of brain network related to emotional responses (Iannilli et al., 2016).

Studies have indicated that Mn can accumulate in brain regions such as the basal ganglia, which has been reported as one of the regions with the highest accumulation of Mn (Dobson et al., 2004). Considering studies with animals chronically exposed to Mn demonstrated an increase of Mn in regions of the hippocampus, frontal cortex, and brainstem (Burton & Guilarte, 2009; Dorman et al., 2001; Guilarte et al., 2006; Schneider et al., 2009), in addition to disrupting dopamine neuron function in the striatum and causing degeneration in the frontal cortex (Guilarte, 2015). These brain regions are related to executive functions and consequently to learning that depend on the frontostriatal circuit and the hippocampus.

The meta-analysis performed by Liu et al. (2020) evaluated the association between IQ and different metrics of exposure to Mn in hair, blood, and drinking water. A tenfold increase of Mn in hair concentrations was associated with a decrease of 2.51 points on Total IQ. When analysis was performed considering the source of exposure, a negative association was found only between Mn in hair and Total IQ related to Mn airborne exposure, but not to water or exposure to Mn mining residues. Considering studies in which the source of exposure of Mn was air, a tenfold increase in hair manganese was associated with a 7.62 point decrease in Total IQ (Liu et al., 2020). In studies with children younger than 6 years, the main effects have been reported by prospective studies and related to neurodevelopment, mainly cognitive, and motor development (Claus Henn et al., 2017; Takser et al., 2003).

Essential and learning-related functions such as working memory (Carvalho et al., 2014; He et al., 1994; Riojas-Rodríguez et al., 2010; Wasserman et al., 2011), verbal memory, and learning (Carvalho et al., 2018; Oulhote et al., 2014; Torres-Agustín et al., 2012; Wright et al., 2006) have been reported in studies such as impaired functions and associated with different markers of exposure to Mn. Moreover, exposure to Mn has been associated with aggressive and disruptive behaviors (Khan et al., 2011; Menezes-Filho et al., 2014; Rodrigues et al., 2018a) and hyperactivity (Bouchard et al., 2007; Ericson et al., 2007).

In Brazil, a series of studies identified high levels of Mn in environmentally exposed children between 7 and 12 years old living close to ferromanganese alloy plant in the region of Simões-Filho Bahia. A negative association was found between Mn in hair levels and IQ, visuospatial organization, and verbal working memory on children (Carvalho et al., 2014). Furthermore, externalizing behavior and hyperactivity were positively related with Mn exposure (Menezes-Filho et al., 2014; Rodrigues et al., 2018b). Further study in this same region, by Menezes-Filho et al. (2018), evaluated the coexposure to Pb, and Mn and the results showed a negative association between IQ and Pb levels on blood only with children with high Mn on toenails (MnTn), but no association was found with children with low MnTn. An executive function intervention Heroes of the Mind was conducted with children from this same region in Brazil (Carvalho, 2017). The results indicated that the group of children exposed to Mn that received the intervention presented more expressive gains with greater effect size compared to the exposure control group in working memory, cognitive flexibility, and phonemic verbal fluency, and there were transferring effects to other cognitive and academic functions such as visual attention and written skills.

In relation to school achievement, the study conducted by Khan et al. (2012) with 840 children between 8 and 11 years old found an association between Mn levels in drinking water and mathematics performance, but not with language scores. Bhang et al. (2013) conducted a study with 1089 children between 8 and 11 years old, and the results revealed that the group with higher levels of Mn showed lower scores in academic performance in relation to thinking, reading, calculations and quotient learning, and a higher number of errors per commission on the continuous performance test (CPT). In Brazil, a study with 63 children aged between 6 and 12 years found a negative association between manganese in the drinking water with writing and language skills, in addition to impaired executive functions (Nascimento et al., 2016).

Lead and Its Compounds

Lead (Pb) is a posttransition toxic metal, soft, malleable, and corrosion resistant (Acharya, 2013). Although it is a natural component of the earth's crust and ubiquitous in the environment, its main source of exposure is due to anthropogenic activities: since antiquity and, mainly, after the industrial revolution which catapulted mining activities and its numerous applications due to its unique physicochemical properties (Acharya, 2013; Burt, 1969; Papanikolaou et al., 2005). As a result of this widespread use, this metal has become abundant in the environment, and despite recent worldwide regulations implemented to control its use, it remains a worrying pollutant for air, soil, water resources, and consequently for agricultural products (Acharya, 2013; ATSDR, 2020).

Human activities that constitute the main sources of exposure to lead include mining, industrial, crafts such as lead glazed ceramics, Pb-based paint, the improper disposal of e-waste, and used batteries (ATSDR, 2020; Bah et al., 2020; Dórea, 2021; O'Connor et al., 2018). Other sources of exposure result from the historical and/or prolonged use of products with considerable lead content: additive gasoline (since the early 1920s), old pipes, and solders (ATSDR, 2020). As Pb is not biodegradable, these previous uses have left a legacy of contaminated sites (ATSDR, 2020; Dórea, 2020; Mohmand et al., 2015; Wong et al., 2003).

The general population is exposed to this metal from inhaled air, food, water, soil, and dust. Several studies have reported significant correlations between Pb levels in these matrices with its biological levels in humans or even in animals: whole blood, hair, and nails (ATSDR, 2020; Brown & Margolis, 2012; Mohmand et al., 2015; Olympio et al., 2018; Rodrigues et al., 2018a). Some factors or situations increase the risk of exposure of populations in general, and particularly in children: socioeconomic level, proximity to contaminated industrial sites, active smoker at home, burning household waste, living near heavy traffic roads or communities having artisanal activities with Pb, such as Oaxaca, Mexico, and Maragogipinho, Brazil (ATSDR, 2020; Azcona-Cruz et al., 2000; Bah et al., 2020; Menezes-Filho et al., 2012).

The route of entry into the human body is mainly by respiratory tract for adults in occupational setting, digestive, especially for children, prenatal exposure, and breast feeding for the newborn. For those, breast milk represents the main source of exposure given Pb similarity to calcium ion (Ca^{2+}) (ATSDR, 2017; ATSDR, 2020; Dórea, 2021). Pb appears to be distributed in the same way, regardless of the route of absorption: in blood it has a half-life of 26–35 days (ATSDR, 2017; Papanikolaou et al., 2005; Ramírez Ortega et al., 2021), in soft tissues such as the brain, liver, and kidneys, it has a half-life of 1–2 months (National Research Council, 1993), and in bones, it has a half-life ranging from 5 to 30 years (ATSDR, 2020). Pb can easily cross the placental and blood–brain barrier (Ramírez Ortega et al., 2021). In general, studies indicate that Pb distribution in the body appears to be similar in children and adults, although most accumulate in bones more in adults (94%) than in children (73%) (ATSDR, 2020; Papanikolaou et al., 2005). Absorbed Pb is excreted mainly in urine and feces, but also in sweat, saliva, hair, nails, and breast milk (ATSDR, 2020; Ramírez Ortega et al., 2021).

Exposure to lead causes damage to almost all vital systems and organs: blood tissue, central and peripheral nervous system, kidneys, cardiovascular system (hypertension), and reproductive system (male infertility). Neurological disorders depend on the degree of exposure, ranging from loss of intellectual performance, mental retardation, maladaptive behavior, to encephalopathy and/or death (Council On Environmental Health, 2016; ATSDR, 2020; Haefliger et al., 2009).

Lead affects the brain through direct and indirect mechanisms. The first is mainly due to its ability to replace the Ca^{2+}, explaining its ability to cross the blood–brain and placental barriers (Dórea, 2019), impact the release of neurotransmitters, induce mitochondrial apoptosis, and affect the activity of second messengers (protein kinases, cyclic AMP, potassium channels) through the activation of calmodulin (Bressler & Goldstein, 1991; Ramírez Ortega et al., 2021). Other mechanisms that should be considered are the sequestration and remobilization of Pb from bone tissue, the effect on various molecules such as synaptotagmin I, NMDA receptors, and protein kinase C (PKC) (Bressler & Goldstein, 1991; Lidsky & Schneider, 2003).

For the fetus, Pb sequestered in maternal bones represents the most important source of exposure; once released, Pb easily crosses the placenta, putting the fetus at risk due to the possibility of being in direct contact with the developing brain (Ramírez Ortega et al., 2021). Pb can indirectly impact the nervous system through other interconnected systems. D-aminolevulinic acid (ALA) suppresses gamma-aminobutyric acid (GABA)-mediated neurotransmission, inhibiting its release and also likely competing with GABA receptors (Lidsky & Schneider, 2003). The ability to reduce iron absorption, in case of its deficiency, from the intestine can aggravate anemia, which is also associated with impaired cognitive neuropsychological development (Kordas, 2010).

Lead has been associated with damage to the intellectual development of children in many studies (Council On Environmental Health, 2016; Dórea, 2021; Ide & Parker, 2005; Koller et al., 2004). Considered without toxic threshold, its exposure among children is associated with developmental problems, including cognitive function impairment, attention deficit, hyperactivity, and short stature (Council On Environmental Health, 2016; Betts, 2012). Haeflinger et al. (Haefliger et al., 2009) reported severe neurological disorders (neuropsychiatric disorders such as irritability, anxiety, sleep disturbances, and aggressiveness) in children exposed to extremely high blood lead levels (BLL) due to car battery recycling without control and protection measures. The study was initiated after the death of 18 children caused by a serious central nervous system disease, with rapid evolution and unexplained origin. BLL circa 129.5 μg/dL was observed in children still alive from the same neighborhood. Low BLLs have also been reported as harmful to children due to cognitive impairments (Betts, 2012; Council on Environmental Health, 2016; Dórea, 2021).

Despite the drop in BLL in the United States (since the 1970s) or in Western richer countries, there is strong evidence in the literature of the occurrence of deficiencies in cognitive and academic skills associated with BLL below 5 μg/dL, current limit recommended by Betts (2012). In Latin America, although little is known about the degree of exposure in children (Olympio et al., 2017), several studies have also reported worrying situations in areas known to have Pb-related activities (Bah et al., 2020; Olympio et al., 2017). For example, in Mexico, Gomaa et al. (2002) reported that Pb levels in the mother's umbilical cord blood and trabecular bone were inversely associated with Mental Development Index (MDI) scores. They concluded that Pb stored in bone could represent a risk factor for neurodevelopment in infants at 24 months of age.

More recently, Horton et al. (2018) showed the relationship between postnatal (dentine) Pb level and the increase in anxiety symptoms in children aged 8–11 years, while Jansen et al. (2019) found association of elevated BLL with greater sleep fragmentation in young adolescents: patients in the highest quartile of BLL in childhood had an average of sleep duration 23 minutes lower than those in the first quartile in adolescence. Although several have found significant inverse associations, others have reported inconclusive results (Dórea, 2020; Buchanan et al., 2011, thus showing the need for further investigation to better understand the mechanisms of Pb neurotoxicity.

It is important to note that over the last few decades, due to the need to consider what happens in real life and social conditions, several investigations have focused on exposure to multiple contaminants. Valeri et al. (2017) estimated that, compared to exposure to a single metal, coexposure to several metals early in life may increase the neurotoxicity resulting from this mixture. Sanders et al. (2015) reviewed studies of the effects of neurotoxic metal mixtures on neurodevelopment and found that the effect associated with Pb was part of most of the studies reviewed.

In Brazil, Menezes-Filho et al. (2018) reported BLLs below 5 μg/dL and a median of 1.8 μg/dL, and schoolchildren living near a ferromanganese alloy plant were significantly associated with loss of IQ points, especially in those with higher exposure to manganese. Most exposures to Pb-containing mixtures occur early in life, during the most vulnerable period, pregnancy, and lactation. Thus, even before birth, exposure to Pb along with other neurotoxic metals can act together to produce adverse effects on normal neurodevelopmental processes, thus causing altered functions, such as cognition and/or behavior later in childhood (Dórea, 2019). In these cases, the central nervous system integrates neurotoxic substances with converging and adverse mechanisms. In addition, environmental exposure to Pb can also interact with other risk factors, such as maternal stress, to negatively affect neurodevelopment (Tamayo et al., 2017).

Final Considerations

Children exposed to metal are at risk for the development of essential functions related to learning. In addition, children and adolescents are usually vulnerable to other social stressors such as low socioeconomic status and low parental education (Lucchini et al., 2019). Studies with populations exposed to metals show that there is a sum of risk factors that are usually present that include low socioeconomic status, low parental education, difficulty in accessing health resources, and low-quality education, thus configuring a context of high vulnerability. The accumulation of risk factors and negative adverse experiences in childhood have been emphasized as being crucial to the formation of brain architecture, consequently leading to negative consequences throughout life for cognitive functioning and expression of adaptive behaviors (National Scientific Council on the Developing Child, 2006).

Studies of some countries of Latin America have showed cognitive and behavioral impairment associated with exposure to metals such as Hg, Mn, and Pb. For example, mercury continues to be used in the extraction of gold in the Amazon region, as well as lead in the countryside of Brazil or manganese in the production of iron alloys, affecting the health of different populations. Considering the risk of continued exposure of metals in the countries and the sociovulnerable conditions of populations, it is important to conduct studies to describe the main effects of exposure, as well as think of ways to minimize risks and enable specific interventions to promote and develop essential cognitive skills for life.

References

Acharya, S. (2013). Lead between the lines. *Nature Chemistry, 5*(10), 894. https://doi.org/10.1038/nchem.1761

Agency for Toxic Substances and Disease Registry (ATSDR). (2020). *Toxicological profile for Lead*. U.S. Department of Health and Human Services, Public Health Service.

ATSDR, U. S. (1999). *Toxicological profile for mercury*. Atlanta – Agency for Toxic Substances and Disease Registry, US Department of Health and Human Services.

ATSDR. (2017). *Lead toxicity; Case studies in environmental medicine (CSEM)*; [accessed 2021 Aug. 27]. https://www.atsdr.cdc.gov/csem/lead/docs/csem-lead_toxicity_508.pdf

Azcona-Cruz, M. I., Rothenberg, S. J., Schnaas, L., Zamora-Muñoz, J. S., & Romero-Placeres, M. (2000). Lead-glazed ceramic ware and blood lead levels of children in the city of Oaxaca, Mexico. *Archives of Environmental Health, 55*(3), 217–222. https://doi.org/10.1080/00039890009603409

Bah, H., Bandeira, M. J., Gomes-Junior, E. A., Anjos, A., Rodrigues, Y., Dos Santos, N. R., Martinez, V. O., Rocha, R., Costa, R. G., Adorno, E. V., & Menezes-Filho, J. A. (2020). Environmental exposure to lead and hematological parameters in Afro-Brazilian children living near artisanal glazed pottery workshops. Journal of environmental science and health. *Part A, Toxic/Hazardous Substances & Environmental Engineering, 55*(8), 964–974. https://doi.org/10.1080/10934529.2020.1761738

Betts, K. S. (2012). CDC updates guidelines for children's lead exposure. *Environmental Health Perspectives, 120*(7), a268. https://doi.org/10.1289/ehp.120-a268

Bhang, S.-Y., Cho, S.-C., Kim, J.-W., Hong, Y.-C., Shin, M.-S., Yoo, H. J., ... Kim, B.-N. (2013). Relationship between blood manganese levels and children's attention, cognition, behavior, and academic performance—A nationwide cross-sectional study. *Environmental Research*, 1–8. https://doi.org/10.1016/j.envres.2013.05.006

Bouchard, M., Laforest, F., Vandelac, L., Bellinger, D., & Mergler, D. (2007). Hair manganese and hyperactive behaviors: Pilot study of school-age children exposed through tap water. *Environmental Health Perspectives, 115*(1), 122–127. https://doi.org/10.1289/ehp.9504

Bouchard, M., Surette, C., Cormier, P., & Foucher, D. (2018). Low level exposure to manganese from drinking water and cognition in school-age children. *Neurotoxicology, 64*, 110–117. https://doi.org/10.1016/j.neuro.2017.07.024

Brenneman, K. A., Wong, B. A., Buccellato, M. A., Costa, E. R., Gross, E. A., & Dorman, D. C. (2000). Direct olfactory transport of inhaled manganese ((54)MnCl(2)) to the rat brain: Toxicokinetic investigations in a unilateral nasal occlusion model. *Toxicology and Applied Pharmacology, 169*(3), 238–248. https://doi.org/10.1006/taap.2000.9073

Bressler, J. P., & Goldstein, G. W. (1991). Mechanisms of lead neurotoxicity. *Biochemical Pharmacology, 41*(4), 479–484. https://doi.org/10.1016/0006-2952(91)90617-e

Brown, M. J., & Margolis, S. (2012). Lead in drinking water and human blood lead levels in the United States. *MMWR Supplements, 61*(4), 1–9.

Buchanan, L. H., Counter, S. A., & Ortega, F. (2011). Environmental lead exposure and otoacoustic emissions in Andean children. *Journal of Toxicology and Environmental Health. Part A, 74*(19), 1280–1293. https://doi.org/10.1080/15287394.2011.587106

Burt, R. (1969). Lead production in England and Wales, 1700-1770. *The Economic History Review, 22*(2), 249–268. https://doi.org/10.2307/2593770

Burton, N. C., & Guilarte, T. R. (2009). Manganese neurotoxicity: Lessons learned from longitudinal studies in nonhuman primates. *Environmental Health Perspectives, 117*(3), 325–332. https://doi.org/10.1289/ehp.0800035

Cano, T. D. M. (2014). *Revista de Medicina e Saúde de Brasília ARTIGO DE REVISÃO Efeitos deletérios e teratogênicos da exposição ao mercúrio – Revisão da literatura*, 288–300.

Carvalho, C. F. (2017). *Programa de estimulação das funções executivas: contribuições para o desenvolvimento cognitivo de crianças em situação de vulnerabilidade e expostas ao manganês*. Tese de Doutorado, Programa de Pós-graduação em Psicologia, Universidade Federal da Bahia.

Carvalho, C. F., Menezes-Filho, J. A., De Matos, V. P., Bessa, J. R., Coelho-Santos, J., Viana, G. F. S., … Abreu, N. (2014). Elevated airborne manganese and low executive function in school-aged children in Brazil. *Neurotoxicology, 45*, 301–308. https://doi.org/10.1016/j.neuro.2013.11.006

Carvalho, C. F., Oulhote, Y., Martorelli, M., de Carvalho, C. O., Menezes-Filho, J. A., Argollo, N., & Abreu, N. (2018). Environmental manganese exposure and associations with memory, executive functions, and hyperactivity in Brazilian children. *Neurotoxicology, 69*, 253–259. https://doi.org/10.1016/j.neuro.2018.02.002

Castoldi, A. F., Onishchenko, N., Johansson, C., Coccini, T., Roda, E., Vahter, M., … Manzo, L. (2008). Neurodevelopmental toxicity of methylmercury: Laboratory animal data and their contribution to human risk assessment. *Regulatory Toxicology and Pharmacology, 51*(2), 215–229. https://doi.org/10.1016/j.yrtph.2008.03.005

Catalán-Vázquez, M., Schilmann, A., & Riojas-Rodríguez, H. (2010). Perceived health risks of manganese in the Molango Mining District, Mexico. *Risk Analysis : An Official Publication of the Society for Risk Analysis, 30*(4), 619–634. https://doi.org/10.1111/j.1539-6924.2010.01377.x

Chevrier, C., Sullivan, K., White, R. F., Comtois, C., Cordier, S., & Grandjean, P. (2009). Qualitative assessment of visuospatial errors in mercury-exposed Amazonian children. *Neurotoxicology, 30*(1), 37–46. https://doi.org/10.1016/j.neuro.2008.09.012

Claus Henn, B., Bellinger, D. C., Hopkins, M. R., Coull, B. A., Ettinger, A. S., Jim, R., … Wright, R. O. (2017). Maternal and cord blood manganese concentrations and early childhood neurodevelopment among residents near a mining-impacted Superfund site. *Environ Health Perspect, 125*(6), 067020. https://doi.org/10.1289/EHP925

Council On Environmental Health. (2016). Prevention of childhood lead toxicity. *Pediatrics, 138*(1), e20161493. https://doi.org/10.1542/peds.2016-1493

Crespo-Lopez, M. E., Augusto-Oliveira, M., Lopes-Araújo, A., Santos-Sacramento, L., Yuki Takeda, P., Macchi, B. M., do Nascimento, J., Maia, C., Lima, R. R., & Arrifano, G. P. (2021). Mercury: What can we learn from the Amazon? *Environment International, 146*, 106223.

Davidson, P. W., Myers, G. J., & Weiss, B. (2004). Mercury exposure and child development outcomes. *Pediatrics, 113*(4 Suppl), 1023–1029. https://doi.org/10.1542/peds.113.4.S1.1023
Debes, F., Budtz-Jørgensen, E., Weihe, P., White, R. F., & Grandjean, P. (2006). Impact of prenatal methylmercury exposure on neurobehavioral function at age 14 years. *Neurotoxicology and Teratology, 28*(3), 363–375. https://doi.org/10.1016/j.ntt.2006.02.004
Debes, F., Weihe, P., & Grandjean, P. (2016). Cognitive deficits at age 22 years associated with prenatal exposure to methylmercury. *Cortex, 74*, 358–369. https://doi.org/10.1016/j.cortex.2015.05.017
Dobson, W. A., Erikson, K. M., & Aschner, M. (2004). Manganese neurotoxicity. *Annals of the New York Academy of Sciences, 1012*, 115–128.
Dórea, J. G. (2019). Environmental exposure to low-level lead (Pb) co-occurring with other neurotoxicants in early life and neurodevelopment of children. *Environmental Research*. https://doi.org/10.1016/j.envres.2019.108641
Dórea, J. G. (2020). Exposure to environmental neurotoxic substances and neurodevelopment in children from Latin America and the Caribbean. *Environmental Research, 192*, 110199. https://doi.org/10.1016/j.envres.2020.110199
Dórea, J. G. (2021). Neurodevelopment and exposure to neurotoxic metal(loid)s in environments polluted by mining, metal scrapping and smelters, and e-waste recycling in low and middle-income countries. *Environmental Research, 197*, 111124. https://doi.org/10.1016/j.envres.2021.111124
Dorman, D. C., Allen, S. L., Byczkowski, J. Z., Claudio, L., Fisher, J. E., Fisher, J. W., … Mileson, B. E. (2001). Methods to identify and characterize developmental neurotoxicity for human health risk assessment. III: Pharmacokinetic and pharmacodynamic considerations. *Environmental Health Perspectives, 109*(1), 101–111. Retrieved from http://www.pubmedcentral.nih.gov/articlerender.fcgi?artid=1240547&tool=pmcentrez&rendertype=abstract
Dorman, D. C., Struve, M. F., Marshall, M. W., Parkinson, C. U., James, R. A., & Wong, B. a. (2006). Tissue manganese concentrations in young male rhesus monkeys following subchronic manganese sulfate inhalation. *Toxicological Sciences, 92*(1), 201–210. https://doi.org/10.1093/toxsci/kfj206
Ericson, J. E., Crinella, F. M., Clarke-Stewart, K. A., Allhusen, V. D., Chan, T., & Robertson, R. T. (2007). Prenatal manganese levels linked to childhood behavioral disinhibition. *Neurotoxicology and Teratology, 29*(2), 181–187. https://doi.org/10.1016/j.ntt.2006.09.020
Fundo das Nações Unidas para a Infância – UNICEF. (2018). *Panorama da distorção idade-série no Brasil*. https://www.unicef.org/brazil/media/461/file/Panorama_da_distorcao_idade-serie_no_Brasil.pdf
Gomaa, A., Hu, H., Bellinger, D., Schwartz, J., Tsaih, S. W., Gonzalez-Cossio, T., Schnaas, L., Peterson, K., Aro, A., & Hernandez-Avila, M. (2002). Maternal bone lead as an independent risk factor for fetal neurotoxicity: a prospective study. *Pediatrics, 110*(1 Pt 1), 110–118. https://doi.org/10.1542/peds.110.1.110
Grandjean, P., & Herz, K. T. (2015). Trace elements as paradigms of developmental neurotoxicants: Lead, methylmercury and arsenic. *Journal of Trace Elements in Medicine and Biology: Organ of the Society for Minerals and Trace Elements (GMS), 31*, 130–134. https://doi.org/10.1016/j.jtemb.2014.07.023
Grandjean, P., & Landrigan, P. J. (2014). Neurobehavioural effects of developmental toxicity. *The Lancet Neurology, 13*(3), 330–338. https://doi.org/10.1016/S1474-4422(13)70278-3
Grandjean, P., Budtz-Jørgensen, E., White, R. F., Jørgensen, P. J., Weihe, P., Debes, F., & Keiding, N. (1999a). Methylmercury exposure biomarkers as indicators of neurotoxicity in children aged 7 years. *American Journal of Epidemiology, 150*(3), 301–305. https://doi.org/10.1093/oxfordjournals.aje.a010002
Grandjean, P., White, R. F., Nielsen, A., Cleary, D., & de Oliveira Santos, E. C. (1999b). Methylmercury neurotoxicity in Amazonian children downstream from gold mining. *Environmental Health Perspectives., 107*(7). https://doi.org/10.1289/ehp.99107587

Guilarte, T. R. (2015). A decade of studies on manganese neurotoxicity in non-human primates: Novel findings and future directions. In L. G. Costa & M. Aschner (Eds.), *Manganese in health and disease* (pp. 459–476). The Royal Society of Chemistry.

Guilarte, T. R., McGlothan, J. L., Degaonkar, M., Chen, M.-K., Barker, P. B., Syversen, T., & Schneider, J. S. (2006). Evidence for cortical dysfunction and widespread manganese accumulation in the nonhuman primate brain following chronic manganese exposure: a 1H-MRS and MRI study. *Toxicological Sciences, 94*(2), 351–358. https://doi.org/10.1093/toxsci/kfl106

Ha, E., Basu, N., Bose-O'Reilly, S., Dórea, J. G., McSorley, E., Sakamoto, M., & Chan, H. M. (2017). Current progress on understanding the impact of mercury on human health. *Environmental Research, 152*, 419–433. https://doi.org/10.1016/j.envres.2016.06.042

Hacon, S. S., Dórea, J. G., de Fonseca, M. F., Oliveira, B. A., Mourão, D. S., Ruiz, C. M. V., ... Bastos, W. R. (2014). The influence of changes in lifestyle and mercury exposure in riverine populations of the Madeira river (Amazon basin) near a hydroelectric project. *International Journal of Environmental Research and Public Health, 11*(3), 2437–2455. https://doi.org/10.3390/ijerph110302437

Haefliger, P., Mathieu-Nolf, M., Lociciro, S., Ndiaye, C., Coly, M., Diouf, A., Faye, A. L., Sow, A., Tempowski, J., Pronczuk, J., Filipe Junior, A. P., Bertollini, R., & Neira, M. (2009). Mass lead intoxication from informal used lead-acid battery recycling in Dakar, Senegal. *Environmental Health Perspectives, 117*(10), 1535–1540. https://doi.org/10.1289/ehp.0900696

He, P., Liu, D. H., & Zhang, G. Q. (1994). Effects of high-level-manganese sewage irrigation on children's neurobehavior. *Zhonghua Yu Fang Yi Xue Za Zhi [Chinese Journal of Preventive Medicine], 28*(4), 216–218. Retrieved from http://scholar.google.com/scholar?hl=en&btnG=Search&q=intitle:Effects+of+high+level+manganese+sewage+irrigation+on+children's+neurobehavior#3

Horton, M. K., Hsu, L., Claus Henn, B., Margolis, A., Austin, C., Svensson, K., Schnaas, L., Gennings, C., Hu, H., Wright, R., Rojo, M., & Arora, M. (2018). Dentine biomarkers of prenatal and early childhood exposure to manganese, zinc and lead and childhood behavior. *Environment International, 121*(Pt 1), 148–158. https://doi.org/10.1016/j.envint.2018.08.045

Iannilli, E., Gasparotti, R., Hummel, T., Zoni, S., Benedetti, C., Fedrighi, C., ... Lucchini, R. G. (2016). Effects of manganese exposure on olfactory functions in teenagers: A pilot study. *PLoS One, 11*(1). https://doi.org/10.1371/journal.pone.0144783

Ide, L. S., & Parker, D. L. (2005). Hazardous child labor: Lead and neurocognitive development. *Public Health Reports (Washington, D.C. : 1974), 120*(6), 607–612. https://doi.org/10.1177/003335490512000608

International Programme on Chemical Safety (IPCS) – Environmental Health Criteria 101, Methylmercury. (1990). World Health Organization.

Jansen, E. C., Dunietz, G. L., Dababneh, A., Peterson, K. E., Chervin, R. D., Baek, J., O'Brien, L., Song, P., Cantoral, A., Hu, H., & Téllez-Rojo, M. M. (2019). Cumulative childhood lead levels in relation to sleep during adolescence. *Journal of Clinical Sleep Medicine : JCSM : Official Publication of the American Academy of Sleep Medicine, 15*(10), 1443–1449. https://doi.org/10.5664/jcsm.7972

Jedrychowski, W., Jankowski, J., Flak, E., Skarupa, A., Mroz, E., Sochacka-Tatara, E., ... Perera, F. (2006). Effects of prenatal exposure to mercury on cognitive and psychomotor function in one-year-old infants: Epidemiologic cohort study in Poland. *Annals of Epidemiology, 16*(6), 439–447. https://doi.org/10.1016/j.annepidem.2005.06.059

Kehrig, H. A., Palermo, E. F. A., Seixas, T. G., Santos, H. S. B., Malm, O., & Akagi, H. (2009). Methyl and total mercury found in two man-made Amazonian reservoirs. *Journal of the Brazilian Chemical Society, 20*(6), 1142–1152. https://doi.org/10.1590/S0103-50532009000600021

Khan, K., Factor-Litvak, P., Wasserman, G. A., Liu, X., Ahmed, E., Parvez, F., ... Graziano, J. H. (2011). Manganese exposure from drinking water and children's classroom behavior in Bangladesh. *Environmental Health Perspectives, 119*(10), 1501–1506. https://doi.org/10.1289/ehp.1003397

Khan, K., Wasserman, G. A., Liu, X., Ahmed, E., Parvez, F., Slavkovich, V., … Factor-Litvak, P. (2012). Manganese exposure from drinking water and children's academic achievement. *Neurotoxicology, 33*(1), 91–97. https://doi.org/10.1016/j.neuro.2011.12.002

Koller, K., Brown, T., Spurgeon, A., & Levy, L. (2004). Recent developments in low-level lead exposure and intellectual impairment in children. *Environmental Health Perspectives, 112*(9), 987–994. https://doi.org/10.1289/ehp.6941

Kordas, K. (2010). Iron, lead, and children's behavior and cognition. *Annual Review of Nutrition, 30*, 123–148. https://doi.org/10.1146/annurev.nutr.012809.104758

Liu, W., Xin, Y., Li, Q., Shang, Y., Ping, Z., Min, J., … Wang, F. (2020). Biomarkers of environmental manganese exposure and associations with childhood neurodevelopment: A systematic review and meta-analysis. *Environmental Health: A Global Access Science Source, 19*(1), 1–22. https://doi.org/10.1186/s12940-020-00659-x

Lidsky, T. I., & Schneider, J. S. (2003). Lead neurotoxicity in children: basic mechanisms and clinical correlates. *Brain: a Journal of Neurology, 126*(Pt 1), 5–19. https://doi.org/10.1093/brain/awg014

Lucchini, R. G., Guazzetti, S., Renzetti, S., Conversano, M., Cagna, G., Fedrighi, C., … Alimonti, A. (2019). Neurocognitive impact of metal exposure and social stressors among schoolchildren in Taranto, Italy. *Environmental Health: A Global Access Science Source, 18*(1), 1–12. https://doi.org/10.1186/s12940-019-0505-3

Menezes-Filho, J. A., Novaes, C. D. O., Moreira, J. C., Sarcinelli, P. N., & Mergler, D. (2011). Elevated manganese and cognitive performance in school-aged children and their mothers. *Environmental Research, 111*(1), 156–163. https://doi.org/10.1016/j.envres.2010.09.006

Menezes-Filho, J. A., Viana, G. F., & Paes, C. R. (2012). Determinants of lead exposure in children on the outskirts of Salvador, Brazil. *Environmental Monitoring and Assessment, 184*(4), 2593–2603. https://doi.org/10.1007/s10661-011-2137-0

Menezes-Filho, J. A., Carvalho, C. F., Viana, G. F. S., Ferreira, J. R. D., Nunes, L. S., Mergler, D., & Abreu, N. (2014). Elevated manganese exposure and school-aged children's behavior: A gender-stratified analysis. *Neurotoxicology, 45*, 293–300. https://doi.org/10.1016/j.neuro.2013.09.006

Menezes-Filho, J. A., Carvalho, C. F., Rodrigues, J., Araújo, C., Dos Santos, N. R., Lima, C. S., Bandeira, M. J., Marques, B., Anjos, A., Bah, H., Abreu, N., Philibert, A., & Mergler, D. (2018). Environmental co-exposure to lead and manganese and intellectual deficit in school-aged children. *International Journal of Environmental Research and Public Health, 15*(11), 2418. https://doi.org/10.3390/ijerph15112418

Mohmand, J., Eqani, S. A., Fasola, M., Alamdar, A., Mustafa, I., Ali, N., Liu, L., Peng, S., & Shen, H. (2015). Human exposure to toxic metals via contaminated dust: Bio-accumulation trends and their potential risk estimation. *Chemosphere, 132*, 142–151. https://doi.org/10.1016/j.chemosphere.2015.03.004

Montgomery, K. S., Mackey, J., Thuett, K., Ginestra, S., Bizon, J. L., & Abbott, L. C. (2008). Chronic, low-dose prenatal exposure to methylmercury impairs motor and mnemonic function in adult C57/B6 mice. *Behavioural Brain Research, 191*(1), 55–61. https://doi.org/10.1016/j.bbr.2008.03.008

Mourão, D. S. (2016). Avaliação da exposição ao mercúrio em comunidades ribeirinhas de Porto Velho, Rondônia. In *Dissertação de Mestrado*. Escola Nacional de Saúde Pública Sérgio Arouca, Fundação Oswaldo Cruz – Rio de Janeiro.

Myers, G. J., Davidson, P. W., Watson, G. E., van Wijngaarden, E., Thurston, S. W., Strain, J. J., … Bovet, P. (2015). Methylmercury exposure and developmental neurotoxicity. *Bulletin of the World Health Organization, 93*(2), 2014–2016. https://doi.org/10.2471/BLT.14.141911

Nascimento, S. N., Baierle, M., Göethel, G., Barth, A., Brucker, N., Charão, M., … Garcia, S. C. (2016). Associations among environmental exposure to manganese, neuropsychological performance, oxidative damage and kidney biomarkers in children. *Environmental Research, 147*, 32–43. https://doi.org/10.1016/j.envres.2016.01.035

National Research Council (US) Committee on Measuring Lead in Critical Populations. (1993). *Measuring lead exposure in infants, children, and other sensitive populations*. National Academies Press (US).

National Scientific Council on the Developing Child. (2006). *Early exposure to toxic substances damages brain architecture*. (Working paper No 4). Retrieved from www.developingchild.harvard.edu

O'Connor, D., Hou, D., Ye, J., Zhang, Y., Ok, Y. S., Song, Y., Coulon, F., Peng, T., & Tian, L. (2018). Lead-based paint remains a major public health concern: A critical review of global production, trade, use, exposure, health risk, and implications. *Environment International, 121*(Pt 1), 85–101. https://doi.org/10.1016/j.envint.2018.08.052

Olympio, K. P., Gonçalves, C. G., Salles, F. J., Ferreira, A. P., Soares, A. S., Buzalaf, M. A., Cardoso, M. R., & Bechara, E. J. (2017). What are the blood lead levels of children living in Latin America and the Caribbean? *Environment International, 101*, 46–58. https://doi.org/10.1016/j.envint.2016.12.022

Olympio, K., Silva, J., Silva, A., Souza, V., Buzalaf, M., Barbosa, F., Jr., & Cardoso, M. (2018). Blood lead and cadmium levels in preschool children and associated risk factors in São Paulo, Brazil. *Environmental Pollution (Barking, Essex : 1987), 240*, 831–838. https://doi.org/10.1016/j.envpol.2018.04.124

Oulhote, Y., Mergler, D., Barbeau, B., Bellinger, D. C., Bouffard, T., Brodeur, M.-È., ... Bouchard, M. (2014). Neurobehavioral function in school-age children exposed to manganese in drinking water. *Environmental Health Perspectives, 122*(12), 1343–1350. https://doi.org/10.1289/ehp.1307918

Papanikolaou, N. C., Hatzidaki, E. G., Belivanis, S., Tzanakakis, G. N., & Tsatsakis, A. M. (2005). Lead toxicity update. A brief review. *Medical Science Monitor: International Medical Journal of Experimental and Clinical Research, 11*(10), RA329–RA336.

Passos, C. J. S., da Silva, D. S., Lemire, D., Fillion, M., Guimarães, J. R. D., Lucotte, M., & Mergler, D. (2008). Daily mercury intake in fish-eating populations in the Brazilian Amazon. *Journal of Exposure Science and Environmental Epidemiology, 18*(1), 76–87. https://doi.org/10.1038/sj.jes.7500599

Perlroth, N. H., & Castelo Branco, C. W. (2017). Current knowledge of environmental exposure in children during the sensitive developmental periods. *Jornal de Pediatria, 93*(1), 17–27. https://doi.org/10.1016/j.jped.2016.07.002

Ramírez Ortega, D., González Esquivel, D. F., Blanco Ayala, T., Pineda, B., Gómez Manzo, S., Marcial Quino, J., Carrillo Mora, P., & Pérez de la Cruz, V. (2021). Cognitive impairment induced by lead exposure during lifespan: Mechanisms of lead neurotoxicity. *Toxics, 9*(2), 23. https://doi.org/10.3390/toxics9020023

Riojas-Rodríguez, H., Solís-Vivanco, R., Schilmann, A., Montes, S., Rodríguez, S., Ríos, C., & Rodríguez-agudelo, Y. (2010). Intellectual function in Mexican children living in a mining area and environmentally exposed to manganese. *Environmental Health Perspectives, 118*(10), 1465–1470. https://doi.org/10.1289/ehp.0901229

Rodrigues, J., Bandeira, M. J., Araújo, C., Dos Santos, N. R., Anjos, A., Koin, N. L., Pereira, L. C., Oliveira, S., Mergler, D., & Menezes-Filho, J. A. (2018a). Manganese and lead levels in settled dust in elementary schools are correlated with biomarkers of exposure in school-aged children. *Environmental Pollution (Barking, Essex: 1987), 236*, 1004–1013. https://doi.org/10.1016/j.envpol.2017.10.132

Rodrigues, J. L. G., Araújo, C. F. S., Nathália, R., Bandeira, M. J., Laura, A., Anjos, S., ... Abreu, S. (2018b). Airborne manganese exposure and neurobehavior in school-aged children living near a ferro-manganese alloy plant. *Environmental Research, 167*(May), 66–77. https://doi.org/10.1016/j.envres.2018.07.007

Sanders, A. P., Claus Henn, B., & Wright, R. O. (2015). Perinatal and childhood exposure to cadmium, manganese, and metal mixtures and effects on cognition and behavior: A review of recent literature. *Current Environmental Health Reports, 2*(3), 284–294. https://doi.org/10.1007/s40572-015-0058-8

Santiago de Vasconcellos, A. C. (2016). Carga de doenças atribuída ao mercúrio no Brasil: estimativas nacional, regional e local. In *Tese de doutorado*. Fundação Oswaldo Cruz – Rio de Janeiro.

Santos-Lima, C. D., Mourão, D. S., Carvalho, C. F., Souza-Marques, B., Vega, C. M., Gonçalves, R. A., Argollo, N., Menezes-Filho, J. A., Abreu, N., & Hacon, S. S. (2020). Neuropsychological effects of mercury exposure in children and adolescents of the Amazon Region, Brazil. *Neurotoxicology, 79*, 48–57. https://doi.org/10.1016/j.neuro.2020.04.004

Santos-Sacramento, L., Arrifano, G. P., Lopes-Araújo, A., Augusto-Oliveira, M., Albuquerque-Santos, R., Takeda, P. Y., Souza-Monteiro, J. R., Macchi, B. M., do Nascimento, J., Lima, R. R., & Crespo-Lopez, M. E. (2021). Human neurotoxicity of mercury in the Amazon: A scoping review with insights and critical considerations. *Ecotoxicology and Environmental Safety, 208*, 111686. https://doi.org/10.1016/j.ecoenv.2020.111686

Schneider, J. S., Decamp, E., Clark, K., Bouquio, C., Syversen, T., & Guilarte, T. R. (2009). Effects of chronic manganese exposure on working memory in non-human primates. *Brain Research, 1258*(215), 86–95. https://doi.org/10.1016/j.brainres.2008.12.035

Sheehan, M., Burke, T. A., Navas-Acien, A., Breysse, P. N., McGready, J., & Fox, M. A. (2014). Global methylmercury exposure from seafood consumption and risk of developmental neurotoxicity: A systematic review. *Bulletin of the World Health Organization, 92*(4), 254–269F. http://doi.org//S0102-311X1993000200006

Takser, L., Mergler, D., Hellier, G., Sahuquillo, J., & Huel, G. (2003). Manganese, monoamine metabolite levels at birth, and child psychomotor development. *Neurotoxicology, 24*(4–5), 667–674. https://doi.org/10.1016/S0161-813X(03)00058-5

Tamayo, Y., Ortiz, M., Téllez-Rojo, M. M., Trejo-Valdivia, B., Schnaas, L., Osorio-Valencia, E., Coull, B., Bellinger, D., Wright, R. J., & Wright, R. O. (2017). Maternal stress modifies the effect of exposure to lead during pregnancy and 24-month old children's neurodevelopment. *Environment International, 98*, 191–197. https://doi.org/10.1016/j.envint.2016.11.005

Torres-Agustín, R., Rodríguez-Agudelo, Y., Schilmann, A., Solís-Vivanco, R., Montes, S., Riojas-Rodríguez, H., … Ríos, C. (2012). Effect of environmental manganese exposure on verbal learning and memory in Mexican children. *Environmental Research, 121*(2013), 39–44. https://doi.org/10.1016/j.envres.2012.10.007

Valeri, L., Mazumdar, M. M., Bobb, J. F., Claus Henn, B., Rodrigues, E., Sharif, O., Kile, M. L., Quamruzzaman, Q., Afroz, S., Golam, M., Amarasiriwardena, C., Bellinger, D. C., Christiani, D. C., Coull, B. A., & Wright, R. O. (2017). The joint effect of prenatal exposure to metal mixtures on neurodevelopmental outcomes at 20-40 months of age: Evidence from rural Bangladesh. *Environmental Health Perspectives, 125*(6), 067015. https://doi.org/10.1289/EHP614

Vieira, S. M., de Almeida, R., Holanda, I. B. B., Mussy, M. H., Galvão, R. C. F., Crispim, P. T. B., … Bastos, W. R. (2013). Total and methyl-mercury in hair and milk of mothers living in the city of Porto Velho and in villages along the Rio Madeira, Amazon, Brazil. *International Journal of Hygiene and Environmental Health, 216*(6), 682–689. https://doi.org/10.1016/j.ijheh.2012.12.011

Wasserman, G. A., Liu, X., Parvez, F., Factor-Litvak, P., Ahsan, H., Levy, D., … Graziano, J. H. (2011). Arsenic and manganese exposure and children's intellectual function. *Neurotoxicology, 32*(4), 450–457. https://doi.org/10.1016/j.neuro.2011.03.009

Wong, C. S. C., Lia, X. D., Zhang, G., Qi, S. H., & Peng, X. Z. (2003). Atmospheric deposition of heavy metals in the Pearl River Delta, China. *Atmospheric Environment, 37*, 767–776. https://doi.org/10.1016/S1352-2310(02)00929-9

Wright, R. O., Amarasiriwardena, C., Woolf, A., Jim, R., & Bellinger, D. (2006). Neuropsychological correlates of hair arsenic, manganese, and cadmium levels in school-age children residing near a ha+zardous waste site. *Neurotoxicology, 27*(2), 210–216. https://doi.org/10.1016/j.neuro.2005.10.001

Part III
Academic Performance

Bilingualism, Foreign Language Learning, and Cognition: Insights for Education

Jose Armando Aguasvivas and Manuel Carreiras

Introduction

The human brain is remarkable in its ability to learn and generalize knowledge. There is ample evidence that different life experiences can serve as the basis for enhanced learning. To cite a few examples, taxi drivers are especially good at navigation and spatial learning tasks (Maguire et al., 2000), musicians are better at learning and discriminating lexical tone categories (Zhao & Kuhl, 2015), and action video games players have increased attentional control and meta-learning (learning to learn) skills (Green & Bavelier, 2012), compared to individuals without these life experiences. It is perhaps not a wild stretch to assume that linguistic experiences, such as experience with two languages (bilingual experience), could also foster language learning skills. This idea summarizes the primary focus of this chapter, which explores how bilingual experience could influence foreign language learning.

One can think of a language as a mental rulebook. Within this book, there are multiple interconnected chapters. The information in each chapter is relevant to other sections of this book, and therefore one cannot read them in isolation. For instance, the first chapter could take the form of a dictionary, where a collection of words (in their written and spoken forms) pair with their concepts to form a language's vocabulary. This dictionary would naturally depend on other book sections detailing how to correctly write, pronounce, and use these words in different

J. A. Aguasvivas (✉)
BCBL, Basque Center on Cognition, Brain and Language, San Sebastian, Spain

Universidad del País Vasco (UPV/EHU), San Sebastian, Spain

M. Carreiras
BCBL, Basque Center on Cognition, Brain and Language, San Sebastian, Spain

Universidad del País Vasco (UPV/EHU), San Sebastian, Spain

Ikerbasque, Basque Foundation for Science, Bilbao, Spain

M. V. Alves et al. (eds.), *Cognitive Sciences and Education in Non-WEIRD Populations*, https://doi.org/10.1007/978-3-031-06908-6_10

contexts. Other chapters in this rulebook would also show how to modify the words, combine them into sentences, and convey information that cannot be transmitted using a single word. Using this analogy, learning a foreign language could be akin to acquiring and memorizing a new rulebook.

The idea that knowing more languages facilitates subsequent language learning spans decades. It is thought that the less predictable linguistic environment faced by bilinguals might induce them to more efficiently explore and acquire new linguistic information than monolinguals (Festman, 2021). In this regard, the experimental work on this idea generally falls into studies that target general or specific aspects of foreign language learning. On the one hand, research focused on measuring the general foreign language proficiency (i.e., writing, reading, speaking, listening) highlights that bilinguals may use different strategies during foreign language learning than monolinguals (Cenoz, 2013). These findings hint toward an additive effect of bilingual experience on foreign language learning, whereby knowing more languages leads to an increased linguistic repertoire available during learning.

On the other hand, some researchers have examined the differences between monolinguals and bilinguals on specific foreign language competencies, the more predominant being vocabulary (Antoniou et al., 2015; Kaushanskaya & Marian, 2009a, b). These studies suggest that bilingual experience improves vocabulary learning outcomes by either strengthening the phonological system, the lexical-semantic network, or inhibitory control mechanisms or making these more flexible to accommodate new information (Kaushanskaya, 2012; Kaushanskaya & Marian, 2009b). Nevertheless, due to these typically ad hoc explanations, the mechanisms underlying the differences remain largely understudied.

To review and integrate all these findings into a common framework, Hirosh and Degani (2018) proposed two broad and complementary routes through which experience with multiple languages could facilitate foreign language learning. The first route involves a *direct* transfer of linguistic information from any (or all) available languages to the foreign language, also known as *cross-linguistic transfer*. Individuals can use the knowledge in any language they know during learning to find similarities to exploit (i.e., positive transfer), but these might also lead to negative (nonfacilitatory) transfer. For example, a Spanish speaker learning English might find the word *vocabulary* easier to learn than the word *country*, even if they both occur frequently. The former has a translation equivalent with a similar meaning and form (vocabulario)—a cognate word.

The second route through which bilingualism could affect foreign language learning according to this framework is *indirect*. Simply put, the multilingual experience might potentiate *linguistic* or *nonlinguistic* cognitive abilities, improving foreign language learning outcomes in turn (Hirosh & Degani, 2018). The linguistic abilities include metalinguistic awareness, verbal working memory, and lexical-semantic network strength, among others. With a few notable exceptions (i.e., verbal working memory), these linguistic abilities tend to be challenging to quantify as they are significantly language-specific (Bialystok et al., 2014; Serratrice et al., 2009). In other words, it would not be possible to measure metalinguistic awareness as a general ability without individuals already knowing a specific foreign language.

As a result, studies have yet to explore (1) how to properly quantify these skills, (2) how multilingualism could enhance these skills, and (3) how such enhancement could modulate foreign language learning.

Compared to the linguistic abilities mentioned above, the nonlinguistic cognitive abilities are perhaps even more controversial. Many studies suggest that bilinguals might possess enhanced cognitive abilities over monolinguals (for reviews, see Adesope et al., 2010; Kroll & Bialystok, 2013; Schroeder & Marian, 2017). The typical narrative is that, due to their constant practice of inhibiting/selecting which language to use, bilinguals might show enhanced nonlinguistic cognitive abilities over monolinguals (such as conflict monitoring, inhibitory, or attentional control). These results have been under severe scrutiny recently due to theoretical and methodological issues (Blanco-Elorrieta & Pylkkänen, 2018; de Bruin et al., 2021). Thus, both the existence and relevance of these findings have been rendered moot.

The logical assumption deriving from these findings is that knowing more languages provides more sources for cross-linguistic transfer than other indirect abilities. That is to say, an individual that knows two languages can potentially transfer more knowledge between these and the foreign language they are learning than someone who only knows one language. However, what is the evidence in favor and against bilingual experience influencing foreign language learning?

Empirical Evidence

This subsection reviews the extant research targeting bilingual and monolingual foreign language learning at different levels of language representation. Since this is currently an active field of research, this review is far from exhaustive (for more comprehensive reviews, see Festman, 2021; Hirosh & Degani, 2018; Montanari, 2019). Instead, we primarily cover the main experimental work addressing differences, or lack thereof, between bilinguals and monolinguals at each linguistic level of representation. Moreover, we primarily focus on experimental studies with bilinguals and monolinguals but avoid, wherever possible, those involving multilingual participants. The research involving multilingual individuals, although informative, has additional confounding factors, such as the interactions between multiple languages and a general lack of control measures (i.e., age of acquisition, proficiency, exposure, general intelligence) for these participants.

Sublexical Phonology: Phonology/Phonetics Learning sublexical phonology can refer to acquiring new words that differ in phonology at the syllable level, learning to discriminate or produce new phonemes, or incidentally (implicitly) learning to segment words with foreign phonology from continuous speech. Here, we will discuss studies covering these three types of tasks. Naturally, if an individual already knows how to pronounce or discriminate specific phonemes (due to their prior linguistic knowledge), they should be better overall in these three tasks. Beyond these direct effects, we also present studies showing differences between bilinguals and monolinguals and arguing for indirect effects.

Antoniou et al. (2015) compared monolinguals (English) and bilinguals (Mandarin-English and Korean-English) across two vocabulary learning experiments. That is, participants learned eight new names (words) for eight objects. The critical manipulations in these experiments were the phonological patterns that comprised the words, which were only one syllable long or monosyllabic. Pairs of syllables in the vocabulary differed in one phoneme, and were either English-like (voiced fricative; e.g., /ɸ/ vs. /β/), Mandarin-like (i.e., retroflex; e.g., /t/ vs. /ʈ/), or Korean-like (i.e., lenition; e.g., /θ/ vs. /θ'/) minimal pairs. In the first experiment, Mandarin-English bilinguals outperformed monolinguals in learning both the English-like and Mandarin-like minimal pairs. In the second experiment, both Mandarin-English and Korean-English bilinguals outperformed the monolinguals in the Mandarin-like minimal pairs. However, only the Korean-English group could distinguish between the more challenging Korean-like minimal pairs. Their results indicated that general bilingual experience (in terms of enhanced phonological network) and phonetic similarity (derived from experience with Korean) could influence learning new phonology over monolinguals.

Other studies have shown minimal or no differences between monolingual and bilingual children and adults when learning to discriminate new phonemic contrasts (Polka et al., 2001; Sundara et al., 2006; Tremblay & Sabourin, 2012). These tasks typically involve determining whether a heard sound belongs to a phonemic category (e.g., Hindi's retroflex t versus dental stop t). For instance, Tremblay and Sabourin (2012) tested the phonemic discrimination performance of English monolingual, English-French bilingual, and multilingual adults using pretraining and posttraining tests. Before training, there were no differences between the groups. After training, only the multilinguals but not the bilinguals performed the discrimination task better than the monolinguals.

Wang and Saffran (2014) compared monolinguals (English and Mandarin) and bilinguals (Mandarin-English and English-Spanish) in a statistical learning task. The task combined syllables and tones to simulate the experience of learning a tonal language as a foreign language. The artificial language contained three trisyllabic words (i.e., words with three syllables) with varying tones. After familiarizing themselves with the artificial language, participants heard a series of word pairs (one target and one foil) and selected which one belonged to the artificial language they had previously heard. Their results indicated that both bilingual groups outperformed the monolingual groups, regardless of their experience with tonal languages. Again, the authors of this study pointed at bilingual experience conferring indirect advantages in terms of the phonological network (Wang & Saffran, 2014). In other words, experience with the different phonology of two languages might confer an advantage when segmenting words from the continuous speech in a foreign language.

Sublexical Orthography: Literacy and Orthotactics Under the sublexical orthography level, we consider studies where individuals learned to read and write in a foreign script (literacy). Some of these studies target bilingual individuals who know two scripts, known as biliteracy. For example, a Spanish-Mandarin bilingual knows an alphabetic (Spanish) and a logographic (Mandarin) script.

Most studies on monolingual and bilingual literacy skills involve infants, children, and adolescents learning new scripts. In brief, some suggest that bilingualism or biliteracy can facilitate spelling and decoding in a new script over monolingualism (Clyne et al., 2004; Kahn-Horwitz et al., 2011; Schwartz et al., 2014; Trapman et al., 2014), whereas others have shown comparable performance between the groups (Van Gelderen et al., 2003). Although these findings imply that bilinguals might possess a more flexible orthographic system than monolinguals, it is essential to consider that other critical sociocultural factors might be involved (Montanari, 2019). For instance, immigrant and immersion bilingual minors might be more willing to communicate and have lower anxiety when facing a foreign language than local monolinguals. Thus, sociocultural factors might have a more considerable effect on ultimate learning achievement than bilingual experience in itself.

The research on adult monolingual and bilingual literacy is very scarce. Of note, Modirkhamene (2006) compared the English reading comprehension of adult bilingual (Turkish-Persian) and monolingual (Persian) speakers three times over two years of their classroom learning studies. Persian possesses a Perso-Arabic script, and the Turkish script contains 29 letters compared to the 26 in the English alphabet. They found that the bilingual participants outperformed their monolingual peers over time, suggesting that bilinguals could better acquire the decoding and comprehension skills to succeed in this test (Modirkhamene, 2006). Similarly, another study found that Turkish-Farsi bilinguals outperformed Farsi monolinguals in their English writing skills (Poorebrahim et al., 2020). Although other mechanisms might be involved, it is easy to see how Turkish knowledge might have directly benefited the outcomes in these studies because of its similar script to English.

Morphosyntax: Grammar Morphosyntax encompasses the variations of words (morphology) and the rules by which multiple words combine to form clauses and sentences (syntax). These aspects are collectively referred to as *grammar*. The research on monolingual and bilingual grammar learning is limited, partly because learning grammar requires prior knowledge of other fine-grained aspects at the sub-lexical, lexical, and even semantic levels (Hirosh & Degani, 2018; Montanari, 2019).

Nation and McLaughlin (1986) tested the grammar learning performance of English monolinguals, bilinguals, and multilinguals (with varying languages) using artificially generated character strings that followed two underlying sets of grammatical rules. A subset in each group was told that the character strings followed specific rules, which they had to discover (explicit condition). Conversely, the remaining participants did not receive any instructions and passively visualized the strings (implicit condition). After seeing the strings, all participants performed a grammaticality judgment test where they had to indicate whether a series of character strings were correct or not. Their results revealed no differences between the groups in the explicit condition. Multilinguals but not bilinguals outperformed the other two groups in the implicit condition. A similar study used English grammar instead of artificial character strings, also indicating a multilingual advantage for grammar learning in adolescents (Klein, 1995). Regardless, there have been no reported differences between monolinguals and bilinguals in these tasks.

A more recent study compared an older population (older than 60) of English monolinguals and English-Spanish bilinguals that explicitly or implicitly learned basic Latin grammar (Cox, 2017). They used four tasks to measure learning outcomes: written and auditory sentence interpretation, grammaticality judgment, and sentence production. Overall, their results showed that bilinguals outperformed monolinguals in both sentence interpretation tasks, irrespective of whether they received explicit or implicit instructions. However, there were no differences between the groups in the other two tasks. They attributed these results to bilinguals having an enhanced metalinguistic awareness than monolinguals. An alternative explanation is that knowledge of Spanish (a Latin-derived language) could have directly influenced the learning of Latin grammar.

Finally, another study compared Mandarin-English bilinguals against English monolinguals in an artificial grammar learning task (Grey et al., 2018). After receiving some instruction in the novel language, the participants performed comprehension and production tasks and a grammaticality judgment task. They then performed the comprehension and production tasks for 20 blocks to practice extensively in the new language. Their results showed that bilinguals and monolinguals did not differ in any of the tasks. This result was consistent when testing the participants at a low or a high proficiency level in the artificial language and comparing them through the comprehension and production learning blocks. In other words, at least at the behavioral level, there was no evidence of bilingual experience conferring advantages over monolingual experience.

Lexical-semantics: Vocabulary This subsection considers studies that target how individuals acquire the mapping between foreign words and their meanings, known as vocabulary learning. Vocabulary learning is the area that has received the most attention from the literature (Festman, 2021; Montanari, 2019). Admittedly, some of the studies we have reviewed thus far fall into the lexical-semantics level. However, their designs and purpose more appropriately addressed other aspects rather than vocabulary specifically.

In two seminal studies, Kaushanskaya and Marian (2009a, b) compared the performance of bilinguals (English-Spanish English-Mandarin) and English monolinguals when learning foreign words paired with their English translations. The words were artificial and had unfamiliar phonology created for these experiments. Additionally, a subset of the participants only heard the new words and saw their English translation (unimodal condition). In contrast, others heard and saw the new words alongside their translation (bimodal condition). Their results indicated that the bilingual group outperformed the monolingual group when recognizing and recalling the words immediately after learning them in both conditions. After a delay, bilinguals were only better than monolinguals in the bimodal condition. To explain this advantage, the authors suggested that bilingual experience either provided a more flexible mapping between orthography and phonology or led to increased phonological working memory (Kaushanskaya & Marian, 2009a, b). An equally plausible interpretation offered by other studies is that bilinguals are better at reducing phonological interference of their known languages by recruiting inhibitory control mechanisms (Bartolotti et al., 2011; Yoshida et al., 2011).

This bilingual advantage in vocabulary learning seems consistent irrespective of whether bilinguals learned their languages since birth, through formal instruction or later in life, or when matching bilinguals and monolinguals on their phonological awareness and verbal working memory (Kan & Sadagopan, 2014; Kaushanskaya, 2012; Nair et al., 2016; Van Hell & Mahn, 1997). Additionally, related studies have shown that the differences between monolinguals and bilinguals are more prominent when learning concrete as compared to abstract meanings for the words (Kaushanskaya & Rechtzigel, 2012). Consequently, it would seem that bilingual experience improves language learning outcomes by either strengthening the phonological network, the lexical-semantic network, or inhibitory control mechanisms (Hirosh & Degani, 2018; Kaushanskaya & Rechtzigel, 2012; Yoshida et al., 2011). All of these comprise the indirect effects mentioned above.

Although the evidence seems overwhelmingly in favor of a bilingual vocabulary learning advantage, other studies employing different designs have shown no differences between monolinguals and bilinguals. For instance, Bakker-Marshall et al. (2021) compared English monolinguals and Spanish-English bilinguals when learning two sets of Swahili-English word pairs over two days. Participants performed four increasingly complex recall tasks and a primed lexical decision task where the Swahili words were used as primes (Bakker-Marshall et al., 2021). The authors included this latter task to measure the lexical integration of the new words. Overall, their results suggested that the groups performed equally on all the recall tasks, and only the monolingual group showed the priming effect indicative of lexical integration. These results could be biased since participants in the bilingual group were learning the words in their second language. Regardless, these results show that it might not always be the case that bilinguals outperform monolinguals.

It seems like the evidence in favor of a bilingual advantage in foreign language learning vastly overpowers the evidence against this claim, particularly at the vocabulary level. These findings imply that learning a third language could differ from learning a second language. Still, no current theoretical account suggests how or why second and third language learning should differ. As an alternative, we summarize the two major fields covering language acquisition from the linguistic perspective and bilingual word processing and learning from the cognitive perspective, respectively.

Linguistic and Cognitive Views on Second and Third Language Learning

Linguistic Perspective

The linguistic perspective on language learning concurs that certain aspects of a known language can transfer to the second or third languages, a concept known as cross-linguistic transfer (Alonso, 2016). A crucial concept in the context of foreign language learning is that of language *typology*, which refers to the structural features or parameters of a language at all analytic levels (VanPatten & Benati, 2015). All languages vary across these parameters, with languages oscillating according to how close or distant they are (i.e., their proximity). Notably, the typological similarity need not be objective, as learners can perceive idiosyncratic and subjective similarities. These similarities are collectively known as psychotypology (Hermas, 2014).

Naturally, the only source for transfer during second language acquisition is the native language (L1). In this case, the transfer will largely depend on the psychotypological proximity between the L1 and the target foreign language. In contrast, in acquiring a third language, the transfer could occur from any or both known languages depending on several factors. Consequently, several linguistic models of third language acquisition have been put forward:

- The *L1 Transfer Model* essentially posits that positive (facilitatory) and negative (interfering) cross-linguistic transfer of linguistic information occurs mainly from the first language (L1), at least during the initial stages of foreign language learning (Hermas, 2014). According to this model, transferring from the L1 incurs the least cognitive effort (i.e., is more efficient), inducing learners to find perceived similarities to their L1 and not other known languages. Under this model, second and third language learning should be approximately equal for individuals with the same L1.
- Contrary to the previous model, the *L2 Status Factor Model* suggests that the second language (L2) can take a more decisive role in cross-linguistic transfer if it is more similar (has a higher status factor) to the foreign language (Bardel & Falk, 2012). The reasoning for this model is that the learner does not actively select which language to use for cross-linguistic transfer. Hence, transfer effects can occur naturally from the L2 if the learner is sufficiently proficient and if the L2 and target foreign language have similar psychotypology.
- The *Cumulative Enhancement Model* assumes that language learning is cumulative, and specific language pairs can enhance learning (Flynn et al., 2004). In other words, the L1 is not exceptional, and either the L1 or the L2 can be used as sources for direct cross-linguistic transfer. This model would predict not only that bilinguals would outperform monolinguals but that trilinguals (speakers of three languages) would, in turn, be better than bilinguals at learning a foreign language. That is, the more languages one knows, the easier it becomes to learn more.

- The *Typological Primacy Model* extended the ideas from the previous model by suggesting that cross-linguistic transfer is selective (Rothman, 2011, 2015). It is not a question of either-or but rather "when" do learners transfer from a specific language. In this model, the more psychotypologically similar language has "primacy" during cross-linguistic transfer. Furthermore, the transfer can occur on a construction-by-construction basis. Therefore, learners can transfer from any language so long as a specific construction is similar to any of them.
- The last model we consider is the *Linguistic Proximity Model* (Westergaard et al., 2017). Compared to the prior models, this account further assumes that transfer could occur from all known languages simultaneously. For instance, there could be a positive transfer from the L1 and a negative influence from the L2 operating upon the same construction (Westergaard et al., 2017). That is, learners find similarities to any language to the foreign language. These similarities range from psychotypology to more abstract and structural similarities.

It is critical to note that this literature does not particularly address the differences between second and third language acquisition (i.e., monolingual versus bilingual foreign language learning). Instead, the differences under these models fundamentally rely on whether bilinguals use their more extensive pool of knowledge to transfer linguistic information during learning. Also, these models provide valuable insights regarding the interplay of languages in the learner's mind. First, according to these models, the information from both languages is active and available for cross-linguistic transfer in bilingual learners. Second, there does not seem to be separate systems for each language an individual acquires, and all languages cumulatively interact during subsequent learning. Third, the cross-linguistic transfer occurs from any or both languages on a construction-by-construction basis. Finally, the similarities do not need to be objective, as learners can transfer based on perceived and subjective similarities.

Cognitive Perspective

The field of cognitive science has focused on the mental processes underlying learning. A general view of the learning process comes from information-processing theory (Schunk, 2012). In simple words, learners first integrate the information in short-term memory, which is then passed to long-term memory using consolidation processes (Walker & Stickgold, 2004). Thus, the knowledge is codified, stored, and retrieved from these memory systems during learning, production, comprehension. The codification process involves creating a mental (or neural) representation that can be efficiently stored and retrieved. Under this information-processing view, third language learners might differ in their codification, storing, and retrieval processes over second language learners (Kaushanskaya & Marian, 2009b).

The majority of the language-related cognitive science literature has focused on the *mental lexicon*, the brain's storage for word forms (orthography and phonology), their meanings, and other aspects (Acha & Carreiras, 2014; Aitchison, 2012). Specifically, the second language processing literature concurs that meaning representations are common to the native and second languages. Recent consensus also suggests that form representations are unified within the multilingual mental lexicon, the so-called *language-nonselective* view of the mental lexicon (Baxter et al., 2021). In other words, there is a common space for words and meanings from all known languages, which could potentially extend to other aspects such as syntax and even pragmatics. Finally, the representations within the mental lexicon are organized according to their similarity, with items more similar in their form or meaning being closer together. This view complements the linguistic perspective in that bilinguals have a more extensive pool to exploit for cross-linguistic transfer during learning.

Computational models of the mental lexicon have not yet focused on the differences between second and third language learning. Nevertheless, some models have targeted word recognition in two languages. Out of these, perhaps the most influential model is the *Bilingual Interactive Activation Plus (BIA+)*. According to the BIA+ model, a visual word will trigger the sublexical orthographic and phonological representations, subsequently activating the lexical representations. These representations activate the meaning representations. Critically, the BIA+ model employs language nodes to determine a word's language after receiving the lexical representations. These nodes aid in selecting the correct meaning of the visualized words. For instance, the word *room* can mean cream in Dutch or a part of a building in English. The main disadvantage of the BIA+ model is that there are no learning mechanisms, and the connections and representations need to be hand-engineered by the modeler. Although some theoretical extensions of this model attempt to verbally account for learning in the BIA+ (Grainger et al., 2008), this model cannot explain how second and third language learning could differ.

To address the gaps in learning mechanisms, other models have used basic learning rules and simple architectures to explain how the organizational principles of the mental lexicon could emerge from the input. The following are some of the most influential models:

- The *Bilingual Simple Recurrent Network* was one of the first models to address bilingual language learning and representation (French, 1998). The architecture was a recurrent connectionist network that received an input word and a copy of the previous state, producing a prediction of the following input (see Fig. 1). This model was trained using simple three-word sentences from two languages with no boundary between languages (e.g., boy lift toy femme touche ballon). This model was the first to show that lexical representations clustered according to each language without the need for explicit language tags, using a language-nonselective lexicon.
- The *Conceptual Feature Mapping model* is a theoretical model of word learning (Lee et al., 1999). According to this model, the bilingual mental lexicon is language-nonselective at the form (conceptual) and meaning levels. An intermediate "lemma" layer maps a word form to its corresponding meaning in a specific

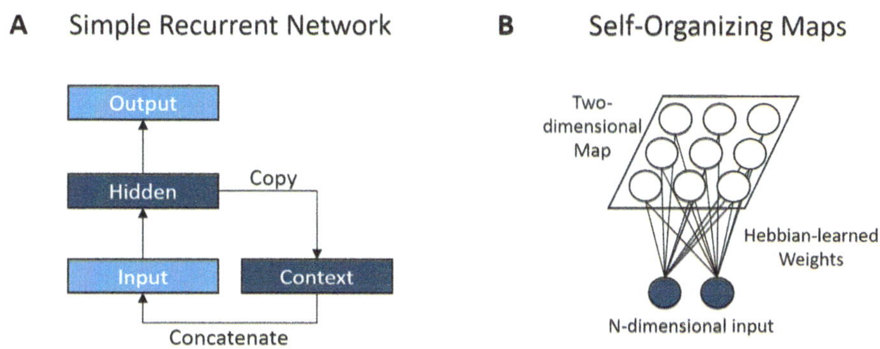

Fig. 1 Models of bilingual word learning. Note (**a**) Depiction of the Simple Recurrent Network model. The model keeps a copy of the prior context to concatenate with the current input and produce the following hidden representation and output. The connections between all layers are learned using backpropagation using the difference between the produced and expected outputs (self-supervised learning). (**b**) Simplified depiction of a self-organizing map, the map learns a topological representation of the input in an unsupervised manner. The connection between input and the map is learned using more biologically plausible mechanisms such as Hebbian learning

language, similar to the language node in the BIA+. Thus, learning a foreign word requires modifying the connections from the form level to the lemma level or from the lemma level to the meaning level, depending on whether the target word's form or meaning overlaps with its native language counterpart.

- These *Self-organizing maps (SOMs)* are a collection of models rather than a single unified account of bilingual word learning. In essence, SOMs are artificial neural networks that map a set of multidimensional inputs onto a two-dimensional map (Fig. 1b) using weights learned through simple learning rules (e.g., Hebbian learning). The *Self-Organizing Map of Bilingual Processing (SOMBIP)* was one of the first models to show the differences in the organization of phonological word forms emerging from early versus late learning of the second language (Li & Farkas, 2002). More recent developments of this model, called *Dev-LEX* and *Dev-LEX II* (Zhao & Li, 2010, 2013), have further argued that distinct language-dependent form representations emerge from these models without any label or language node. Furthermore, these representations are structured depending on the onset of L2 learning, with segregate representations for early bilinguals, and parasitic L2 representations for late learners (Zhao & Li, 2010).

As with the linguistic perspective, the cognitive perspective does not strictly differentiate between second and third language learning. Instead, the insights from these models and theories rely on the mental organization of information within the mental lexicon. The exposure to foreign input might shape the language-nonselective mental lexicon in several ways. First, there are more potentially distinct representations to facilitate the cross-linguistic transfer in a third than a second language. Second, the novel representations might lie closer to already existing representations, facilitating their acquisition and production. Last, the onset of foreign language exposure can shape these representations, making them overlapping or distinct in the bilingual mind.

Bilingualism and Language Learning in Non-WEIRD Populations

Although some of the literature reviewed here already highlights bilingualism research with diverse and non-WEIRD populations, the purpose of this segment is to consider the main ideas of this chapter in the context of Latin America. Bilingualism is ubiquitous in Latin America, ranging from countries where two languages coexist (e.g., Haiti, Puerto Rico) to countries where knowing two languages provides an advantage in opportunities (e.g., Brazil, Mexico). But how do the ideas discussed in this chapter relate to the academic achievement of these particular populations? We consider two main points regarding immigration in non-WEIRD populations:

Immigration and Language Learning Migrant individuals from bilingual contexts might find themselves better suited to learn the country's language in which they are migrating, compared to monolingual individuals with similar background. Research shows that exposure to diverse linguistic contexts (common in Latin America) could facilitate learning novel vocabulary (Bice & Kroll, 2019). Because of this, both monolingual and bilingual migrants might benefit from their knowledge in acquiring a novel language. Admittedly, a vocabulary learning advantage does not imply learning all aspects of a language, but vocabulary does lie at the heart of these other aspects. It is essential for countries to put forward fair language policies that facilitate migrant children and adult academic, social, and cultural integration.

Immigration and Academic Achievement Language comprehension is at the core of academic achievement (Parmis et al., 2020). Children and adolescents of migrant parents might already be exposed to one or two languages and receive education in an additional language. While research shows that bilingualism itself might not predict academic achievement in other areas such as math (Han, 2012), the ability to understand mathematical instruction and prompts could ultimately affect their academic achievement. It is well known that immigrant learners (particularly Latin American immigrants) often struggle in adapting to education in a foreign language (Martone, 2014), but there is not much research regarding the influence of bilingualism on their performance. It would be interesting for future studies to explore the influence of language background on ultimate academic achievement in migrant children and adolescents. This is particularly important for countries to develop language and educational policies that are inclusive and fair for students from diverse populations.

Insights for Education

How do the ideas mentioned above impact the language and educational practice? We offer a few points:

- *It is essential to consider the learner's linguistic background.* The particularities of known languages, and similarities and differences with the target foreign language, can implicitly or explicitly be exploited to maximize language learning achievement.
- *Experience with two languages facilitates vocabulary learning.* This statement does not mean to discourage monolingual learners. Instead, the essential aspect to consider is that learning a foreign language can facilitate further vocabulary acquisition. Surprisingly, even in monolinguals, passive exposure to a nonnative language seems to confer vocabulary learning advantages (Bice & Kroll, 2019).
- *Objective and perceived similarity can facilitate and hinder language learning.* As shown throughout this chapter, objective similarities serve an important purpose in language learning, aiding individuals to quickly recognize and produce foreign constructions. Perceived similarities, on the other hand, are idiosyncratic and will largely depend on individual experiences and their state of mind. Still, it is beneficial to find similarities during vocabulary learning regardless of a specific word's meaning.
- *Consistent practice ultimately yields the best results.* Even though at the initial levels of learning, bilinguals might outperform monolinguals, learning achievement ultimately depends on consistent practice and exposure. The more opportunities an individual has to perceive or produce different constructions, the more likely they are to learn them.
- *Linguistic experiences could also influence learning of other skills.* Although we did not explore this idea in this chapter, it is essential to consider that language mediates the way we learn and perform in different areas (Han, 2012). It is always vital to consider linguistic diversity in educational practice.

References

Acha, J., & Carreiras, M. (2014). Exploring the mental lexicon: A methodological approach to understanding how printed words are represented in our minds. *The Mental Lexicon*. https://doi.org/10.1075/ml.9.2.03ach

Adesope, O. O., Lavin, T., Thompson, T., & Ungerleider, C. (2010). A systematic review and meta-analysis of the cognitive correlates of bilingualism. *Review of Educational Research*. https://doi.org/10.3102/0034654310368803

Aitchison, J. (2012). Words in the mind: An introduction to the mental lexicon. In *Language*. Wiley. https://doi.org/10.2307/416662

Alonso, R. A. (2016). Cross-linguistic influence in second language acquisition. *Multilingual Matters*. https://doi.org/10.1002/9781405198431.wbeal0292

Antoniou, M., Liang, E., Ettlinger, M., & Wong, P. C. M. (2015). The bilingual advantage in phonetic learning. *Bilingualism, 18*(4), 683. https://doi.org/10.1017/S1366728914000777

Bakker-Marshall, I., Takashima, A., Fernandez, C. B., Janzen, G., McQueen, J. M., & Van Hell, J. G. (2021). Overlapping and distinct neural networks supporting novel word learning in bilinguals and monolinguals. *Bilingualism, 24*(3), 524–536. https://doi.org/10.1017/S1366728920000589

Bardel, C., & Falk, Y. (2012). Behind the L2 status factor : A neurolinguistic framework for L3 research. In *Third language acquisition in adulthood* (pp. 61–78). John Benjamins Pub. Co.

Bartolotti, J., Marian, V., Schroeder, S. R., & Shook, A. (2011). Bilingualism and inhibitory control influence statistical learning of novel word forms. *Frontiers in Psychology, 2*, 324. https://doi.org/10.3389/fpsyg.2011.00324

Baxter, P., Bekkering, H., Dijkstra, T., Droop, M., van den Hurk, M., & Léoné, F. (2021). Grounding second language vocabulary instruction in cognitive science. *Mind, Brain, and Education*, 1–11. https://doi.org/10.1111/mbe.12278

Bialystok, E., Peets, K. F., & Moreno, S. (2014). Producing bilinguals through immersion education: Development of metalinguistic awareness. *Applied PsychoLinguistics, 35*, 177–191. https://doi.org/10.1017/S0142716412000288

Bice, K., & Kroll, J. F. (2019). English only? Monolinguals in linguistically diverse contexts have an edge in language learning. *Brain and Language*. https://doi.org/10.1016/j.bandl.2019.104644

Blanco-Elorrieta, E., & Pylkkänen, L. (2018). Ecological validity in bilingualism research and the bilingual advantage. *Trends in Cognitive Sciences, 22*(12), 1117–1126. https://doi.org/10.1016/j.tics.2018.10.001

Cenoz, J. (2013). The influence of bilingualism on third language acquisition: Focus on multilingualism. *Language Teaching*. https://doi.org/10.1017/S0261444811000218

Clyne, M., Hunt, C. R., & Isaakidis, T. (2004). Learning a community language as a third language. *International Journal of Multilingualism, 1*(1), 33–52. https://doi.org/10.1080/14790710408668177

Cox, J. G. (2017). Explicit instruction, bilingualism, and the older adult learner. *Studies in Second Language Acquisition, 39*(1), 29–58. https://doi.org/10.1017/S0272263115000364

de Bruin, A., Dick, A. S., & Carreiras, M. (2021). Clear theories are needed to interpret differences: Perspectives on the bilingual advantage debate. *Neurobiology of Language*, 1–46. https://doi.org/10.1162/nol_a_00038

Festman, J. (2021). Learning and processing multiple languages: The more the easier? *Language Learning, 71*(S1), 121–162. https://doi.org/10.1111/lang.12437

Flynn, S., Foley, C., & Vinnitskaya, I. (2004). The cumulative-enhancement model for language acquisition: Comparing adults' and children's patterns of development in first, second and third language acquisition of relative clauses. *International Journal of Multilingualism, 1*(1), 3–16. https://doi.org/10.1080/14790710408668175

French, R. M. (1998). A simple recurrent network model of bilingual memory. In *Proceedings of the twentieth annual conference of the cognitive science society* (pp. 368–373). Routledge.

Grainger, J., Manjaly, K., & Holcomb, P. J. (2008). Re-thinking the bilingual interactive-activation model from a developmental perspective (BIA-d). *Langauge Acquisition across Linguistic and Cognitive Systems, 52*, 267–283.

Green, C. S., & Bavelier, D. (2012). Learning, attentional control, and action video games. *Current Biology, 22*(6), 197–206. https://doi.org/10.1016/j.cub.2012.02.012

Grey, S., Sanz, C., Morgan-Short, K., & Ullman, M. T. (2018). Bilingual and monolingual adults learning an additional language: ERPs reveal differences in syntactic processing. *Bilingualism, 21*(5), 970–994. https://doi.org/10.1017/S1366728917000426

Han, W. J. (2012). Bilingualism and academic achievement. *Child Development, 83*(1), 300–321. https://doi.org/10.1111/j.1467-8624.2011.01686.x

Hermas, A. (2014). Restrictive relatives in L3 english: L1 transfer and ultimate attainment convergence. *Australian Journal of Linguistics, 34*(3), 361–387. https://doi.org/10.1080/07268602.2014.898227

Hirosh, Z., & Degani, T. (2018). Direct and indirect effects of multilingualism on novel language learning: An integrative review. *Psychonomic Bulletin and Review, 25*(3), 892–916. https://doi.org/10.3758/s13423-017-1315-7

Kahn-Horwitz, J., Schwartz, M., & Share, D. (2011). Acquiring the complex English orthography: A triliteracy advantage? *Journal of Research in Reading, 34*(1), 136–156. https://doi.org/10.1111/j.1467-9817.2010.01485.x

Kan, P. F., & Sadagopan, N. (2014). Novel word retention in bilingual and monolingual speakers. *Frontiers in Psychology*. https://doi.org/10.3389/fpsyg.2014.01024

Kaushanskaya, M. (2012). Cognitive mechanisms of word learning in bilingual and monolingual adults: The role of phonological memory. *Bilingualism*. https://doi.org/10.1017/S1366728911000472

Kaushanskaya, M., & Marian, V. (2009a). Bilingualism reduces native-language interference during novel-word learning. *Journal of Experimental Psychology: Learning Memory and Cognition*. https://doi.org/10.1037/a0015275

Kaushanskaya, M., & Marian, V. (2009b). The bilingual advantage in novel word learning. *Psychonomic Bulletin and Review, 16*(4), 705–710. https://doi.org/10.3758/PBR.16.4.705

Kaushanskaya, M., & Rechtzigel, K. (2012). Concreteness effects in bilingual and monolingual word learning. *Psychonomic Bulletin and Review*. https://doi.org/10.3758/s13423-012-0271-5

Klein, E. C. (1995). Second versus third language acquisition: Is there a difference? *Language Learning, 45*(3), 419–466. https://doi.org/10.1111/j.1467-1770.1995.tb00448.x

Kroll, J. F., & Bialystok, E. (2013). Understanding the consequences of bilingualism for language processing and cognition. *Journal of Cognitive Psychology*. https://doi.org/10.1080/20445911.2013.799170

Lee, M.-W., de Groot, A. M. B., & Kroll, J. F. (1999). Tutorials in bilingualism: Psycholinguistic perspectives. *Language*. https://doi.org/10.2307/417531

Li, P., & Farkas, I. (2002). A self-organizing connectionist model of bilingual processing. *Advances in Psychology*. https://doi.org/10.1016/S0166-4115(02)80006-1

Maguire, E. A., Gadian, D. G., Johnsrude, I. S., Good, C. D., Ashburner, J., Frackowiak, R. S. J., & Frith, C. D. (2000). Navigation-related structural change in the hippocampi of taxi drivers. *Proceedings of the National Academy of Sciences of the United States of America, 97*(8), 4398–4403. https://doi.org/10.1073/pnas.070039597

Martone, J. (2014). *Latino academic achievement: Impact of individual, family, school, community and immigration factors*. Dissertation Abstracts International Section A: Humanities and Social Sciences.

Modirkhamene, S. (2006). The reading achievement of third language versus second language learners of english in relation to the interdependence hypothesis. *International Journal of Multilingualism, 3*(4), 280–295. https://doi.org/10.2167/ijm043.0

Montanari, S. (2019). Facilitated language learning in multilinguals. In *Multidisciplinary perspectives on multilingualism* (pp. 302–324). https://doi.org/10.1515/9781501507984-014

Nair, V. K. K., Biedermann, B., & Nickels, L. (2016). Consequences of late bilingualism for novel word learning: Evidence from Tamil–English bilingual speakers. *International Journal of Bilingualism, 20*(4), 473–487. https://doi.org/10.1177/1367006914567005

Nation, R., & McLaughlin, B. (1986). Novices and experts: An information processing approach to the "good language learner" problem. *Applied Psycholinguistics, 7*(1), 41–55.

Parmis, A. A., Bandalan, R. B., & Clerigo, J. C. (2020). Language learning: Reading comprehension, motivation, strategies, and academic achievement. *Archives of Business Research*. https://doi.org/10.14738/abr.86.8324

Polka, L., Colantonio, C., & Sundara, M. (2001). A cross-language comparison of /d /–/ð / perception: Evidence for a new developmental pattern. *The Journal of the Acoustical Society of America, 109*(5), 2190–2201. https://doi.org/10.1121/1.1362689

Poorebrahim, F., Afsharrad, M., & Ghonsooly, B. (2020). Bilingualism, monoliteracy, and third language writing: A case from Turkish-Persian context in Iran. *IIUM Medical Journal Malaysia, 10*(2), 369–381. https://doi.org/10.17509/ijal.v10i2.28608

Rothman, J. (2011). L3 syntactic transfer selectivity and typological determinacy: The typological primacy model. *Second Language Research, 27*(1), 107–127. https://doi.org/10.1177/0267658310386439

Rothman, J. (2015). Linguistic and cognitive motivations for the Typological Primacy Model (TPM) of third language (L3) transfer: Timing of acquisition and proficiency considered. *Bilingualism*. https://doi.org/10.1017/S136672891300059X

Schroeder, S. R., & Marian, V. (2017). Cognitive consequences of trilingualism. *International Journal of Bilingualism*. https://doi.org/10.1177/1367006916637288

Schunk, D. H. (2012). Learning theories: An educational perspective. In *Reading*. Pearson.

Schwartz, M., Kahn-Horwitz, J., & Share, D. L. (2014). Orthographic learning and self-teaching in a bilingual and biliterate context. *Journal of Experimental Child Psychology, 117*, 45–58. https://doi.org/10.1016/j.jecp.2013.08.008

Serratrice, L., Sorace, A., Filiaci, F., & Baldo, M. (2009). Bilingual children's sensitivity to specificity and genericity: Evidence from metalinguistic awareness. *Bilingualism, 12*(2), 239–257. https://doi.org/10.1017/S1366728909004027

Sundara, M., Polka, L., & Genesee, F. (2006). Language-experience facilitates discrimination of /d-{a figure is presented}/ in monolingual and bilingual acquisition of English. *Cognition, 100*(2), 369–388. https://doi.org/10.1016/j.cognition.2005.04.007

Trapman, M., van Gelderen, A., van Steensel, R., van Schooten, E., & Hulstijn, J. (2014). Linguistic knowledge, fluency and meta-cognitive knowledge as components of reading comprehension in adolescent low achievers: Differences between monolinguals and bilinguals. *Journal of Research in Reading, 37*(S1), S3–S21. https://doi.org/10.1111/j.1467-9817.2012.01539.x

Tremblay, M.-C., & Sabourin, L. (2012). Comparing behavioral discrimination and learning abilities in monolinguals, bilinguals and multilinguals. *The Journal of the Acoustical Society of America, 132*(5), 3465–3474. https://doi.org/10.1121/1.4756955

Van Gelderen, A., Schoonen, R., de Glopper, K., Hulstijn, J., Snellings, P., Simis, A., & Stevenson, M. (2003). Roles of linguistic knowledge, metacognitive knowledge and processing speed in L3, L2 and L1 reading comprehension: A structural equation modeling approach. *International Journal of Bilingualism, 7*(1), 7–25. https://doi.org/10.1177/13670069030070010201

Van Hell, J. G., & Mahn, A. C. (1997). Keyword mnemonics versus rote rehearsal: Learning concrete and abstract foreign words by experienced and inexperienced learners. *Language Learning*. https://doi.org/10.1111/0023-8333.00018

VanPatten, B., & Benati, A. G. (2015). *Key terms in second language acquisition*. Bloomsbury Publishing.

Walker, M. P., & Stickgold, R. (2004). Sleep-dependent learning and memory consolidation. *Neuron, 44*(1), 121–133. https://doi.org/10.1016/j.neuron.2004.08.031

Wang, T., & Saffran, J. R. (2014). Statistical learning of a tonal language: The influence of bilingualism and previous linguistic experience. *Frontiers in Psychology, 5*, 953. https://doi.org/10.3389/fpsyg.2014.00953

Westergaard, M., Mitrofanova, N., Mykhaylyk, R., & Rodina, Y. (2017). Cross-linguistic influence in the acquisition of a third language: The linguistic proximity model. *International Journal of Bilingualism, 21*(6), 666–682. https://doi.org/10.1177/1367006916648859

Yoshida, H., Tran, D. N., Benitez, V., & Kuwabara, M. (2011). Inhibition and adjective learning in bilingual and monolingual children. *Frontiers in Psychology, 2*, 210. https://doi.org/10.3389/fpsyg.2011.00210

Zhao, T. C., & Kuhl, P. K. (2015). Effect of musical experience on learning lexical tone categories. *The Journal of the Acoustical Society of America, 137*(33), 1452–1463. https://doi.org/10.1121/1.4913457

Zhao, X., & Li, P. (2010). Bilingual lexical interactions in an unsupervised neural network model. *International Journal of Bilingual Education and Bilingualism*. https://doi.org/10.1080/13670050.2010.488284

Zhao, X., & Li, P. (2013). Simulating cross-language priming with a dynamic computational model of the lexicon. *Bilingualism*. https://doi.org/10.1017/S1366728912000624

Development of Mathematical Cognition: The Role of Technology in Low-SES Populations

Francisco Martín López, Dinorah de León, Nadir Díaz-Simón, and Alejandro Maiche

Introduction

There is a cultural problem with mathematics. Many children, and also some adults, show an aversion toward dealing with numbers and arithmetic operations. In extreme cases, this behavior leads to stress and fear akin to a phobia, which receives the name of mathematical anxiety (Dowker et al., 2016). Unlike the social and parental pressure regarding language difficulties, which are usually corrected due to the need to read and write in order to communicate with others, struggles in math are often overlooked and accepted, particularly in underprivileged contexts (DeFlorio & Beliakoff, 2015). The consequences of poor mathematical performance are damaging and long-lasting, both for the individual and for their communities.

Data from large-scale surveys reveal significant differences in mathematical abilities across socioeconomic status (SES) (OECD, 2019). SES refers to the position of an individual or a group within society and is usually measured using household income, occupation, and degree of education of parents. Low-SES populations

---

F. M. López
Interdisciplinary Center in Cognition for Education and Learning (CICEA), Universidad de la República, Montevideo, Uruguay

Frankfurt Institute for Advanced Studies (FIAS), Frankfurt, Germany

D. de León
Interdisciplinary Center in Cognition for Education and Learning (CICEA), Universidad de la República, Montevideo, Uruguay

Faculty of Psychology, Universidad Católica del Uruguay, Montevideo, Uruguay

N. Díaz-Simón · A. Maiche (✉)
Interdisciplinary Center in Cognition for Education and Learning (CICEA), Universidad de la República, Montevideo, Uruguay
e-mail: amaiche@psico.edu.uy

M. V. Alves et al. (eds.), *Cognitive Sciences and Education in Non-WEIRD Populations*, https://doi.org/10.1007/978-3-031-06908-6_11

present more health problems, stress, malnutrition, and domestic violence, as well as less neurocognitive and socioemotional development (Bradley & Corwyn, 2002; Noble et al., 2005) which usually results in lower academic achievement.

Traditional educational approaches have failed to reduce this academic achievement gap (Ryan et al., 2006). The recent emergence of new information and communication technologies (ICTs) can provide innovative and exciting opportunities to balance the disparities. These opportunities depend not only on giving students a real chance to interact with the technology but also on the type of activity that the new technologies support in the school environment (Alderete & Formichella, 2016). While policy makers and technology investors tend to quickly align themselves behind new seemingly magical solutions, the academic community needs to make critical revisions of past and present studies and thus provide guidelines for future research and educational policies. Such is the approach of this work.

In this chapter, we discuss the role that educational technologies may play in supporting the development of mathematical cognitive skills in children, specifically those from low-SES populations in Latin America. Educational technologies encompass all alternatives to traditional education based on ICTs, such as audio and video lessons, computer-assisted instruction, and digital and hybrid games. First, we review the existing evidence that shows that differences in mathematics performance across SES levels are already present in early childhood. Then, we explore some technological solutions that have been implemented in educational settings aiming to boost the math knowledge of underprivileged children. The main section of this chapter focuses on research done in Latin America around this topic. In this context, we also present the work that our group has been carrying out with technology-based interventions in Uruguayan schools, including the main results and insights from recent years, as well as ongoing studies and future directions.

From the revision of the positive and negative outcomes of the inclusion of educational technology in Latin American schools, we shall conclude that the discussion about whether or not to use ICTs must be replaced with a discussion about how to use them. Technological tools can provide children in underprivileged contexts with some of the specific stimuli, questions, and answers that their peers in middle- and high-SES contexts are usually exposed to in their households.

Early Math Performance and SES

Mathematical concepts emerge from an early age. Although children come to the world with some innate abilities to approximately quantify entities and events (Feigenson et al., 2004; Izard et al., 2009), these abilities are further developed throughout the first stages of life (Odic et al., 2013) and later solidified by formal education (Halberda et al., 2012). Several longitudinal studies have shown strong correlations between early numerical skills and later mathematical performance (Starr et al., 2013; Gilmore et al., 2010). It is therefore essential to identify children with underdeveloped mathematical abilities and focus on boosting their competencies in those first years of life.

One source of disparity in academic achievement is SES. Children from low-SES households underperform their peers from high-SES contexts in basic mathematical knowledge as early as kindergarten (Starkey et al., 2004) which results in a "school entry gap" (Janus & Duku, 2007). Crucially, the disparities between low-SES and high-SES individuals in math are greater than those in language and other areas of knowledge (Cross et al., 2009; Duncan et al., 2007). This gap increases across the lifespan and might influence long-term academic and professional achievement.

The underlying origins for the lower math achievement in children from underprivileged contexts may be manifold. From a developmental point of view, growing up in poverty is linked with hindered physical, mental, and emotional health (Bradley & Corwyn, 2002; Yoshikawa et al., 2012). Children from low-SES families show different brain structure (Noble et al., 2005; Ursache et al., 2016). Research has made it possible to corroborate the influence of living in poverty on basic cognitive processes, such as self-regulation, and to detect its traces in adults with a past of child poverty (Lipina, 2016). Low SES is associated with reduced vocabulary and language processing skills (Fernald et al., 2013) and lower numerical abilities (Jordan et al., 1994) compared to their peers from middle and high SES even before the start of formal schooling. These differences subsequently increase throughout childhood and adolescence (Duncan et al., 2007) and result in higher dropout rates (Alexander et al., 1997). Furthermore, although parents from low-SES households tend to get involved in home math activities as much as parents from middle- and high-SES households, the latter show higher expectations which might explain these disparities (Susperreguy et al., 2021). Parents from different contexts not only act but also think differently about the importance of math (Elliott & Bachman, 2018). Moreover, differences in math achievement might also emerge from disparate quality of home activities, suggesting that not any kind of parental involvement impacts learning equally.

However, while poverty is associated with lower school performance, differences can be observed within children from low-SES households. Halle et al. (1997) conducted interviews with low-income minority children and parents and found that mothers with higher education had higher expectations for their children's academic achievement which influenced their children's subsequent achievement in math and reading. Thus, growing up in a low-SES household does not necessarily imply low academic achievement. Consequently, and independently of the need to eradicate poverty, finding the most efficient ways to help children from underprivileged contexts learn mathematics is an effort worth making.

Can Technology Be the Modern Great Equalizer?

Education is often perceived as the main tool for reducing social disparities. In the nineteenth century, Horace Mann defined education as the "great equalizer of conditions of men." Under this view, if all children enter a schooling system at the same age, learn the same contents, and are held to the same standards, they should finish their education at the same level and have the same opportunities, regardless of their SES. As we know, this traditional view is far from the truth. On the contrary, social context routinely influences the quality of education, which widens rather than closes the academic gap between children living in poverty and their peers from middle and high SES. The educational policies implemented in the last century have failed to change this reality, making education a reflection of social hierarchy rather than a contributor to social mobility.

An alternative view suggests that ICTs might replace traditional educational methodologies (Bando et al., 2016; Pea, 1987), in particular as a solution for hindered development in children from low-SES backgrounds. Technological tools might provide children with the specific educational content they need and thus balance social differences. Children from underprivileged contexts can have access, through the use of computers, smartphones, and other ICTs, to a myriad of stimuli, questions, and answers that their families and schools might not be able to provide at the time they need.

In this sense, the inclusion of educational technologies in school settings or at home can balance math performance differences by compensating for specific areas in low-achieving children. In fact, technological tools can also be used to perform fast and detailed math evaluations in order to obtain detailed assessments of a child's abilities and potential areas of deficiencies. This assessment can in turn be used to improve their performance before those children fall behind their peers. Frequent testing also facilitates the retention of information (Roediger & Karpicke, 2006), but pen-and-paper tests can take too long for educators to correct. Using ICTs can not only accelerate this process, but also provide educators with opportunities for adaptive testing and result in valuable feedback about their students and what areas they should work on. Computer-based assessments are already in use in most countries for standardized tests such as PISA (OECD, 2019) and will likely be a key component in the future of education.

However, the evidence supporting the benefits of the systematic use of tablets and mobile devices in schools is still scarce (Haßler et al., 2016) with significant yet moderate effects on learning (Cheung & Slavin, 2013; Sung et al., 2016). A review of 74 studies that include the use of educational technology applications in mathematical learning shows an overall small effect size compared to traditional methods (Cheung & Slavin, 2013). The authors highlight supplemental computer-assisted instruction (CAI) as being the methodology that produces the largest effects on mathematics achievement and conclude that incorporating these educational technologies into the classroom curriculum appears to be the best approach.

Interesting observations can be made from studies comparing technology-based and traditional teaching methodologies. Outhwaite et al. (2017) report significant learning gains from tablet-based interventions in UK schools. Their results indicate that low-achieving children, in particular those with weak short-term memory skills, show the largest improvements. The authors attribute this outcome to the reduced cognitive demands of a tablet game in comparison to traditional classroom learning. When controlling for previous mathematical knowledge, the authors find no significant effect of SES level. Previous works support the idea that technological tools are useful for children with lower numerical skills (Räsänen et al., 2009; Shin et al., 2012).

Educational technologies can also be used outside school settings, particularly at home. Parental expectations, practices, and attitudes toward the learning of math are known to impact a child's mathematical performance (Elliott & Bachman, 2018; Halle et al., 1997). Low-achieving students are likely not getting high-quality interactions and support at home. This may be due to parents' lack of knowledge about how to help their children or the (misleading) idea that their interactions will have no effect on their children's learning. In such cases, ICTs can provide an effective and accessible solution. Berkowitz et al. (2015) carried out an intervention to promote home math interactions with 587 children and their parents in the United States. A tablet app called *Bedtime Learning Together* presented a short story with numerical content and related questions. The story was read by parents who also answered as many questions as they wished together with their children, thus promoting the expected interaction. The questions included counting, geometry, and arithmetic. Parents used the app several times per week throughout the school year. Results show that children in the experimental group improved significantly compared to children in the control group who used a similar tool with reading rather than math content.

Insights from Latin America

Given the cultural and educational differences, it is unclear how the results from studies conducted in developed countries translate to the Latin American context. One relevant distinction between children in western, educated, industrialized, rich, and democratic (WEIRD) and Latin American countries is their access to internet-based technological tools. Many studies have shown the correlation between availability of Internet connection and various household or individual socioeconomic characteristics, such as income and education (see Grazzi & Vergara, 2014). For this reason, it is important to understand how the use of educational technologies impacts the learning in Latin American children, in particular those from peripheral and low-SES contexts.

One rudimentary yet powerful educational technology is audio. Radio lessons were popular in the 1980s, mainly as a widespread and inexpensive teaching mechanism for children in rural areas in Central America (Jamison et al., 1981), and again during the COVID-19 pandemic in regions where children lacked the internet connectivity to attend classes remotely (Dreesen et al., 2020). Naslund-Hadley et al. (2014) explored the benefits of a pilot audio version of Tikichuela, a Paraguayan mathematical instruction program for preschool children adapted from the Big Math for Little Kids program used in low-income schools in New York. These audio lessons included games, songs, and interactive activities intended to capture the interest of children. The results from this study show an improvement in test scores for children participating in the pilot program in comparison to controls following a five-month period with significantly larger improvements in rural versus urban schools but no differences on SES. The authors highlight audio lessons as an encouraging tool to reduce student learning disparities that emerge from educators lacking specific training or resources. Audio-based mathematical instruction was also used to improve the memory and math abilities of blind children in Chile (Sánchez & Flores, 2005). Cuba's video-based reading and writing program "Yo, sí puedo" has been implemented in 28 countries world-wide and has helped eradicate adult illiteracy in Venezuela and Bolivia among others (Canavire, 2011). In sum, audiovisual tools such as these are good examples of the potential for ICTs in critical situations, when other alternatives are not available.

More recently, the wide access to computers has inspired researchers to explore the educational benefits of using computer games. Goldin et al. (2014) implemented a 10-week-long intervention using the *Mate Marote* software with 111 low-SES first graders in Argentina. Children played three different computer games involving executive functions which resulted in significant improvements in inhibitory control and attention, but not planning, in pretest–posttest measures compared to a control group that played commercial video games. What is more, training their executive functions resulted in a far transfer to math and language abilities, as measured by school grades. The researchers also split the children who used Mate Marote into two categories according to their attendance to school. Following the intervention, children with low attendance, who initially had lower school grades than their high attendance peers, showed the largest improvements in school grades even though they were not directly trained in language or mathematics. Other uses of educational software in Mexico (Zaldívar-Colado et al., 2017) and Argentina (Furman et al., 2019) also support the use of ICTs, although the latter showed no significant differences between groups of children that learned science with and without tablets.

Motivated by these possibilities, several Latin American governments resorted to the One Laptop Per Child (OLPC) program as a remedy for their ailing educational systems. The outcomes are ambiguous. The OLPC initiative was created in 2005 with the idea that distributing low-cost hardware that children could take home would reduce the emerging digital gap between developed and developing countries. As a side effect, the inclusion of ICTs in classrooms could have a positive impact on academic performance. Data from the educational benefits of using laptops at schools are controversial. While OLPC advertises their products with test scores from Nicaragua that showed an increase in math and reading skills, more rigorous studies conducted in Latin America do not show positive outcomes. For example, a large-scale randomized study with data from 15 months of use of OLPC devices in 318 primary schools in Perú resulted in no effects on math and language test scores, and a nonsignificant increase in cognitive skills (Cristia et al., 2017). Other smaller-scale studies conducted in Perú revealed that computer skills improved in children that received OLPC devices but with no significant differences on academic achievement (Beuermann et al., 2015; Malamud et al., 2019; Severín & Capota, 2011). Similarly disappointing results were obtained from OLPC programs in Colombia (Severín & Capota, 2011), Costa Rica (Meza-Cordero, 2017), Paraguay (Ames, 2013), and Uruguay (de Melo et al., 2014; Yanguas, 2020). While it is undoubtedly beneficial that children improve their computer skills, the lack of transfer to math and language proficiency reveals that the naive approach of the OLPC program is insufficient to produce profound changes in education.

In sum, educational technologies have been used in Latin American schools with varying degrees of success. Besides the extreme cases where alternatives to traditional lessons had to be implemented because in-person classes were nonviable, positive outcomes were obtained mainly through the use of software especially designed by researchers and educators to train basic cognitive abilities. On the other hand, the unrestricted inclusion of laptops and tablets from the OLPC program showed no positive effects when it was not supported by a well-structured educational platform. This could show that it is not the technological tool in itself, but rather the use that is given to it, that allows for enhanced learning.

In this sense, we think focus must be shifted from a comparison between tech-based and traditional learning to a comparison between the contents that can be taught with each of these methodologies. Technology presents innovative, widespread, and efficient ways to include contents based on the principles of educational neuroscience in school settings which can result in improvements in academic performance (Goldin et al., 2014). The shift in focus is further supported by data about the massive and unrestricted inclusion of ICTs in Latin American schools. To illustrate this, in the next section, we present a detailed account of the work our research group has made over the last decade in Uruguayan schools.

Learnings from a Decade of Technology-Based School Interventions in Uruguay

In 2007, the Uruguayan government and OLPC partnered to establish the Plan Ceibal, a program that provides every school-aged student with an Android-based tablet that they can use at school and at home. The students have free internet connection in schools and multiple public areas around the country. This makes Uruguay an ideal site for studying the usefulness of ICTs in educational contexts.

Our research group has been working in collaboration with the Plan Ceibal for nearly ten years. We have worked on a number of projects involving technology-based interventions in classrooms that can be clustered into two main categories. On the one hand, we have designed and tested early assessment tools that can be used directly from the Plan Ceibal tablets available in Uruguayan schools. On the other hand, we have created several games aimed at reinforcing the basic mathematical abilities of young children. All the software developed by our group is free and available to everyone.[1] These tools are targeted to Uruguayan children, involving characters, places, stories, and vocabulary that they can identify with (see Outhwaite et al., 2020). From the personal reports of educators, parents, and the children themselves, this makes our games more entertaining and engaging than other educational software available in Plan Ceibal. The simultaneous use of both testing and training learning tools has shown beneficial results, particularly in low-SES contexts. Our long-term goal is to incorporate a set of applications into the Plan Ceibal tablets that the teachers themselves can use to facilitate mathematical learning in the classroom.

To the best of our knowledge, there is a lack of digital math assessment tools for children under the age of 7. One of the most used tests for assessing young children in math is TEMA-3 (Ginsburg & Baroody, 2003), which takes around 30 minutes to complete and requires one-on-one interactions between a trained evaluator and each child. Although TEMA-3 is a powerful tool, it requires an amount of time and training that makes it unreasonable for large-scale interventions and regular use in schools. Our group has always been concerned with the need for tools that make testing fast and scalable, and that can ultimately give teachers useful feedback about the particular topics that require attention. For this reason, in 2013 we began the development of the Prueba Uruguaya de Matemática (PUMA), a digital tool where children are tested on their knowledge of number symbols, ordering, number composition and decomposition, number line placing, and number words (see Odic et al., 2016) for a description of the first version of PUMA). PUMA provides self-administered assessment that a single researcher or teacher can apply simultaneously to an entire class. A second version of PUMA was developed in 2020[2] with improved design and usability (Fig. 1). An assessment of 187 Uruguayan preschoolers and first graders with the updated version of PUMA and TEMA-3 showed a strong correlation ($r = 0.803$, $p < 0.001$; De León et al., in prep). These positive outcomes show that ICTs have great potential regarding assessments. However, so far we have failed to convince educators and policymakers about the benefit of incorporating PUMA into regular school programs. We attribute this to a failure on our part to translate the quantitative outputs of PUMA into easily understandable qualitative feedback. This must be a key point to address in the near future.

The first large-scale intervention that we conducted using Plan Ceibal tablets took place in 2013. We tested 503 Uruguayan first graders from 10 schools across SES levels on their abilities to discriminate approximate quantities, using the Panamath software (Halberda et al., 2008), and approximate time (Odic et al., 2016). We also assessed their formal mathematical performance with the first version of PUMA. Our results showed correlations between math performance and approximate number discrimination as well as approximate time discrimination.

[1] For more information, go to: http://www.cognicionnumerica.psico.edu.uy/en/#recursos

[2] A demo version is available here: http://math.psico.edu.uy/puma/

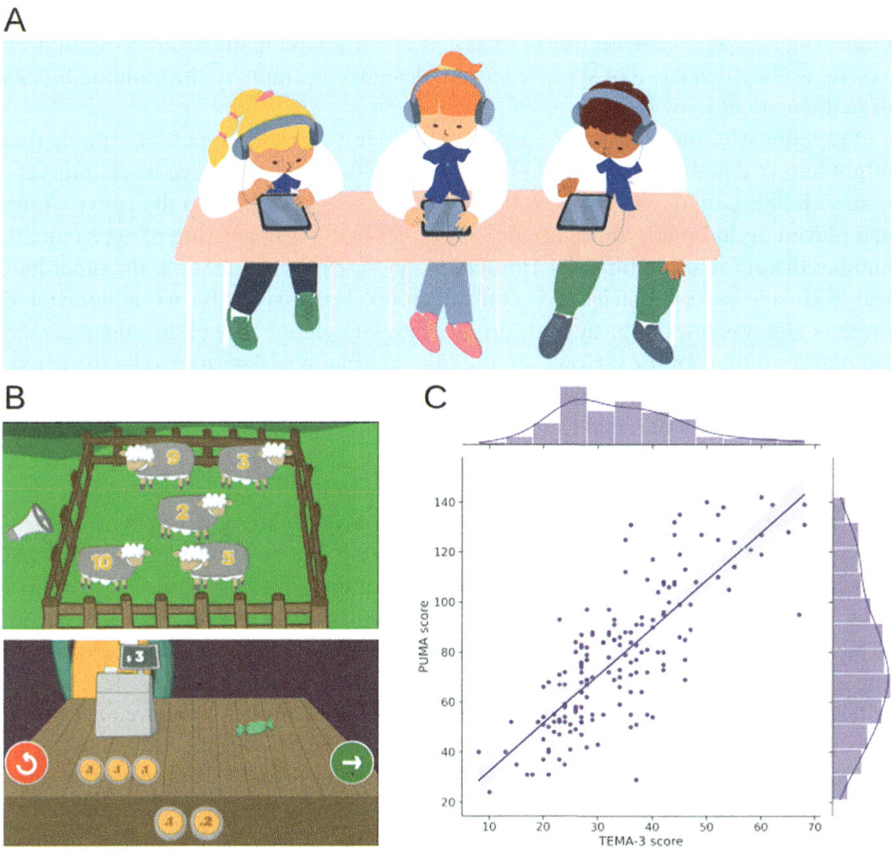

Fig. 1 Second version of the Prueba Uruguaya de Matemática (PUMA). (**a**) Three Uruguayan children taking the PUMA test with their Plan Ceibal tablets. (**b**) Screenshots of two PUMA tasks (top: verbal-numeral to Arabic transcoding; bottom: composition and decomposition). (**c**) Correlation of the 2nd version of PUMA and TEMA-3 (Ginsburg & Baroody, 2003) with $N = 187$, $r = 0.803$ ($p < 0.001$). Both tests take a similar time to complete, but PUMA is a tablet-based assessment that does not require one-on-one interactions between each child and one evaluator. It can therefore be applied simultaneously to a large number of children, e.g., an entire class could be assessed in detail in less than one hour

Most importantly, this first large-scale use of the Plan Ceibal tablets for doing research in school settings confirmed the validity of ICTs as assessment tools. A total of 454 of those children participated in a short intervention consisting of four six-minute sessions of approximate number system (ANS) training (Valle-Lisboa et al., 2016). Pretest–posttest differences revealed that children across all SES levels improved their ANS and formal math skills without significant differences on SES. However, only low-SES children showed a significant correlation between the number of times they played and the PUMA scores improvement. This could show that cognitive stimulation plays a higher role in underprivileged contexts. The learnings from these initial studies were in line with other studies conducted in Latin America. ICTs can be useful to assess and train the math abilities of children as long as they make use of specific stimuli designed to improve the building blocks of mathematical knowledge.

Subsequently, our emphasis veered to studying different types of stimuli that might improve math abilities. Langfus et al. (2019) conducted a five week intervention with 386 Uruguayan children. Half of them were assigned to the active group and played digital mini-games on the Plan Ceibal tablets consisting of approximate number discrimination, time discrimination, and area discrimination; the other half was a Business-As-Usual (BAU) control group. For this study, we administered pretests and posttests of basic and formal math knowledge as well as language and general cognitive abilities. However, the playing time was determined by the teachers, who agreed to use the games during their classes, and by the children themselves, who were encouraged to continue playing at home. No significant differences were found between the active and the control groups. A detailed analysis of the data indicates that this was likely due to the low number of plays: on average, children played the time, number and area games only 10, 16, and 26 times, respectively, over a five-week period. The children had positive comments about the games, but they were not interested enough to continue playing in their spare time. This highlights a particular difficulty with tech-based learning. If children are expected to engage in educational games voluntarily, then these games must be as engaging and attractive as other commercial games that they may be used to. Alternatively, tech-based learning must be framed within well-structured activities. Educators should be trained on how to make the most of educational technologies so that children can be cognitively stimulated with stimuli specifically designed for them.

Currently, we are also interested in exploring the differences between individual and social learning. Tablets and computers are most useful for single users, but the peer interactions that usually emerge in group settings have powerful influences on early cognition and learning (de la Hera et al., 2019; Dillon et al., 2017). To this end, we designed a new game based on the cards used by Dillon et al. (2017), with a digital version that can be played individually on tablets and a hybrid version that can be played individually or collectively with a Magic Box (see Fig. 2). That latter consists of playing cards equipped with radiofrequency identification (RFID) tags and an Arduino-powered smart box with RFID readers. This hybrid version of the game incorporates ICTs without isolating the children and, since it promotes peer interaction to advance in the game, is predicted to be most useful in school environments, whereas individual games such as the digital version of the game can be used at home. A comparative study of 266 children with two active groups, one playing with the Magic Box and the other one with tablets, and a BAU control group, is scheduled for late 2021.

In sum, the trajectory followed by our research group is representative of the general approach to educational technologies in Latin America and in the rest of the world. Our initial enthusiasm for the use of ICTs in schools, backed by large governmental investments in tech-based learning programs such as OLPC programs

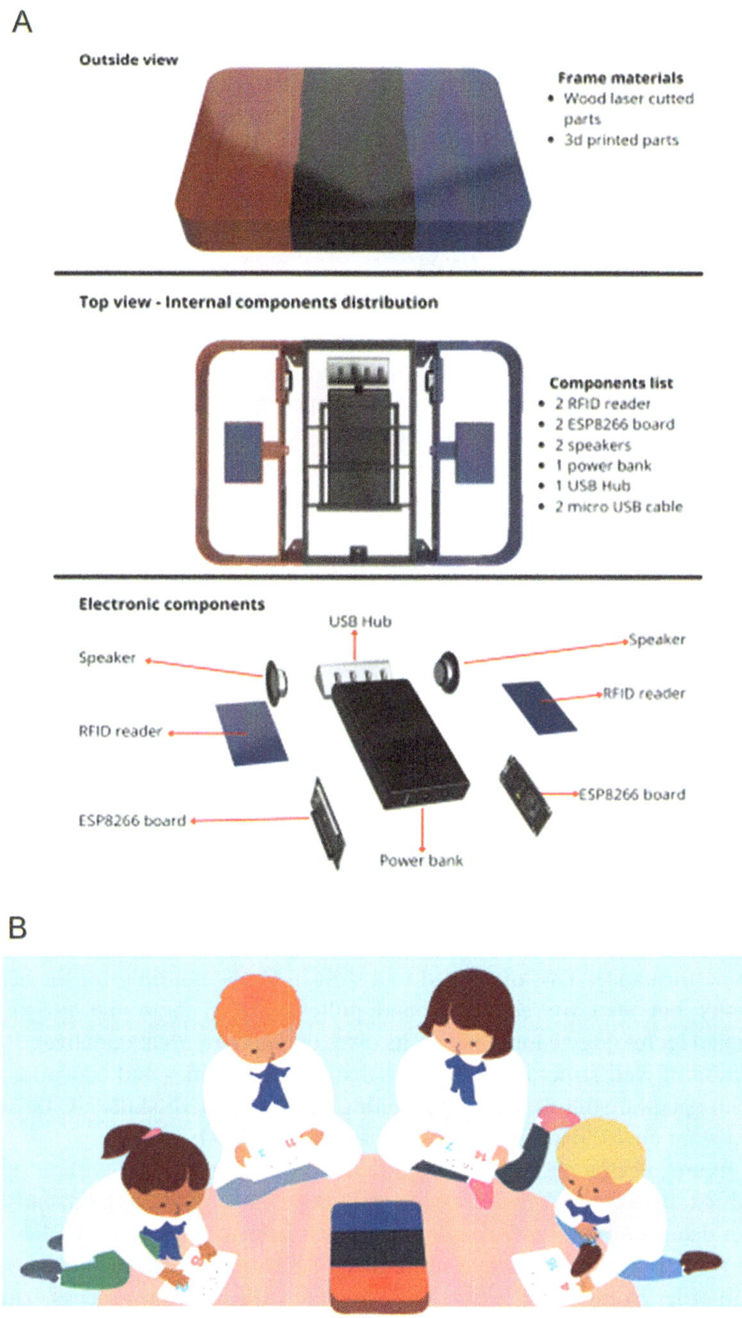

Fig. 2 The Magic Box. (**a**) outside and inside views of the Magic Box and its components. Note that the box is closed while playing so children do not have access to any of the internal components. (**b**) A typical situation of four children playing in the collective version of the game with the Magic Box. Children take turns to decide whether each of their cards should go on the red or the blue half of the box. After four cards (one for each child), the Magic Box gives a positive feedback sound (if all four cards are correct) or a negative feedback sound (if at least one of the cards is wrong). Children are encouraged to interact and help each other in order to obtain a correct response of the Magic Box and advance in the game

like Plan Ceibal in Uruguay, morphed into a critical examination of the differences between educational technologies and other methodologies, with a particular focus on characterizing the cognitive stimuli which produce the most efficient learning.

Conclusions and Perspectives

The evidence presented in this chapter reveals that, while SES strongly correlates with academic achievement, it is possible to boost low-SES children's math learning by improving the cognitive abilities that support math development. Educational technologies were initially thought to be a possible solution for these disparities, but most research shows no SES-related differences between ICTs and traditional methodologies when initial performance is controlled for. That is, educational technologies are most useful for low-achieving children, regardless of their SES. By way of illustration, it is not a households' income that determines the child's academic performance but rather the parental expectations and quality of home learning activities. In our view, this is because it is not poverty itself, but rather the individual, parental, and social consequences of poverty that are at the core of the disparity in math achievement. Even though technology cannot directly change the contexts where children from low-SES live and grow, it might compensate for the lack of math-related home activities by providing the stimuli needed for their cognitive development. Thus, ICTs are likely most useful in such contexts.

Positive and negative outcomes have been observed from the use of educational technologies in Latin America. The unrestricted inclusion of ICTs in classrooms, as is the case with the OLPC program, resulted in no measurable differences in achievement other than computer and internet skills. Conversely, numerous school interventions carried out by research groups did produce significant improvements in math performance. This must lead to a shift in focus, not only for the academic community but also for educators and policymakers, about the use of ICTs. Educational technology cannot be, on its own, the modern great equalizer. But with the addition of well-structured programs that result in games and educational software with specific stimuli aimed at training basic cognitive skills, ICTs can help low-achieving children and particularly those from low SES.

The future of educational technologies is promising but not without its challenges. First, researchers must continue studying basic cognition in the early stages of life to determine the best specific stimuli to improve learning. We also need to connect the use of these stimuli with realistic educational practices in schools and homes in order to involve teachers and parents in the learning process. This must include studies about how to include ICTs in classrooms and homes. The involvement of teachers as researchers vastly broadens the inputs of information and new ideas needed to advance knowledge. Put together with the versatility of ICTs, this joint work can allow for the exploration of multiple possibilities and can provide high-quality feedback about the impact of educational technologies in the classroom. On a similar note, researchers should aim to increase and improve the involvement of parents and other family members through the use of ICTs. This requires the development of tools that are suitable for use at home and promote the natural interaction with children about everyday math. Finally, all the software must be engaging and relatable to the children's reality, which increases the requirements and potentially the costs of developing new tools. Collaborative work with countries that share a common culture, as is the case in Latin America, is a much needed approach.

References

Alderete, M. V., & Formichella, M. M. (2016). The effect of ICTs on academic achievement: The conectar igualdad programme in Argentina. *CEPAL Review, 2016*(119), 83–100. https://doi.org/10.18356/f23c6662-en

Alexander, K. L., Entwisle, D. R., & Horsey, C. S. (1997). From first grade forward: Early foundations of high school dropout. *Sociology of Education, 70*(2), 87. https://doi.org/10.2307/2673158

Ames, M. G. (2013). *From MIT to Paraguay: A critical historical and ethnographic analysis of one laptop per Child.* Stanford University.

Bando, R., Gallego, F., Gertler, P., & Romero, D. (2016). *Books or laptops? The cost-effectiveness of shifting from printed to digital delivery of educational content* (Working Paper No. 22928; Working Paper Series). National Bureau of Economic Research. https://doi.org/10.3386/w22928

Berkowitz, T., Schaeffer, M. W., Maloney, E. A., Peterson, L., Gregor, C., Levine, S. C., & Beilock, S. L. (2015). Math at home adds up to achievement in school. *Science (New York, N.Y.), 350*(6257), 196–198. https://doi.org/10.1126/science.aac7427

Beuermann, D. W., Cristia, J., Cueto, S., Malamud, O., & Cruz-Aguayo, Y. (2015). One laptop per child at home: Short-term impacts from a randomized experiment in Peru. *American Economic Journal: Applied Economics, 7*(2), 53–80. https://doi.org/10.1257/app.20130267

Bradley, R. H., & Corwyn, R. F. (2002). Socioeconomic status and child development. *Annual Review of Psychology, 53*(1), 371–399. https://doi.org/10.1146/annurev.psych.53.100901.135233

Canavire, L., V. B. (2011). Educación para adultos en América Latina: Programa de alfabetización "Yo, sí puedo". *CPU-e, Revista de Investigación Educativa, 13*, 128–143. https://doi.org/10.25009/cpue.v0i13.40

Cheung, A. C. K., & Slavin, R. E. (2013). The effectiveness of educational technology applications for enhancing mathematics achievement in K-12 classrooms: A meta-analysis. *Educational Research Review, 9*, 88–113. https://doi.org/10.1016/j.edurev.2013.01.001

Cristia, J., Ibarrarán, P., Cueto, S., Santiago, A., & Severín, E. (2017). Technology and child development: Evidence from the one laptop per child program. *American Economic Journal: Applied Economics, 9*(3), 295–320. https://doi.org/10.1257/app.20150385

Cross, C. T., Woods, T. A., Schweingruber, H. A., & National Research Council (U.S.) (Eds.). (2009). *Mathematics learning in early childhood: Paths toward excellence and equity.* National Academies Press.

de la Hera, D. P., Sigman, M., & Calero, C. I. (2019). Social interaction and conceptual change pave the way away from children's misconceptions about the Earth. *Npj Science of Learning, 4*(1), 1–12. https://doi.org/10.1038/s41539-019-0051-3

De León, D., López-Guzmán, F., Díaz-Simón, N., Puyol, L., González, M., Sierna, C., & Maiche, A. (in prep). *Self-administered digital math assessment test.* Universidad de La República.

de Melo, G., Machado, A., & Miranda, A. (2014). *The impact of a one laptop per child program on learning: Evidence from Uruguay.* Banco de México. https://doi.org/10.36095/banxico/di.2014.22

DeFlorio, L., & Beliakoff, A. (2015). Socioeconomic status and preschoolers' mathematical knowledge: The contribution of home activities and parent beliefs. *Early Education and Development, 26*(3), 319–341. https://doi.org/10.1080/10409289.2015.968239

Dillon, M. R., Kannan, H., Dean, J. T., Spelke, E. S., & Duflo, E. (2017). Cognitive science in the field: A preschool intervention durably enhances intuitive but not formal mathematics. *Science, 357*(6346), 47–55. https://doi.org/10.1126/science.aal4724

Dowker, A., Sarkar, A., & Looi, C. Y. (2016). Mathematics anxiety: What have we learned in 60 years? *Frontiers in psychology, 7*. https://doi.org/10.3389/fpsyg.2016.00508

Dreesen, T., Akseer, S., Brossard, M., Dewan, P., Giraldo, J.-P., Kamei, A., Mizunoya, S., & Ortiz, J. S. (2020). Promising practices for equitable remote learning: Emerging lessons from COVID-19 education responses in 127 countries. *Unicef Innocenti Research Briefs, 10*.

Duncan, G. J., Dowsett, C. J., Claessens, A., Magnuson, K., Huston, A. C., Klebanov, P., Pagani, L. S., Feinstein, L., Engel, M., Brooks-Gunn, J., Sexton, H., Duckworth, K., & Japel, C. (2007). School readiness and later achievement. *Developmental Psychology, 43*(6), 1428–1446. https://doi.org/10.1037/0012-1649.43.6.1428

Elliott, L., & Bachman, H. J. (2018). SES disparities in early math abilities: The contributions of parents' math cognitions, practices to support math, and math talk. *Developmental Review, 49*, 1–15. https://doi.org/10.1016/j.dr.2018.08.001

Feigenson, L., Dehaene, S., & Spelke, E. (2004). Core systems of number. *Trends in Cognitive Sciences, 8*(7), 307–314. https://doi.org/10.1016/j.tics.2004.05.002

Fernald, A., Marchman, V. A., & Weisleder, A. (2013). SES differences in language processing skill and vocabulary are evident at 18 months. *Developmental Science, 16*(2), 234–248. https://doi.org/10.1111/desc.12019

Furman, M., De Angelis, S., Dominguez Prost, E., & Taylor, I. (2019). Tablets as an educational tool for enhancing preschool science. *International Journal of Early Years Education, 27*(1), 6–19. https://doi.org/10.1080/09669760.2018.1439368

Gilmore, C. K., McCarthy, S. E., & Spelke, E. S. (2010). Non-symbolic arithmetic abilities and mathematics achievement in the first year of formal schooling. *Cognition, 115*(3), 394–406. https://doi.org/10.1016/j.cognition.2010.02.002

Ginsburg, H., & Baroody, A. (2003). *TEMA-3 examiners manual*. Pro-Ed.

Goldin, A. P., Hermida, M. J., Shalom, D. E., Elias Costa, M., Lopez-Rosenfeld, M., Segretin, M. S., Fernandez-Slezak, D., Lipina, S. J., & Sigman, M. (2014). Far transfer to language and math of a short software-based gaming intervention. *Proceedings of the National Academy of Sciences, 111*(17), 6443–6448. https://doi.org/10.1073/pnas.1320217111

Grazzi, M., & Vergara, S. (2014). Internet in Latin America: Who uses it? … and for what? *Economics of Innovation and New Technology, 23*(4), 327–352. https://doi.org/10.1080/10438599.2013.854513

Halberda, J., Mazzocco, M. M. M., & Feigenson, L. (2008). Individual differences in non-verbal number acuity correlate with maths achievement. *Nature, 455*(7213), 665–668. https://doi.org/10.1038/nature07246

Halberda, J., Ly, R., Wilmer, J. B., Naiman, D. Q., & Germine, L. (2012). Number sense across the lifespan as revealed by a massive Internet-based sample. *Proceedings of the National Academy of Sciences, 109*(28), 11116–11120. https://doi.org/10.1073/pnas.1200196109

Halle, T. G., Kurtz-Costes, B., & Mahoney, J. L. (1997). Family influences on school achievement in low-income, African American children. *Journal of Educational Psychology, 89*(3), 527–537. https://doi.org/10.1037/0022-0663.89.3.527

Haßler, B., Major, L., & Hennessy, S. (2016). Tablet use in schools: A critical review of the evidence for learning outcomes: Tablet use in schools: A critical review. *Journal of Computer Assisted Learning, 32*(2), 139–156. https://doi.org/10.1111/jcal.12123

Izard, V., Sann, C., Spelke, E. S., & Streri, A. (2009). Newborn infants perceive abstract numbers. *Proceedings of the National Academy of Sciences, 106*(25), 10382–10385. https://doi.org/10.1073/pnas.0812142106

Jamison, D. T., Searle, B., Galda, K., & Heyneman, S. P. (1981). Improving elementary mathematics education in Nicaragua: An experimental study of the impact of textbooks and radio on achievement. *Journal of Educational Psychology, 73*(4), 556–567. https://doi.org/10.1037/0022-0663.73.4.556

Janus, M., & Duku, E. (2007). The school entry gap: Socioeconomic, family, and health factors associated with children's school readiness to learn. *Early Education & Development, 18*(3), 375–403. https://doi.org/10.1080/10409280701610796a

Jordan, N. C., Huttenlocher, J., & Levine, S. C. (1994). Assessing early arithmetic abilities: Effects of verbal and nonverbal response types on the calculation performance of middle- and low-income children. *Learning and Individual Differences, 6*(4), 413–432. https://doi.org/10.1016/1041-6080(94)90003-5

Langfus, J., Maiche, A., De León, D., Fitipalde, D., Mailhos, Á., & Halberda, J. (2019). Chapter 2—The effects of SES, grade-repeating, and IQ in a game-based approximate math intervention. In D. C. Geary, D. B. Berch, & K. Mann Koepke (Eds.), *Cognitive foundations for improving mathematical learning* (Vol. 5, pp. 37–67). Academic Press. https://doi.org/10.1016/B978-0-12-815952-1.00002-5

Lipina, S. J. (2016). *Pobre cerebro: Lo que la neurociencia nos propone pensar y hacer acerca de los efectos de la pobreza sobre el desarrollo cognitivo y emocional*. Siglo Veintiuno Editores.

Malamud, O., Cueto, S., Cristia, J., & Beuermann, D. W. (2019). Do children benefit from internet access? Experimental evidence from Peru. *Journal of Development of Economics, 138*, 41–56. https://doi.org/10.1016/j.jdeveco.2018.11.005

Meza-Cordero, J. A. (2017). Learn to play and play to learn: Evaluation of the one laptop per child program in Costa Rica: Learn to play and play to learn. *Journal of International Development, 29*(1), 3–31. https://doi.org/10.1002/jid.3267

Naslund-Hadley, E., Parker, S. W., & Hernandez-Agramonte, J. M. (2014). Fostering early math comprehension: Experimental evidence from Paraguay. *Global Education Review, 1*(4).

Noble, K. G., Norman, M. F., & Farah, M. J. (2005). Neurocognitive correlates of socioeconomic status in kindergarten children. *Developmental Science, 8*(1), 74–87. https://doi.org/10.1111/j.1467-7687.2005.00394.x

Odic, D., Libertus, M. E., Feigenson, L., & Halberda, J. (2013). Developmental change in the acuity of approximate number and area representations. *Developmental Psychology, 49*(6), 1103–1112. https://doi.org/10.1037/a0029472

Odic, D., Lisboa, J. V., Eisinger, R., Olivera, M. G., Maiche, A., & Halberda, J. (2016). Approximate number and approximate time discrimination each correlate with school math abilities in young children. *Acta Psychologica, 163*, 17–26. https://doi.org/10.1016/j.actpsy.2015.10.010

OECD. (2019). *PISA 2018 results (volume I): What students know and can do*. OECD. https://doi.org/10.1787/5f07c754-en

Outhwaite, L. A., Gulliford, A., & Pitchford, N. J. (2017). Closing the gap: Efficacy of a tablet intervention to support the development of early mathematical skills in UK primary school children. *Computers & Education, 108*, 43–58. https://doi.org/10.1016/j.compedu.2017.01.011

Outhwaite, L. A., Gulliford, A., & Pitchford, N. J. (2020). Language counts when learning mathematics with interactive apps. *British Journal of Educational Technology, 51*(6), 2326–2339. https://doi.org/10.1111/bjet.12912

Pea, R. D. (1987). Cognitive technologies for mathematics education. In *Cognitive science and mathematics education* (pp. 89–122). Routledge.

Räsänen, P., Salminen, J., Wilson, A. J., Aunio, P., & Dehaene, S. (2009). Computer-assisted intervention for children with low numeracy skills. *Cognitive Development, 24*(4), 450–472. https://doi.org/10.1016/j.cogdev.2009.09.003

Roediger, H. L., & Karpicke, J. D. (2006). Test-enhanced learning: Taking memory tests improves long-term retention. *Psychological Science, 17*(3), 249–255. https://doi.org/10.1111/j.1467-9280.2006.01693.x

Ryan, R. M., Fauth, R. C., & Brooks-Gunn, J. (2006). Childhood poverty: Implications for school readiness and early childhood education. In *Handbook of research on the education of young children* (2nd ed., pp. 323–346). Lawrence Erlbaum Associates Publishers.

Sánchez, J., & Flores, H. (2005). AudioMath: Blind children learning mathematics through audio. *International Journal on Disability and Human Development, 4*(4). https://doi.org/10.1515/IJDHD.2005.4.4.311

Severín, E., & Capota, C. (2011). *One-to-one laptop programs in Latin America and the Caribbean: Panorama and perspectives*. Inter-American Development Bank. https://publications.iadb.org/publications/english/document/One-to-One-Laptop-Programs-in-Latin-America-and-the-Caribbean-Panorama-and-Perspectives.pdf

Shin, N., Sutherland, L. M., Norris, C. A., & Soloway, E. (2012). Effects of game technology on elementary student learning in mathematics: The effects of game technology on student learning. *British Journal of Educational Technology, 43*(4), 540–560. https://doi.org/10.1111/j.1467-8535.2011.01197.x

Starkey, P., Klein, A., & Wakeley, A. (2004). Enhancing young children's mathematical knowledge through a pre-kindergarten mathematics intervention. *Early Childhood Research Quarterly, 19*(1), 99–120. https://doi.org/10.1016/j.ecresq.2004.01.002

Starr, A., Libertus, M. E., & Brannon, E. M. (2013). Number sense in infancy predicts mathematical abilities in childhood. *Proceedings of the National Academy of Sciences, 110*(45), 18116–18120. https://doi.org/10.1073/pnas.1302751110

Sung, Y.-T., Chang, K.-E., & Liu, T.-C. (2016). The effects of integrating mobile devices with teaching and learning on students' learning performance: A meta-analysis and research synthesis. *Computers & Education, 94*, 252–275. https://doi.org/10.1016/j.compedu.2015.11.008

Susperreguy, M. I., Jiménez Lira, C., Xu, C., LeFevre, J.-A., Blanco Vega, H., Benavides Pando, E. V., & Ornelas Contreras, M. (2021). Home learning environments of children in Mexico in relation to socioeconomic status. *Frontiers in Psychology, 12*, 756. https://doi.org/10.3389/fpsyg.2021.626159

Ursache, A., Noble, K. G., & the Pediatric Imaging, Neurocognition and Genetics Study. (2016). Socioeconomic status, white matter, and executive function in children. *Brain and Behavior, 6*(10), e00531. https://doi.org/10.1002/brb3.531

Valle-Lisboa, J., Cabana, Á., Eisinger, R., Mailhos, Á., Luzardo, M., Halberda, J., & Maiche, A. (2016). Cognitive abilities that mediate SES's effect on elementary mathematics learning: The Uruguayan tablet-based intervention. *Prospects, 46*(2), 301–315. https://doi.org/10.1007/s11125-017-9392-y

Yanguas, M. L. (2020). Technology and educational choices: Evidence from a one-laptop-per-child program. *Economics of Education Review, 76*(C). https://ideas.repec.org/a/eee/ecoedu/v76y2020ics0272775719302729.html

Yoshikawa, H., Aber, J. L., & Beardslee, W. R. (2012). The effects of poverty on the mental, emotional, and behavioral health of children and youth: Implications for prevention. *American Psychologist, 67*(4), 272–284. https://doi.org/10.1037/a0028015

Zaldívar-Colado, A., Alvarado-Vázquez, R. I., & Rubio-Patrón, D. E. (2017). Evaluation of using mathematics educational software for the learning of first-year primary school students. *Education Sciences, 7*(4), 79. https://doi.org/10.3390/educsci7040079

Number Processing and Low Arithmetic Achievement in Cuban and Chilean Children: From Neurocognitive Theories to Educational Practice

Nancy Estévez Pérez, Danilka Castro Cañizares, and Miladys Orraca Castillo

Introduction

Adequate mathematical competences are indispensable in professional and social life as the role of technology in contemporary society increases. However, difficulties to represent and compare quantities and perform arithmetic operations are very common. Severe and persistent difficulties in number processing and calculation have been classified as a specific learning disorder for math (or developmental dyscalculia, DD), a neurodevelopmental disorder caused by an impairment in the brain's ability to perceive and process numerical information efficiently and accurately (APA, 2013). Neuroimaging studies focusing on the nature of the causal mechanisms of DD provided evidence for alterations in the structure, function, and connectivity of the brain, affecting the parietal as well as the temporal and prefrontal brain regions (Kucian, 2016). Note, however, that DD children show spared-to-high intellectual capacity and usually cope well with the rest of the academic subjects.

Prevalence of DD ranges between 2.5% and 6.4% in school-age population (Koumoula et al., 2004; Gross-Tsur et al., 2008; Lewis et al., 1994; Ramaa & Gowramma, 2002; Reigosa-Crespo et al., 2012). DD significantly interferes with

N. Estévez Pérez (✉)
Neurodevelopment Department, Cuban Neurosciences Center, Havana, Cuba
e-mail: nancy.estevez@cneuro.edu.cu

D. Castro Cañizares
School of Psychology, Universidad Mayor, Providencia, Santiago, Chile

Center for Advanced Research in Education, Institute of Education, Universidad de Chile, Santiago de Chile, Chile

M. Orraca Castillo
Cuban Society for the Development of the Family, Ministry of Health, Havana, Cuba

M. V. Alves et al. (eds.), *Cognitive Sciences and Education in Non-WEIRD Populations*, https://doi.org/10.1007/978-3-031-06908-6_12

academic achievement and professional and daily life activities involving numerical skills (APA, 2013); affects individuals' social success and access to employment opportunities; and, in general, hinders personal and professional fulfillment. Interestingly, a high co-occurrence of DD has been reported with dyslexia, the specific learning disability for reading (Wilson et al., 2015), and with attention-deficit hyperactivity disorder (Czamara et al., 2013). Additionally, a recent meta-analysis showed that children with math difficulties, including DD, show deficits in processing speed, short-term memory, phonological processing, working memory, attention, visuospatial skills, and executive functions (Peng et al., 2018). Altogether, the state of the art suggests that multiple brain networks and cognitive processes involved in different aspects of information processing contribute to proficiency in math and atypical numerical cognition.

Neural Underpinnings of Mathematical Cognition

Most of the studies addressing the etiology of learning difficulties for math have focused in describing the association between the individual variability in morphometric and functional descriptors of brain regions and specific/basic cognitive/numerical processes (Barnea-Goraly et al., 2005; Rykhlevskaia et al., 2009). Others have focused in identifying differences in morphometric and functional properties in key brain areas between typically developing populations and subjects with low academic achievement in math (LAM), with developmental dyscalculia (DD), and children with genetic syndromes that exhibit LAM/DD as part of their cognitive profile (Arsalidou & Taylor, 2011; Dehaene et al., 2003; Isaacs et al., 2001; Kaufmann et al., 2011; Molko et al., 2004; Ranpura et al., 2013; Rotzer et al., 2008; Rykhlevskaia et al., 2009).

Neuroimaging studies of children with neurodevelopmental disorders and genetic syndromes exhibiting learning disorders are particularly relevant for translational neuroscience. They constitute natural experiments that allow to test theoretical models regarding the genetic foundations, neural underpinnings, and cognitive mechanisms implicated in typical development. Its findings have the potential to provide neurobiological justification for the design of inexpensive behavioral screening tools for the large-scale detection of children at risk of LAM/DD: the behavioral measures that are significantly associated with characteristic anatomic-functional features of the "dyscalculic brain" could be included as risk indicators in universal screening tools.

However, these studies are expensive and technologically challenging or inaccessible for most of our Latin American contexts. Hence, most of the literature comes from developed countries with historically high investments dedicated to basic science and technology. Here, we present two studies conducted in Cuba, aimed at assessing the neural correlates of key cognitive processes that have been reported to covary or predict typical/atypical performance in math, or are significantly impaired in LAM/DD subjects, in order to incorporate them in the assessment and diagnostic protocols used in our country.

A seminal study conducted in Havana city by Reigosa-Crespo et al. (2012) reported that 4.5% of the school-age population (11,652 2nd–9th graders) showed deficits in basic numerical abilities such as the instantaneous and precise estimation of small quantities (or "subitizing"), number counting, and numerical comparison (Kaufman et al., 1949; Mandler & Shebo, 1982; Moyer & Landauer, 1967; Piazza et al., 2002). These deficits affected more boys than girls (2.4:1). The estimated prevalence of DD, defined as a significant deficits in both basic numerical capacities and arithmetic dysfluency (AD), was 3.4%, and the male:female ratio was 4:1. In contrast, the prevalence of AD was almost 3 times as high (9.3%), and no gender differences were found (male:female ratio = 1.07:1). These contrastive findings, suggested that DD, defined as a defective sense of numerosity, could be a distinctive disorder that affects only a portion of the children with AD. These basic numerical processes were also reported as significant predictors of mathematical performance by a longitudinal study conducted in Havana (Reigosa-Crespo et al., 2013). Hence, we conducted the first Cuban study aiming to describe the association in individual variability of morphometric properties of the brain to these measures of basic numerical processing and mathematical attainment.

This study explored the association between the length, depth, volume, and surface area of the horizontal segment of the Intraparietal Sulcus (IPS) and the central sulcus (a region included as a reference for its lack of known association to numerical processing) and the three aforementioned basic numerical abilities: subitizing, counting, and magnitude comparison, in DD children with specific deficits in subitizing. This study was inspired by previous manuscripts by Bruandet et al. (2004) and Molko et al. (2003) in Turner Syndrome (TS), a genetic condition caused by the complete or partial deletion of one of the two X chromosomes in subjects with a female phenotype that includes DD in its neurocognitive profile. Their results showed significant impairments in the subitizing effect (the slope of reaction time per item when exactly estimating the numerosity of small sets of objects: 1 to 4 items), in presence of spared counting and magnitude comparison effects in TS girls presented with DD (Bruandet et al., 2004). Also, they described DD female subjects exhibited significantly lower values of depth and length of the IPS, suggesting these might be morphometric features associated to subitizing deficits in DD (Molko et al., 2003).

Sulcal morphometry was conducted in five Cuban DD children (four boys, replicating the male:female ratio described by the Havana City Survey, Reigosa-Crespo et al., 2012) and controls matched by grade and gender ($N = 8$, 1 girl) (Estévez et al., 2016). Statistically significant nonparametric Spearman rank correlations were confirmed, among all three morphometric descriptors of the horizontal segment of IPS and the subitizing effect in the left hemisphere (subitizing effect X depth: $r = .62$, $p < .05$; subitizing effect X volume: $r = .79$, $p < .01$; and subitizing effect X surface area: $r = .81$, $p < .01$). No significant correlations between the morphometric descriptors and counting or comparison abilities were found. A significant correlation among the depth of the left Central Sulcus and subitizing ($r = .61$, $p < .05$) was also found.

In contrast to Molko et al. (2003), here, the DD children showed values significantly higher in the morphometric descriptors compared to the controls. It is necessary to point out that the sample of the present study is composed mainly of male subjects (only one girl in each group), while the sample in Molko et al. (2003) included only female subjects with Turner Syndrome. The gender characteristics of the sample may influence the results, since genes that regulate brain gyrification during embryonic development are found on the X chromosome. The IPS is one of the deep sulci present in primates, which is visible during 29th week of gestation in humans (Armstrong et al., 1995; Chi et al., 1977). The presence of anomalies in the IPS in TS may be related to damage to the function of X chromosome genes related to the brain gyrification process. It has been described that mutations in the Xq22.3 gene, q23, on the X chromosome, cause patterns of simplified gyrification and cortical disorganization due to atypical neuronal migration (Olson & Walsh, 2002). Additionally, experiments in monkeys indicate that damage to afferent pathways during pregnancy may affect the characteristics of cortical sulci and gyri (Dehay et al., 1996; Rakic, 1988).

On the other hand, according to Molko et al. (2004), there are no clear predictions regarding the direction of the changes to be found in developmental disorders. Previous results in TS have revealed that gray matter density and volume may decrease or increase in different brain regions. This has been also described in other disorders such as autism, fetal alcohol syndrome (Molko et al., 2004), and even in DD (Kaufmann et al., 2011).

In children with dyslexia (DL), better reading skills have been reported to be associated to lower values of cortical surface area. It has been suggested that smaller values of surface area reflect less space between gyri, and this, in turn, suggests the existence of stronger connections within and between cortical areas (Frye et al., 2010) pointing to a reinforcement in the connectivity of the region. The significantly larger values of cortical surface area found in the DD group suggest deficits in the IPS connectivity with the rest of the network involved in numerical processing (in particular, in the process of subitizing). This suggest that DD, similar to DL, may be, in essence, a disconnection syndrome (Rykhlevskaia et al., 2009) and advocates for analyzing the contribution of other brain regions to numerical cognition.

The second exploratory study was conducted in children with Neurofibromatosis 1(NF1), exhibiting specific learning disorders in their cognitive profile (NF1_LD). This study aimed to characterize brain morphometric properties associated to poor academic achievement in NF1 following up on previous research dedicated to determining the prevalence of specific learning disabilities in this genetic disorder in Cuba (Orraca-Castillo et al., 2014). NF1 is an autosomal-dominant disease caused by mutations of the NF1 gene, a tumor-suppressor gene on chromosome 17. In addition to a highly variable clinical profile, learning, cognitive, and neurobehavioral deficits are highly prevalent in this condition (Acosta et al., 2012). As a single-gene disorder, NF1 provides a unique genetic model to identify and analyze the molecular and cellular bases underlying cognitive dysfunction. Thus, this disorder has received the full attention of the scientific community dedicated to the translation of basic science to clinical practice or dissemination to population-based community interventions (Acosta et al., 2012).

Orraca-Castillo et al. (2014) studied 32 children with NF1 (7–14 year olds) using neurocognitive tests dedicated to assess basic capacities which are involved in reading and mathematical achievement. The estimated prevalence of DD (defined as the presence of arithmetic dysfluency and/or basic numeric deficits) was 18.8%, and the male:female ratio was 5:1. However, deficits in core numeric capacities were not found in the sample, suggesting that is not the cause of AD. On the other hand, the prevalence of developmental dyslexia (defined as the presence of deficits in phonological and/or lexical deficits) was almost three times as high (50%), and no gender differences were found (male:female ratio = 1:1). Deficits in lexical and phonological strategies were both underlying the reading disorder. Additionally, efficiencies in lexical/phonological strategies and mental arithmetic were significant predictors of individual performance in reading and math attainment tests.

Ten NF1_LD children diagnosed using the criteria described in Orraca-Castillo et al. (2014) gave consent to participate in a voxel-based morphometry (VBM) study. Two of the children showed only DD (one of them exhibited deficits in basic numerical capacities and the other, AD). Eight of them showed comorbidity between DD and DL (three children showed DD due to deficits in basic numerical capacities and mixed DL, four children showed AD and mixed DL, and one child showed AD and DL due to phonological deficits). They were compared with seven matched controls matched in age and gender.

The NF1_LD sample showed significantly lower cognitive function compared to the control group: significantly lower scores on Raven's CPMT intellectual capacity's test (NF1_LD: $M = 23.7$, $SD = 5.12$; control: $M = 30$, $SD = 2.49$; $p < .01$), the WAIS-III Digit Span (backward) working memory test (NF1_LD: $M = 3.4$, $SD = .96$; control: $M = 5$, $SD = .57$; $p < .01$), and the Peabody vocabulary test (NF1_LD: $M = 8.79$, $SD = 3.08$; control: $M = 11.49$, $SD = 3.21$; $p < .01$). Also, they exhibited significantly lower mathematical attainment in the test used by the Cuban Ministry of Education (MAT, Reigosa-Crespo et al., 2012) (NF1_LD: $M = 2.75$, $SD = 1.83$; control: $M = 8$, $SD = 0$; $p < .01$) and lower basic numerical capacities (exact enumeration test, NF1_LD: $M = 2600.9$, $SD = 792.03$; control: $M = 1940.85$, $SD = 459.6$; $p < .05$; digits comparison test, NF1_LD: $M = 1697.3$, $SD = 751.9$; control: $M = 857.7$, $SD = 283.9$; $p < .01$) and arithmetic fluency (basic calculation test, NF1_LD: $M = 5885.5$, $SD = 2417.02$; control: $M = 2034.1$, $SD = 522.32$; $p < .001$), assessed using the Basic Numeric Battery (Reigosa-Crespo et al., 2012).

High-resolution anatomic images (MP-RAGE) of the whole brain were taken using a 1.5 T Siemens Symphony scanner (Erlangen, Germany) and a single-volume sequential EPI (160 planes of 1 mm thickness in sagittal orientation; 1 × 1 mm^2 resolution, ET/TR = 3.93 ms/3000 ms). Gray matter probability density maps were calculated and plotted in standard space. A VBM analysis was carried out on the gray matter probability density, using version 5.0 of the Statistic Parametric Mapping processing suite. NF1-LD group showed a significant reduction ($p < .01$) of the gray matter of the left temporal lobe in the left hemisphere, at the level of the

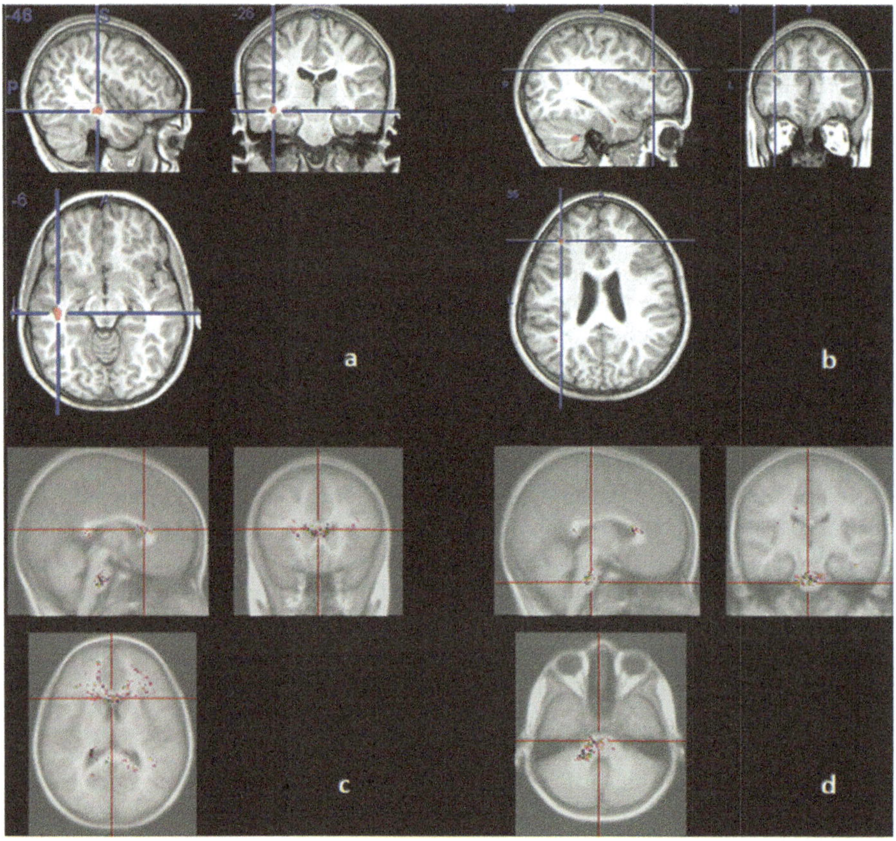

Fig. 1 (**a**, **b**) Difference in gray matter density between NF1-LD children and controls: (**a**) Crosshairs show the superior Temporal Gyrus, left Temporal Lobe. (**b**) Crosshairs show the Superior Frontal Gyrus, Frontal Lobe, and Cerebellum, left Hemisphere. (**c**, **d**) Significant differences in the white matter architecture of the NF1_LD children compared to the control group: (**c**) Crosshairs show the Genus of the Corpus Callosum (rostral section). Also note the Splenium, in the caudal section. (**d**) Crosshairs show the Corticospinal tract and adjacent fibers that connect to the left Cerebellum

superior temporal gyrus (Fig. 1a), in the left frontal lobe, at the level of the superior frontal gyrus, and in the left cerebellum ($p < .01$) (Fig. 1b).

Additionally, 12 diffusion-weighted anatomical images and an additional image ($b = 0$) with the following parameters were taken per participant: 50 slices of 3 mm thickness, $b = 1200$ s/mm^2 for weighted images, spatial resolution of 2×2 mm^2 in plane; ET/RT = 160 ms/7000 ms. The procedure was repeated five times to improve on the signal/noise ratio. Phase and magnitude images of a T2 echo gradient sequence with ET = 7.7 ms and 12.47 msm were also recorded and used to improve the quality of the diffusion-weighted images. Fractional anisotropy maps were calculated and plotted in standard space. Fractional anisotropy (FA) is an indirect measure of white matter integrity, inferred from the direction of water diffusion. High

FA values (anisotropic diffusion) indicate the presence of intact and well-organized fibers. Low FA values indicate loss of nerve fibers or demyelination, and to some extent, suggest a less efficient axonal conduction (Rotzer et al., 2008). A permutation test was carried out (based on a bootstrap procedure) to evaluate the possible differences between FA maps. The results suggest significant differences between the NF1-LD and control groups in the architecture of the white matter of the Corpus Callosum, at the level of rostral section, in the Genus, and in the Splenium (Fig. 1c) and also, in the corticospinal tract ($p > .05$) at the level of the Pons (in an adjacent region through the fibers that connect it with the left Cerebellum) (Fig. 1d).

The results of the VBM coincide with the results of the neurocognitive evaluation of the NF1_LD children. The regions showing significant differences in gray matter density in this sample have been associated to altered cognitive functions: vocabulary (temporal lobe), working memory, and intellectual capacity (frontal and temporal lobes; Rotzer et al., 2008). On the other hand, the white matter findings are in line with brain atypicalities systematically reported in the NF1 syndrome: The Corpus Callosum, the Cerebellum, and the Brain Stem (Corticospinal tract) (Wignall et al., 2010). Again, the results from the second study presented suggest large brain networks, and the corresponding cognitive processing resources contribute to mathematical cognition.

Contribution of Domain-Specific and Domain-General Cognitive Processes to Math Proficiency

Over the last decades, multiple studies have been carried out to identify the cognitive origin of difficulties in mathematics. In this regard, two fundamental hypotheses have been put forward regarding the origin of these difficulties. The first hypothesis proposes that mathematical competence is based on the development of foundational numerical abilities (Butterworth, 2005; Dehaene, 1997) including informal numerical abilities (e.g., "subitizing," enumeration, and comparing sets of objects) and knowledge of number symbols (e.g., to understand the name of numbers and the quantities they represent). This last capacity suggests the existence of an interface responsible for the mapping between the nonsymbolic representation system and the verbal numbering system (Rousselle & Noël, 2007). Consequently, deficits in these domain-specific capacities could be the origin of future difficulties in academic performance in mathematics, and even cause DD (Butterworth, 2005; Castro Cañizares et al., 2012, 2021a; De Smedt & Gilmore, 2011; Inglis et al., 2011; Kolkman et al., 2013; Landerl et al., 2004; Libertus et al., 2011; Rousselle & Noël, 2007). The second hypothesis proposes that there is a close relationship between the development of general-domain processes and mathematical skills, so that deficits in mathematical tasks could be explained by difficulties in nonnumerical processes, such as executive functions, verbal reasoning, intellectual capacity, attention, etc. (Castro et al., 2017; Castro Cañizares et al., 2021b; Fuchs et al., 2010; Geary, 2011;

Geary et al., 2000; Geary et al., 2012; Kaufmann et al., 2013; LeFevre et al., 2013; Swanson, 2011; Szucs et al., 2013). To test these hypotheses, researchers from Chile and Cuba conducted several studies with samples of Chilean school-age children.

Is Low Mathematical Achievement the Result of Domain-Specific Deficits?

Recently, 85 Chilean school-age children (3rd–6th grades) were assessed using numerical nonsymbolic, symbolic, and mapping tasks and classified as children with very low achievement in arithmetic (VLA, $N = 23$), children with low achievement in arithmetic (LA, $N = 31$), and children with typical arithmetic achievement (TA, $N = 31$) (Castro Cañizares et al., 2021a). In this study, intellectual capacity and working memory were controlled for. Children with VLA showed deficits, both in the nonsymbolic processing of numbers and in the mapping skills, required to interphase between numerical symbols and the analog magnitudes they represent, supporting the idea that a damage in nonsymbolic numerical representations underlies their severe difficulties in arithmetic. In contrast, LA children showed deficits in mapping skills only; hence, their low achievement in arithmetic seems to be better explained by deficits in the access to the nonsymbolic representation of quantities, through the symbols they represent (see Fig. 2). These results support the hypothesis of deficits in basic numerical abilities as the origin of difficulties in mathematics and offer new evidence regarding the existence of different cognitive mechanisms that underlie the behavioral profiles of children with different degrees of difficulties in arithmetic.

Do General-Domain Processing Deficits Contribute to Low Mathematical Achievement?

To test the second hypothesis that proposes that there is a close relationship between the general-domain processes and the development of mathematical skills, two studies were carried out, the first aimed at exploring the contribution of the components of working memory (WM) to mathematic attainment and the second to explore the contribution of attentional networks.

In the first of these studies (Castro et al., 2017), the specific contribution of each WM component (phonological loop, visuospatial sketchpad, and central executive) to the explanation of individual variability in efficiency of basic arithmetic was assessed, in different moments of school development. Two hundred eighty-five Chilean school children (1st–6th grades) with and without difficulties in basic arithmetic were evaluated. Hierarchical regression analyzes were performed controlling for basic numerical capacities (symbolic and nonsymbolic numeric comparison).

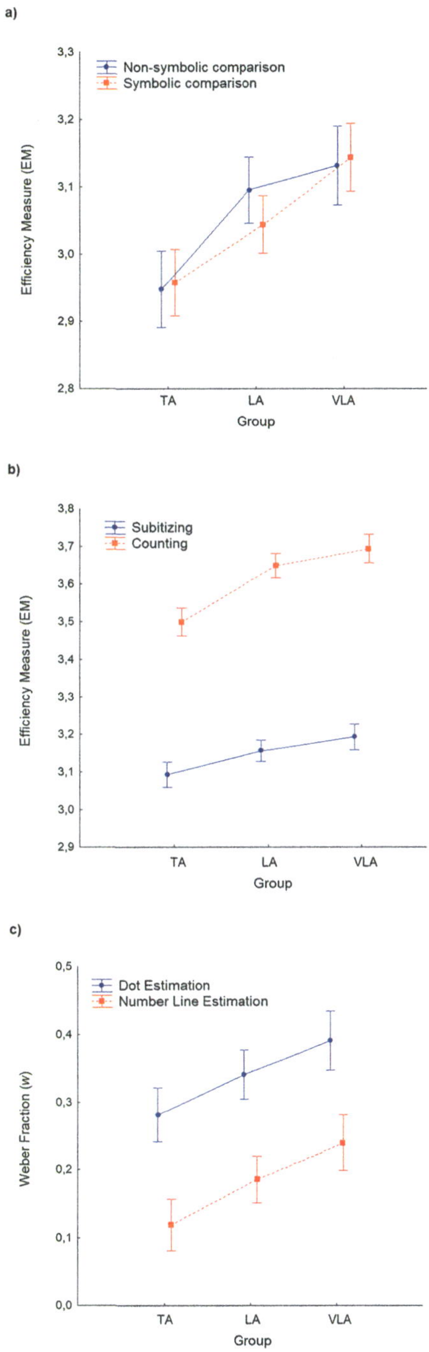

Fig. 2 Group differences in numerical skills tasks. (**a**) Efficiency measures by groups for numerical comparison tasks: nonsymbolic and symbolic. (**b**) Efficiency measures by groups for object counting task for small (subitizing) and large (counting) numerosities. (**c**) Weber fraction by groups for numerical estimation tasks: dot estimation and number line estimation. TA Typical arithmetic achievement, LA Low arithmetic achievement, VLA Very low arithmetic achievement. Efficiency measure is an inverse measure: higher values indicate worse performance. The errors bars represent the *SD*. (Reproduced from Castro Cañizares et al., 2021a)

The results showed that the WM components have a significant contribution to the explanation of the variance in the efficiency in basic arithmetic, and that this varies during school development, but only in the children without difficulties in basic arithmetic. In younger children, a contribution of the visuospatial component ($R^2 = .035$) for 1st and 2nd grades and for 3rd and 4th grades ($R^2 = .032$) was found. In 5th and 6th graders, a contribution of the visuospatial component was also found ($R^2 = .055$), but individual variability in arithmetic performance was also explained by the verbal component ($R^2 = .038$). In line with these results, previous studies have described that the arithmetic performance of children aged 6–8 years is affected by difficulties in the visuospatial and central executive components of WM, when the processing of visuospatial information is required (Berg, 2008; Holmes & Adams, 2006; McKenzie et al., 2003). Around age ten, children usually already perform basic arithmetic tasks through retrieval of numerical facts directly from long-term memory, which involves the phonological loop (Dehaene & Cohen, 1995) and reflects mastery of symbolic-linguistic arithmetic strategies, reliant on verbal codes (Berg, 2008; Dehaene & Cohen, 1995; Houdé, 1997). This could explain the fact that toward the end of elementary school (5th and 6th grades), as a result of the internalization of number symbols and the automation of addition and subtraction operations as numerical facts, the central executive component (which includes the storage carried out by the phonological loop and the simultaneous processing of verbal information) has a greater contribution to the development of arithmetic. In contrast, in children with arithmetic difficulties, no contribution of any of the WM components assessed was found in efficiency in basic arithmetic. Repeated measures' ANOVAs were also performed to explore possible differences between groups in performance on tasks exploring WM. This analysis showed that the performance of the group with arithmetic difficulties was lower than that of their control pairs, particularly in those components that showed a specific contribution to the efficiency in basic arithmetic in the latter group (see Fig. 3a, b). As a whole, the findings of this study suggest that a lower development of WM (mainly visuospatial components) could be causing difficulties in the maintenance of mental images related to the problem space of arithmetic operations and, therefore, could generate difficulties in this domain. In addition, lower WM operability could also interfere with the quality of the coding and retrieval processes of numerical facts, processes closely linked to verbal WM (Dehaene & Cohen, 1995; Fuchs et al., 2006).

In the second study, to test the hypothesis of the influence of different general-domain processes on the origin of difficulties in mathematics, the specific contributions of executive attention and attentional orientation networks to performance in basic arithmetic were explored (Castro Cañizares et al., 2021b). The executive control network (executive attention) is a brain network related to the detection of errors and the ability to resolve conflicts between response alternatives (Botvinick et al., 2001; Dehaene et al., 1994). The attentional orientation network is responsible for attentional shifting, that is, the ability to change the attentional focus between sets, tasks, or strategies. It allows the disconnection of a set of irrelevant tasks and the subsequent selection of a new, more appropriate set (Fan et al., 2002).

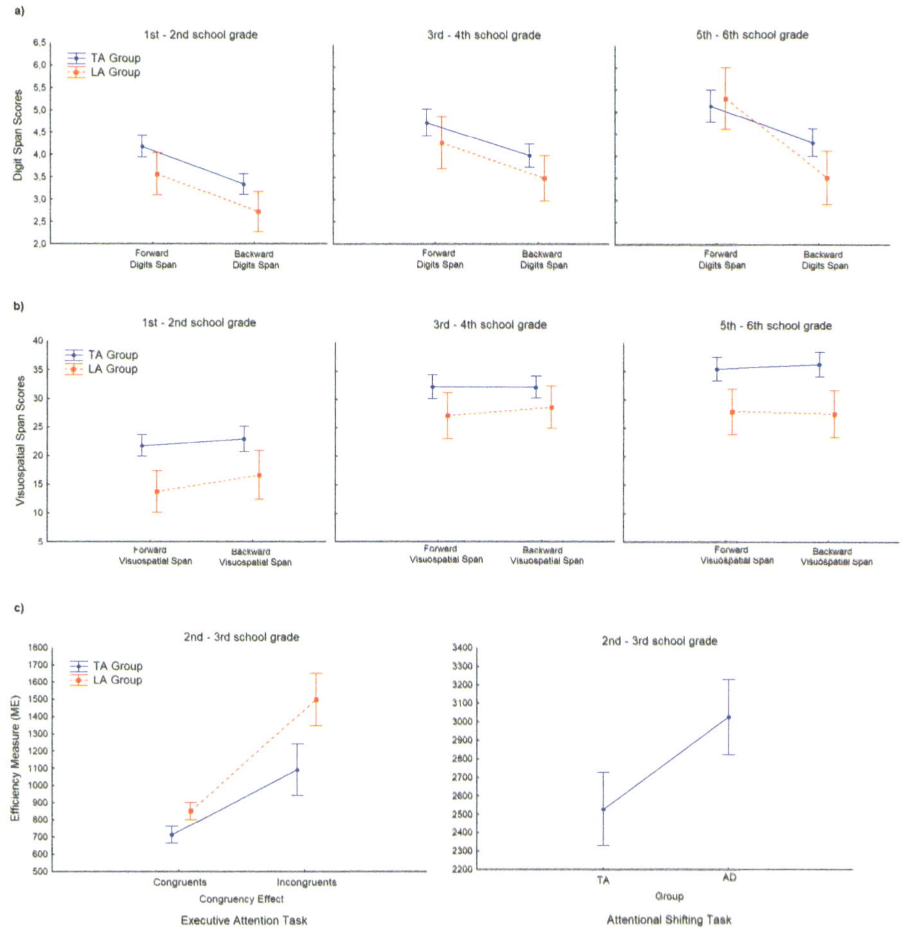

Fig. 3 Group differences in working memory and attentional tasks. (**a**) Scores in verbal working memory tasks by school grade. (**b**) Scores in visuospatial working memory tasks by school grade. (**c**) Efficiency measures by group for attentional tasks. TA Typical arithmetic achievement, LA Low arithmetic achievement. Efficiency measure is an inverse measure: higher values indicate worse performance. The errors bars represent the *SD*. (Reproduced from Castro et al., 2017; Castro Cañizares et al., 2021b)

In this study, 158 Chilean school-age children (2nd–3th grades) with and without difficulties in basic arithmetic were evaluated in tasks that explored the performance of these attentional networks, controlling for the effect of basic numerical abilities. The group of children with difficulties in arithmetic (low arithmetic achievement group, LA) showed a significantly lower performance compared to the control group, both in their efficiency in attentional shifting and in executive attention (see Fig. 3c).

These data support the hypothesis of a contribution of general-domain cognitive processes to arithmetic achievement, and are consistent with previous studies reporting less ability to shift attention (Bull et al., 1999; Bull & Scerif, 2001; van der Sluis et al., 2007) and poor attentional control (LeFevre et al., 2013; LeFevre & Kulak, 1994; Siegler & Shrager, 1984) in children with arithmetic difficulties. The analysis of the specific contribution of attentional shifting and executive attention to the explanation of variance in efficiency arithmetic showed significant correlations with basic arithmetic in both attentional networks (attentional orientation and executive attention). Furthermore, in this group, executive attention showed a significant contribution to the explanation of the variance of efficiency in basic arithmetic ($R^2 = .11, p < .01$). No significant contribution of attentional networks to arithmetic efficiency was found in the control group.

The previous results suggest children who do not have difficulties in basic arithmetic have achieved a correct automation of numerical facts and the manipulation of numerical symbols, which leads to a fluent resolution of arithmetic tasks, with little demand on attentional resources. However, children with low performance in arithmetic have been described as having persistent difficulties in retrieving numerical facts, which affects their arithmetic fluency (Geary, 2010; Geary et al., 2007; Kaufmann et al., 2004; Temple & Sherwood, 2002); therefore, these children could be relying on much more immature strategies for solving arithmetic problems (such as using their fingers to count, counting-all, etc.). These strategies would demand much more attentional control and could explain that in these children the attentional networks are much more involved in arithmetic than in their peers without difficulties. The previous relationship between high levels of executive attention and the development of arithmetic fluency has been previously reported by LeFevre et al. (2013).

The findings of the studies presented above suggest that, although basic numerical skills explain a great variety of difficulties in learning mathematics, the origin of these difficulties does not stem only from a deficit in the development of core numerical skills (nonsymbolic, symbolic, and symbolic-to-nonsymbolic mapping skills) as indicated by some of the current theories regarding the origin of disorders in numerical cognition. Rather different general-domain processes seem to play a part in the origin of difficulties in learning mathematics. Following the results described, we hypothesize that such general-domain cognitive processes, including WM and attentional processes, could support the development of a range of higher level strategies that leverage the computations subserved by the core numerical architecture, and afford the cognitive system efficient shortcuts to deal with numerical information. In contrast, undeveloped basic numerical abilities would impose greater demands to general-domain processing resources and use them to compensate for the inefficient numerical strategies. Alternatively, children with typical development of nonsymbolic representations but atypical development of symbolic and mapping skills and/or atypical development of general-domain cognitive processes could face an inability to take advantage of additional information processing resources and would have to continue to resort to less sophisticated strategies that eventually would no longer match the task demands and lead to difficulties in learning mathematics. In short, our findings suggest that difficulties in mathematics could be associated with a heterogeneous combination of deficits, including both domain-specific and general-domain deficits.

Implications of the Evidence for Educational Practice

It has been rightfully pointed out that no classroom-ready knowledge from neuroscience is ever likely to exist (Howard-Jones, 2010; Thomas et al., 2018). Nevertheless, the relevance of neuroscience for education is accepted by policymakers and teachers worldwide (see Bishop, 2014; Bowers, 2016; Bruer, 1997, 2016; Howard-Jones et al., 2016; Thomas et al., 2018 for alternative views within the academic community).

The results of the Fourth Regional Comparative and Explanatory Study ERCE-2019 (LLECE, 2020) show that the contents with highest presence (in average) in the mathematics curricula in Latin America and the Caribbean are: numbers and operations, magnitudes and measures, and geometry. Figure 4 shows the comparison between the frequency of these contents in the curricula (for third and sixth grades) in Cuba and Chile, with respect to the region. The contents are presented within the frame of the problem-solving approach, which encourages the construction of knowledge in a process that involves analyzing, reflecting, and discovering strategies to solve concrete problems in a creative way. The acquisition of such content, as shown in multiple studies, depends upon both the pedagogical strategies used and the development of cognitive skills involved in learning. It is at this point that neuroscientific knowledge acquires high practical relevance. Teachers aim to use scientific knowledge about the neural foundations of learning to facilitate the children's access to curriculum. Also, it is relevant for policymakers as a foundation

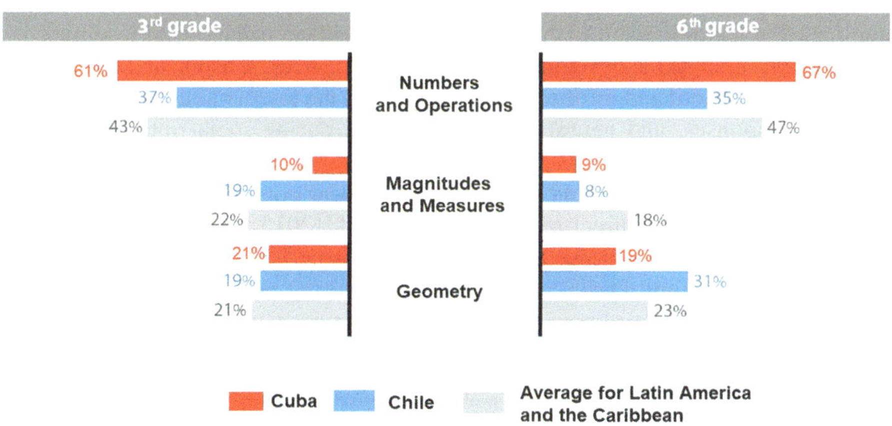

Fig. 4 Average frequency of contents included in the mathematics curricula for third and sixth grades in Cuba and Chile, compared to the average for Latin America and the Caribbean. (Adapted from LLECE, 2020)

for evidence-based decision making regarding curriculum design/selection and the implementation of remediation programs or the introduction of technologies in the schools, at a local or national scale.

Scientific knowledge regarding the neurocognitive architecture supporting math proficiency and the cognitive (and emotional) challenges faced by children at risk of low mathematical achievement could be capitalized to the advantage of educators via two complementary (direct and indirect) routes. The direct route of action is in line with the "Precision Education Initiative" (Hart, 2016). This initiative extends the goals of precision medicine regarding the individualization of treatment and prevention based on genes, environment, and lifestyle; to the classification and treatment of learning disorders, including the design of educational interventions. Concrete actions encompassed by this approach may include: (1) the design of universal screening tools for the early detection of children with special educational needs regarding math education (children at risk of specific learning disorder or highly able and talented children at Math), including basic behavioral indicators validated by neuroimaging studies; (2) the design of highly sensitive and specific diagnostic and cognitive profiling tools for neuropsychologists and school psychologists to assess children at risk of specific learning disorders; (3) the design of personalized stimulation and intervention programs for children with low academic attainment in math; (4) the assessment of efficacy of interventions programs targeting specific neural networks and cognitive components implicated in mathematical achievement.

The indirect route for neuroscience findings to make an impact include: (1) the translation of information regarding the sociodemographic, cultural, and individual predictors of the typical development of numerical cognition and its dissemination to parents, teachers, and policymakers; (2) the inclusion of the corresponding theoretical models and supporting scientific findings in the teachers' curriculum design and in other postgraduate professional learning opportunities available to them; (3) incorporating the knowledge regarding the milestones and cognitive processes implicated in learning math to the design of personalized pedagogical resources by the own teachers; (4) the development of wearable devices, sensors, and data analysis methods to record and inform teachers regarding the levels of engagement, physiological responses, and neural states of students during social interactions and learning in educational settings.

Several initiatives exemplify these alternative and complementary routes. For instance, regarding the direct route, Iuculano et al. (2015) demonstrated that 1:1 cognitive tutoring during 8 weeks remediates poor performance in children with Mathematical Learning Disability (MLD). The tutoring program focused on conceptual aspects of number knowledge, speeded practice on efficient counting strategies, and systematic learning of number families (e.g., all the problems that add to 10, and the corresponding subtraction problems). The intervention induced widespread neuroplasticity changes in brain activity, leading to a normalization of aberrant functional responses in a distributed network including parietal, prefrontal, and ventral temporal–occipital areas supporting successful numerical problem solving. Remarkably, brain activity was correlated with performance gains and machine learning algorithms that successfully discriminated brain activity patterns in children with MLD from those of typically developing peers before the interventions were no longer able to correctly classify them after tutoring. This suggests that behavioral gains were not due to compensatory mechanisms: after performance deficits were successfully remediated, children with MLD stopped showing atypical responses in the brain areas they did before tutoring; and additional and distinct (compensatory) brain areas were not recruited. Additionally, this study provided novel metrics to assess the response to intervention in the neural level.

Regarding the indirect approach, LaPsyDÉ (Laboratoire de Psychologie du Développement et de l'Éducation de l'enfant; CNRS, University Paris Descartes) undertook in 2017 the first participatory science initiative including teachers. Citizen or participatory science, a methodology traditionally used in scientific domains demanding massive data collection, was recently employed to bring the communities of neuroscientists and educators closer together. This approach is characterized by collaboration between researchers and nonresearchers on a project. Levels of participation of nonresearchers in the projects range from data collection to definition of research problems and methodological design and materials, in coconstructed projects (Miller-Rushing et al., 2012).

The project involved some 3600 students in all and 130 teachers running a coconstructed study in their classrooms (Houdé & Borst, 2018). They address the errors children make frequently in class that can be explained by a lack of inhibitory control over erroneous automatisms and deceptive heuristics (Houdé, 2019) and at the same time contrast the validity of neurosciences findings regarding the relevance of cognitive inhibition (a prefrontal cortex function allowing to ignore irrelevant information or automatic responses) to learning. Also, they assess the feasibility of involving teachers in the design of the pedagogical interventions and compare the effect sizes obtained with those of studies carried out under traditional laboratory conditions (Houdé & Borst, 2018). Teachers received professional training on inhibitory control and scientific methodology. They codesigned the study with the researchers, randomly assigned children to experimental and control groups, identified errors to be addressed, and selected and validated interventions proposed by the research team. Also, they collected the data before (pretest) and after the intervention (posttest) and ran the daily training activities in the classroom. The results showed a significant increase on inhibitory control in the experimental group at the end of the pedagogical intervention. The effect size of the intervention was similar to that of interventions run under classical experimental conditions.

This result shows it is possible to translate scientific knowledge into pedagogical resources, use them to improve cognition and academic attainment, test theoretical models without the direct intervention of researchers, and assess the impact of the interventions at the neural level, with the collaboration of neuroscientists. Participatory science is a very promising alternative that can be adapted to any context, including Latin America. It promotes close communication and collaboration between scientists and teachers and relies fundamentally in the teacher's assessment regarding the topics their children find particularly challenging and the errors they frequently commit. The design of interventions and pedagogical resources is also in the hands of teachers. Hence, the relevance of evidence-based teacher's curriculum is apparent.

Currently, we use the basic knowledge gathered in our countries and successful initiatives as the ones previously described to promote an implementation of these two routes through the collaboration of the human and technologic resources of the ministries of education and those at universities and research institutions dedicated to promote neurocognitive development and academic success. In Cuba, a program inspired in the Precision Education approach was already implemented. It has been designed around the organizational architecture of the Cuban Ministry of Education, which includes a network of Diagnosis and Orientation Centers, responsible for coordinating the attention of children with special educational needs in their municipality's schools and offer professional development opportunities for the teachers. A package of tools designed by the Cuban Neurosciences Center allows the detection of children at risk of learning disorders, the cognitive characterization of them, and the design of personalized stimulation and intervention strategies at these centers. Additionally, it includes an e-learning environment for teachers training in educational neuroscience (Reigosa-Crespo, 2019).

General Conclusions

- Studies conducted in Cuba and Chile suggest difficulties in mathematics stem from a heterogeneous combination of domain-specific and general-domain processing deficits.
- Core numerical abilities and executive functions can be used as early predictors of the development of numerical cognition and included in universal screening instruments to detect children at risk of DD.
- Significant translational neuroscience work is required for the evidence to be accessible for teachers and families, incorporated into policy design and decision-making processes and included in teachers' curriculum design and training.

References

Acosta, M. T., Bearden, C. E., Castellanos, X. F., Cutting, L., Elgersma, Y., Gioia, G., et al. (2012). The learning disabilities network (LeaDNet): Using neurofibromatosis type1 (NF1) as a paradigm for translational research. *American Journal of Medical Genetics. Part A, 158A*, 2225–2232. https://doi.org/10.1002/ajmg.a.35535

APA. (2013). *American Psychiatric Association: Diagnostisches und Statistisches Manual Psychischer Störungen – DSM-V*. American Psychiatric Association.

Armstrong, E., Schleicher, A., Omran, H., Curtis, M., & Zilles, K. (1995). The ontogeny of human gyrification Cereb. *Cortex, 5*, 56–63.

Arsalidou, M., & Taylor, M. J. (2011). NeuroImage is 2 + 2 = 4? Meta-analyses of brain areas needed for numbers and calculations. *NeuroImage, 54*, 2382–2393. https://doi.org/10.1016/j.neuroimage.2010.10.009

Barnea-Goraly, N., Menon, V., Eckert, M., Tamm, L., Bammer, R., Karchemskiy, A., et al. (2005). White matter development during childhood and adolescence: A cross-sectional diffusion tensor imaging study. *Cerebral Cortex, 15*, 1848–1854. https://doi.org/10.1093/cercor/bhi062

Berg, D. H. (2008). Working memory and arithmetic calculation in children: The contributory roles of processing speed, short-term memory, and reading. *Journal of Experimental Child Psychology, 99*(4), 288–308. https://doi.org/10.1016/j.jecp.2007.12.002

Bishop, D. V. M. (2014). *What is educational neuroscience?* Available from: https://figshare.com/articles/What_is_educational_neuroscience_/1030405

Botvinick, M. M., Braver, T. S., Barch, D. M., Carter, C. S., & Cohen, J. D. (2001). Conflict monitoring and cognitive control. *Psychological Review, 108*(3), 624–652. https://doi.org/10.1037/0033-295X.108.3.624

Bowers, J. S. (2016). The practical and principled problems with educational neuroscience. *Psychological Review, 123*, 600–612.

Bruandet, M., Molko, N., Cohen, L., & Dehaene, S. (2004). A cognitive characterization of dyscalculia in Turner syndrome. *Neuropsychologia, 42*(3), 288–298. https://doi.org/10.1016/j.neuropsychologia.2003.08.007

Bruer, J. T. (1997). Education and the brain: A bridge too far. *Educational Researcher, 26*, 4–16.

Bruer, J. T. (2016). Where is educational neuroscience? *Educational Neuroscience, 1*, 1–12. https://doi.org/10.1177/2377616115618036

Bull, R., & Scerif, G. (2001). Executive functioning as a predictor of children's mathematics ability: Inhibition, switching and working memory. *Developmental Neuropsychology, 19*(3), 273–293. https://doi.org/10.1207/S15326942DN1903_3

Bull, R., Johnston, R. S., & Roy, J. A. (1999). Exploring the roles of the visual–spatial sketch pad and central executive in children's arithmetical skills: Views from cognition and developmental neuropsychology. *Developmental Neuropsychology, 15*(3), 421–442. https://doi.org/10.1080/87565649909540759

Butterworth, B. (2005). The development of arithmetical abilities. *Journal of Child Psychology and Psychiatry, 46*(1), 3–18. https://doi.org/10.1111/j.1469-7610.2005.00374.x

Castro, D., Amor, V., Gómez, D., & Dartnell, P. (2017). Contribución de los componentes de la memoria de trabajo a la eficiencia en aritmética básica durante la edad escolar. *Psykhe, 26*(2), 1–17. https://doi.org/10.7764/psykhe.26.2.1141

Castro Cañizares, D., Crespo, V. R., & Alemañy, E. G. (2012). Symbolic and non-symbolic number magnitude processing in children with developmental dyscalculia. *The Spanish Journal of Psychology, 15*(3), 952–966. https://doi.org/10.5209/rev_SJOP.2012.v15.n3.39387

Castro Cañizares, D., Dartnell, P., & Pérez, N. E. (2021a). Exploring basic numerical capacities in children with difficulties in simple arithmetical achievement. *Suma Psicológica, 28*(1), 1–9. https://doi.org/10.14349/sumapsi.2021.v28.n1.1

Castro Cañizares, D., Kettlun Poblete, R., & Estevez Perez, N. (2021b). Contribution of attentional networks to basic arithmetic achievement in school-age children. *Psicologia Educativa*. Ahead of print. https://doi.org/10.5093/psed2021a20

Chi, J. G., Dooling, E. C., & Gilles, F. H. (1977). Gyral development of the human brain. *Annals of Neurology, 1*, 86–93.

Czamara, D., et al. (2013). Children with ADHD symptoms have a higher risk for reading, spelling and math difficulties in the GINIplus and LISAplus cohort studies. *PLoS One, 8*(5), e63859.

Dehaene, S. (1997). *The number sense: How the mind creates mathematics*. Oxford University Press.

Dehaene, S., & Cohen, L. (1995). Towards an anatomical and functional model of number processing. *Mathematical Cognition, 1*, 83–120.

Dehaene, S., Posner, M. I., & Tucker, D. M. (1994). Localization of a neural system for error detection and compensation. *Psychological Science, 5*(5), 303–305. https://doi.org/10.1111/j.1467-9280.1994.tb00630.x

Dehaene, S., Piazza, M., Pinel, P., & Cohen, L. (2003). Three parietal circuits for number processing. *Cognitive Neuropsychology, 20*, 487–506. https://doi.org/10.1080/02643290244000239

Dehay, C., Giroud, P., Berland, M., Killackey, H., & Kennedy, H. (1996). Contribution of thalamic input to the specification of cytoarhitectonic cortical fields in the primate: Efects of bilateral enucleation in the fetal monkey on the boundaries, dimensions, and gyrification of striate and extrastriate cortex. *Journal of Comparative Neurology, 367*, 70–89. https://doi.org/10.1002/(SICI)1096-9861(19960325)367:1<70::AIDCNE6>3.0.CO;2-G

De Smedt, B., & Gilmore, C. K. (2011). Defective number module or impaired access? Numerical magnitude processing in first graders with mathematical difficulties. *Journal of Experimental Child Psychology, 108*(2), 278–292. https://doi.org/10.1016/j.jecp.2010.09.003

Estévez Pérez, N., Alemán Gómez, Y., Mosquera Suárez, Recio Morales, B., Rodríguez Cápiro, M., Marine Isangue, R., López Inguanzo, J. M., Valdés Sosa, M. & Reigosa Crespo, V. (2016). Estudio de las propiedades morfométricas del Surco Intraparietal en niños con Discalculia del Desarrollo con déficits específicos en el proceso de subitización. Revista Neuropsicología, Neuropsiquiatría y Neurociencias, Julio-Diciembre 2016, Volumen 16, Número 3, pp. 53–74, ISSN: 0124-1265.

Fan, J., McCandliss, B. D., Sommer, T., Raz, A., & Posner, M. I. (2002). Testing the efficiency and independence of attentional networks. *Journal of Cognitive Neuroscience, 14*(3), 340–347. https://doi.org/10.1162/089892902317361886

Frye, R. E., Liederman, J., Malmberg, B., Mclean, J., Strickland, D., & Beauchamp, M. S. (2010). Surface area accounts for the relation of gray matter volume to reading related skills and history of dyslexia. *Cerebral Cortex, 20*, 2625–2635. https://doi.org/10.1093/cercor/bhq010

Fuchs, L. S., Fuchs, D., Compton, D. L., Powell, S. R., Seethaler, P. M., Capizzi, A. M., et al. (2006). The cognitive correlates of third-grade skill in arithmetic, algorithmic computation, and arithmetic word problems. *Journal of Educational Psychology, 98*, 29–43. https://doi.org/10.1037/0022-0663.98.1.29

Fuchs, L. S., Geary, D. C., Compton, D. L., Fuchs, D., Hamlett, C. L., & Bryant, J. D. (2010). The contributions of numerosity and domain-general abilities to school readiness. *Child Development, 81*, 1520–1533. https://doi.org/10.1111/j.1467-8624.2010.01489.x

Geary, D. C. (2010). Mathematical disabilities: Reflections on cognitive, neuropsychological, and genetic components. *Learning and Individual Differences, 20*(2), 130–133. https://doi.org/10.1016/j.lindif.2009.10.008

Geary, D. C. (2011). Consequences, characteristics, and causes of mathematical learning disabilities and persistent low achievement in mathematics. *Journal of Developmental & Behavioral Pediatrics, 32*, 250–263. https://doi.org/10.1097/DBP.0b013e318209edef

Geary, D. C., Hamson, C. O., & Hoard, M. K. (2000). Numerical and arithmetical cognition: A longitudinal study of process and concept deficits in children with learning disability. *Journal of Experimental Child Psychology, 77*(3), 236–263. https://doi.org/10.1006/jecp.2000.2561

Geary, D. C., Hoard, M. K., Byrd-Craven, J., Nugent, L., & Numtee, C. (2007). Cognitive mechanisms underlying achievement deficits in children with mathematical learning disability. *Child Development, 78*(4), 1343–1359. https://doi.org/10.1111/j.1467-8624.2007.01069.x

Geary, D. C., Hoard, M. K., Nugent, L., & Bailey, D. H. (2012). Mathematical cognition deficits in children with learning disabilities and persistent low achievement: A five-year prospective study. *Journal of Educational Psychology, 104*(1), 206–223. https://doi.org/10.1037/a0025398

Gross-Tsur, V., Manor, O., & Shalev, R. S. (2008). Developmental dyscalculia: Prevalence and demographic features. *Developmental Medicine & Child Neurology, 38*(1), 25–33. https://doi.org/10.1111/j.1469-8749.1996.tb15029.x

Hart, S. A. (2016). Precision education initiative: Moving toward personalized education. *Mind, Brain, and Education, 10*(4), 209–211. https://doi.org/10.1111/mbe.12109

Holmes, J., & Adams, J. W. (2006). Working memory and children's mathematical skills: Implications for mathematical development and mathematics curricula. *Educational Psychology, 26*, 339–366. https://doi.org/10.1080/01443410500341056

Houdé, O. (1997). Numerical development: From the infant to the child. Wynn's (1992) paradigm in 2- and 3-year olds. *Cognitive Development, 12*, 373–391. https://doi.org/10.1016/S0885-2014(97)90009-5

Houdé, O. (2019). *3-system theory of the cognitive brain: A post-Piagetian approach to cognitive development* (p. 134). Routledge. ISBN 9781138069701.

Houdé, O. & Borst, G. (2018). Le Cerveau et les apprentissages. NATHAN FERNAND, Collection les Repères Pédagogiques, pp. 335. ISBN 9782091246406 (2091246409).

Howard-Jones, P. A. (2010). *Introducing neuroeducational research: Neuroscience, education and the brain from contexts to practice*. Routledge.

Howard-Jones, P., Varma, S., Ansari, D., Butterworth, B., De Smedt, B., Goswami, U., et al. (2016). The principles and practices of educational neuroscience: Commentary on Bowers. *Psychological Review, 123*, 620–627.

Inglis, M., Attridge, N., Batchelor, S., & Gilmore, C. (2011). Non-verbal number acuity correlates with symbolic mathematics achievement: But only in children. *Psychonomic Bulletin & Review, 18*(6), 1222–1229. https://doi.org/10.3758/s13423-011-0154-1

Isaacs, E. B., Edmonds, C. J., Lucas, A., & Gadian, D. G. (2001). Calculation difficulties in children of very low birthweight: A neural correlate. *Brain, 124*, 1701–1707. https://doi.org/10.1093/brain/124.9.1701

Iuculano, T., Rosenberg-Lee, M., Richardson, J., Tenison, C., Fuchs, L., Supekar, K., & Menon, V. (2015). Cognitive tutoring induces widespread neuroplasticity and remediates brain function in children with mathematical learning disabilities. *Nature Communications, 6*(1). https://doi.org/10.1038/ncomms9453

Kaufman, E. L., Lord, M. W., Reese, T. W., & Volkmann, J. (1949). The discrimination of visual number. *The American Journal of Psychology, 62*, 498–525.

Kaufmann, L., Lochy, A., Drexler, A., & Semenza, C. (2004). Deficient arithmetic fact retrieval—storage or access problem?. A case study. *Neuropsychologia, 42*(4), 482–496. https://doi.org/10.1016/j.neuropsychologia.2003.09.004

Kaufmann, L., Wood, G., & Rubinsten, O. (2011). Meta-analyses of developmental fMRI studies investigating typical and atypical trajectories of number processing and calculation. *Developmental Neuropsychology, 36*, 763–787. https://doi.org/10.1080/87565641.2010.549884

Kaufmann, L., Mazzocco, M. M., Dowker, A., von Aster, M., Goebel, S. M., Grabner, R. H., Henik, A., Jordan, N. C., Karmiloff-Smith, A. D., Kucian, K., Rubinsten, O., Szucs, D., Shalev, R., & Nuerk, H.-C. (2013). Dyscalculia from a developmental and differential perspective. *Frontiers in Psychology, 4*, 516. https://doi.org/10.3389/fpsyg.2013.00516

Kolkman, M. E., Kroesbergen, E. H., & Leseman, P. P. (2013). Early numerical development and the role of non-symbolic and symbolic skills. *Learning and Instruction, 25*, 95–103. https://doi.org/10.1016/j.learninstruc.2012.12.001

Koumoula, A., Tsironi, V., Stamouli, V., Bardani, I., Siapati, S., Graham, A., et al. (2004). An epidemiological study of number processing and mental calculation in Greek schoolchildren. *Journal of Learning Disabilities, 37*(5), 377–388. https://doi.org/10.1177/00222194040370050201

Kucian, K. (2016). Developmental dyscalculia and the brain. In D. B. Berch et al. (Eds.), *Development of mathematical cognition: Neural substrates and genetic influences* (pp. 165–193). Elsevier.

Laboratorio Latinoamericano de Evaluación de la Calidad de la Educación (LLECE). (2020). Análisis curricular Estudio Regional Comparativo y Explicativo (ERCE-2019). Reporte Técnico. OREALC-UNESCO, Santiago de Chile.

Landerl, K., Bevan, A., & Butterworth, B. (2004). Developmen¬tal dyscalculia and basic numerical capacities: A study of 8-9 years-old students. *Cognition, 93*(2), 99–125. https://doi.org/10.1016/j.cognition.2003.11.004

LeFevre, J. A., & Kulak, A. G. (1994). Individual differences in the obligatory activation of addition facts. *Memory & Cognition, 22*(2), 188–200. https://doi.org/10.3758/bf03208890

LeFevre, J. A., Berrigan, L., Vendetti, C., Kamawar, D., Bisanz, J., Skwarchuk, S. L., & Smith-Chant, B. L. (2013). The role of executive attention in the acquisition of mathematical skills for children in grades 2 through 4. *Journal of Experimental Child Psychology, 114*(2), 243–261. https://doi.org/10.1016/j.jecp.2012.10.005

Lewis, C., Hitch, G. J., & Walker, P. (1994). The prevalence of specific arithmetic difficulties and specific Reading difficulties in 9- to 10-year-old boys and girls. *Journal of Child Psychology and Psychiatry, 35*(2), 283–292. https://doi.org/10.1111/j.1469-7610.1994.tb01162.x

Libertus, M. E., Feigenson, L., & Halberda, J. (2011). Preschool acuity of the approximate number system correlates with school math ability. *Developmental Science, 14*(6), 1292–1300. https://doi.org/10.1111/j.1467-7687.2011.01080.x

Mandler, G., & Shebo, B. J. (1982). Subitizing: An analysis of its component processes. *Journal of Experimental Psychology: General, 111*, 1–21. https://doi.org/10.1037/0096-3445.111.1.1

McKenzie, B., Bull, R., & Gray, C. (2003). The effects of phonological and visual-spatial interference on children's arithmetical performance. *Educational & Child Psychology, 20*(3), 93–108.

Miller-Rushing, A., Primack, R., & Bonney, R. (2012). The history of public participation in ecological research. *Frontiers in Ecology and the Environment, 10*(6), 285–290. https://doi.org/10.1890/110278

Molko, N., Cachia, A., Bruandet, M., Bihan, D. L., Cohen, L., & Dehaene, S. (2003). Functional and structural alterations of the intraparietal sulcus in a developmental dyscalculia of genetic origin. *Neuron, 40*, 847–858. https://doi.org/10.1016/S0896-6273(03)00670-6

Molko, N., Cachia, A., Riviere, D., Mangin, J. F., Bruandet, M., LeBihan, D., & Dehaene, S. (2004). Brain anatomy in Turner syndrome: Evidence for impaired social and spatial-numerical networks. *Cerebral Cortex, 14*, 840–850. https://doi.org/10.1093/cercor/bhh042

Moyer, R., & Landauer, T. (1967). Time required for judgments of numerical inequality. *Nature, 215*, 1519–1520. https://doi.org/10.1038/2151519a0

Olson, E. C., & Walsh, C. A. (2002). Smooth, rough and upside- down neocortical development. *Current Opinion in Genetics & Development, 12*, 320–327. https://doi.org/10.1016/S0959-437X(02)00305-2

Orraca-Castillo, M., Estévez-Pérez, N., & Reigosa-Crespo, V. (2014). Neurocognitive profiles of learning disabled children with neurofibromatosis type 1. *Frontiers in Human Neuroscience, 8*, 386. https://doi.org/10.3389/fnhum.2014.00386

Peng, P., Wang, C., & Namkung, J. (2018). Understanding the cognition related to mathematics difficulties: A meta-analysis on the cognitive deficit profiles and the bottleneck theory. *Review of Educational Research, 88*(3), 434–476.

Piazza, M., Mechelli, A., Butterworth, B., & Price, C. J. (2002). Are subitizing and counting implemented as separate or functionally overlapping processes? *NeuroImage, 15*(2), 435–446. https://doi.org/10.1006/nimg.2001.0980

Rakic, P. (1988). Specification of cerebral cortical areas. *Science, 241*, 170–176. https://doi.org/10.1126/science.3291116

Ramaa, S., & Gowramma, I. P. (2002). A systematic procedure for identifying and classifying children with dyscalculia among primary school children in India. *Dyslexia, 8*(2), 67–85. https://doi.org/10.1002/dys.214

Ranpura, A., Isaacs, E., Edmonds, C., Rogers, M., Lanigan, J., Singhal, A., & Butterworth, B. (2013). Trends in neuroscience and education developmental trajectories of grey and white matter in dyscalculia. *Trends in Neuroscience and Education, 1-9*. https://doi.org/10.1016/j.tine.2013.06.007

Reigosa-Crespo, V. (2019). Beyond the "third method" for the assessment of developmental dyscalculia. Implications for research and practice. In *The international handbook of math learning difficulties: from the lab to the classroom*. Springer International Publishing AG.

Reigosa-Crespo, V., Valdés-Sosa, M., Butterworth, B., Estévez, N., Rodríguez, M., Santos, E., et al. (2012). Basic numerical capacities and prevalence of developmental dyscalculia: The Havana survey. *Developmental Psychology, 48*(1), 123–135. https://doi.org/10.1037/a0025356

Reigosa-Crespo, V., González-Alemañy, E., León, T., Torres, R., Mosquera, R., & Valdés-Sosa, M. (2013). Numerical capacities as domain-specific predictors beyond early mathematics learning: A longitudinal study. *PLoS One, 8*, 1–11. https://doi.org/10.1371/journal.pone.0079711

Rotzer, S., Kucian, K., Martin, E., Aster, M. V., Klaver, P., & Loenneker, T. (2008). Optimized voxel-based morphometry in children with developmental dyscalculia. *NeuroImage, 39*, 417–422. https://doi.org/10.1016/j.neuroimage.2007.08.045

Rousselle, L., & Noël, M. -P. (2007). Basic numerical skills in children with mathematics learning disabilities: A comparison of symbolic vs non-symbolic number magnitude processing. *Cognition, 102*, 361–395. https://doi.org/10.1016/j.cognition.2006.01.005

Rykhlevskaia, E., Uddin, L. Q., Kondos, L., & Menon, V. (2009). Neuroanatomical correlates of developmental dyscalculia: Combined evidence from morphometry and tractography. *Frontiers in Human Neuroscience, 3*, 51. https://doi.org/10.3389/neuro.09.051.2009

Siegler, R. S., & Shrager, J. (1984). Strategy choice in addition and subtraction: How do children know what to do? In C. Sophian (Ed.), *Origins of cognitive skills* (pp. 229–293). Erlbaum.

Swanson, H. L. (2011). Working memory, attention, and mathematical problem solving: A longitudinal study of elementary school children. *Journal of Educational Psychology, 103*(4), 821–837. https://doi.org/10.1037/a0025114

Szucs, D., Devine, A., Soltesz, F., Nobes, A., & Gabriel, F. (2013). Developmental dyscalculia is related to visuo-spatial memory and inhibition impairment. *Cortex, 49*, 2674–2688. https://doi.org/10.1016/j.cortex.2013.06.007

Temple, C. M., & Sherwood, S. (2002). Representation and retrieval of arithmetical facts: Developmental difficulties. *The Quarterly Journal of Experimental Psychology: Section A, 55*(3), 733–752. https://doi.org/10.1080/02724980143000550

Thomas, M. S. C., Ansari, D., & Knowland, V. C. P. (2018). Annual research review: Educational neuroscience: Progress and prospects. *Journal of Child Psychology and Psychiatry.* https://doi.org/10.1111/jcpp.12973

Van der Sluis, S., de Jong, P. F., & van der Leij, A. (2007). Executive functioning in children, and its relations with reasoning, reading, and arithmetic. *Intelligence, 35*(5), 427–449. https://doi.org/10.1016/j.intell.2006.09.001

Wignall, E. L., Griffiths, P. D., Papadakis, N. G., Wilkinson, I. D., Wallis, L. I., Bandmann, O., et al. (2010). Corpus callosum morphology and microstructure assessed using structural MR imaging and diffusion tensor imaging: Initial findings in adults with neurofibromatosis type 1. *American Journal of Neuroradiology, 31*(5), 856–861. https://doi.org/10.3174/ajnr.a2005

Wilson, A. J., Andrewes, S. G., Struthers, H., Rowe, V. M., Bogdanovic, R., & Waldie, K. E. (2015). Dyscalculia and dyslexia in adults: Cognitive bases of comorbidity. *Learning and Individual Differences, 37*, 118–132. https://doi.org/10.1016/j.lindif.2014.11.017

The Perfect Hurricane in Latin America: School Start Time, Chronotype, Sleep, and Academic Performance During Adolescence

Guadalupe Rodríguez Ferrante, Andrea Paula Goldin, and María Juliana Leone

Chronotype, Social Jetlag, and Sleep Duration

Human physiology and behavior fluctuate along the day. As a diurnal species, humans tend to be active during the day and rest at night (Roenneberg et al., 2003). However, sleep–wake cycles show a wide variability among individuals in several characteristics, including the sleep timing (Roenneberg, Kumar, & Merrow, 2007; Wittmann et al., 2006). For example, while some individuals usually go to sleep around 04:00, others spontaneously wake up at the same time. From a biological standpoint, chronotype is the expression of an individual's internal timing in a specific environment or their "temporal phenotype" (Adan et al., 2012; Roenneberg, 2015; Roenneberg, Pilz, et al., 2019). People who go to bed very late at night would be categorized as "owls," as they probably exhibit late chronotypes, and those who wake up very early in the morning as "larks," presenting early chronotypes. Independently of how it is measured, chronotype ranges on a continuum between early and late types, with most people exhibiting intermediate values. Even though chronotype has a genetic basis (Casiraghi et al., 2010; Hirano et al., 2016; Patke et al., 2017), it is modulated by several factors, such as the light-dark cycle (Roenneberg, Kumar, & Merrow, 2007; Stothard et al., 2017; Wright et al., 2013), age (Kuula et al., 2018; Randler et al., 2017; Roenneberg et al., 2004), culture

G. Rodríguez Ferrante (✉) · M. J. Leone
Universidad Torcuato Di Tella, CONICET, Laboratorio de Neurociencia, Buenos Aires, Argentina
e-mail: guadarodriguezf@gmail.com

Universidad Nacional de Quilmes, CONICET, Laboratorio de Cronobiología, Departamento de Ciencia y Tecnología, Buenos Aires, Argentina

A. P. Goldin
Universidad Torcuato Di Tella, CONICET, Laboratorio de Neurociencia, Buenos Aires, Argentina

M. V. Alves et al. (eds.), *Cognitive Sciences and Education in Non-WEIRD Populations*, https://doi.org/10.1007/978-3-031-06908-6_13

(Randler, 2008a; Short et al., 2013), and social cues (e.g., school or work timing) (Goldin et al., 2020; Leone et al., 2020; Skeldon et al., 2017). Thus, chronotype can be considered as a construct that aims to characterize a very complex system (Roenneberg, Pilz, et al., 2019), which comprises biological rhythms (cyclic oscillations) at the behavioral (Mecacci & Zani, 1983; Roenneberg et al., 2015), physiological (Baehr et al., 2000; Duffy et al., 1999), and even molecular levels (Piggins, 2002; Zanello et al., 2000).

Chronotype can be assessed using several methodologies, which measure the timing of specific biological rhythmic processes. The rise in the circulating levels of melatonin (a hormone associated with nighttime and sleep) (Pandi-Perumal et al., 2007) and the minimum body temperature (Refinetti & Menaker, 1992; Waterhouse et al., 2005) are some examples of physiological markers that can be evaluated to estimate chronotype. Other estimates are based on behavioral rhythms. A widely used one is the Midpoint of Sleep in Free days (MSF) (Fig. 1), which is a measure of sleep timing that takes into account both sleep onset and offset. MSF was originally proposed as a chronotype index, based on the idea that, during free days, sleep timing approaches the individuals' endogenous tendencies because they are not influenced by work and/or school schedules as they are during weekdays (Roenneberg et al., 2015; Wittmann et al., 2006). More recently, researchers found that people tend to sleep more on free days than on weekdays, a phenomenon which is even more evident in people with later MSF (Roenneberg, Kuehnle, et al., 2007; Roepke & Duffy, 2010). Consequently, an individual correction for this oversleep on free days was included, constituting an alternative version of MSF, named MSF sleep-corrected (MSFsc), which was proposed as a new chronotype index (Fig. 1) (Roenneberg et al., 2004). Both MSF and MSFsc are local times and can be obtained through the Munich ChronoType Questionnaire (MCTQ), an easy and reliable standardized questionnaire about sleep habits (Roenneberg et al., 2003). MSFsc correlates highly with the timing of physiological rhythms, like melatonin or temperature (Kantermann et al., 2015; Kitamura et al., 2014), and markers of sleep–rest activity rhythms obtained from sleep diaries and actigraphy (Roenneberg et al., 2003, 2015; Ryu et al., 2018; Santisteban et al., 2018).

It is worth noticing that chronotype has also been conceptualized as a psychological trait (Kandeger et al., 2019; Randler, 2008b; Roenneberg, Pilz, et al., 2019). In line with this idea, several questionnaires evaluating individuals' diurnal preferences have been developed, including the Morningness-Eveningness Scale for Children (MESC) (Carskadon et al., 1993; Estevan, 2020) and the Morning-Eveningness Questionnaire (MEQ) (Horne & Östberg, 1976). Although the biological and psychological perspectives of chronotype conceptually differ, the outcomes of questionnaires such as MEQ highly correlate with MSFsc (Goldin et al., 2020; Kitamura et al., 2014; Zavada et al., 2005) as well as with physiological (Baehr et al., 2000; Duffy et al., 1999; Kantermann et al., 2015) and behavioral rhythms (Carrier et al., 1997; Mecacci & Zani, 1983).

Individuals' internal or biological timing is daily aligned (i.e., synchronized) to the external time by different factors, including light exposure. In fact, the external light-dark cycle is the strongest "time giver" or synchronizer of human endogenous clocks (Roenneberg, Kumar, & Merrow, 2007; Stothard et al., 2017;

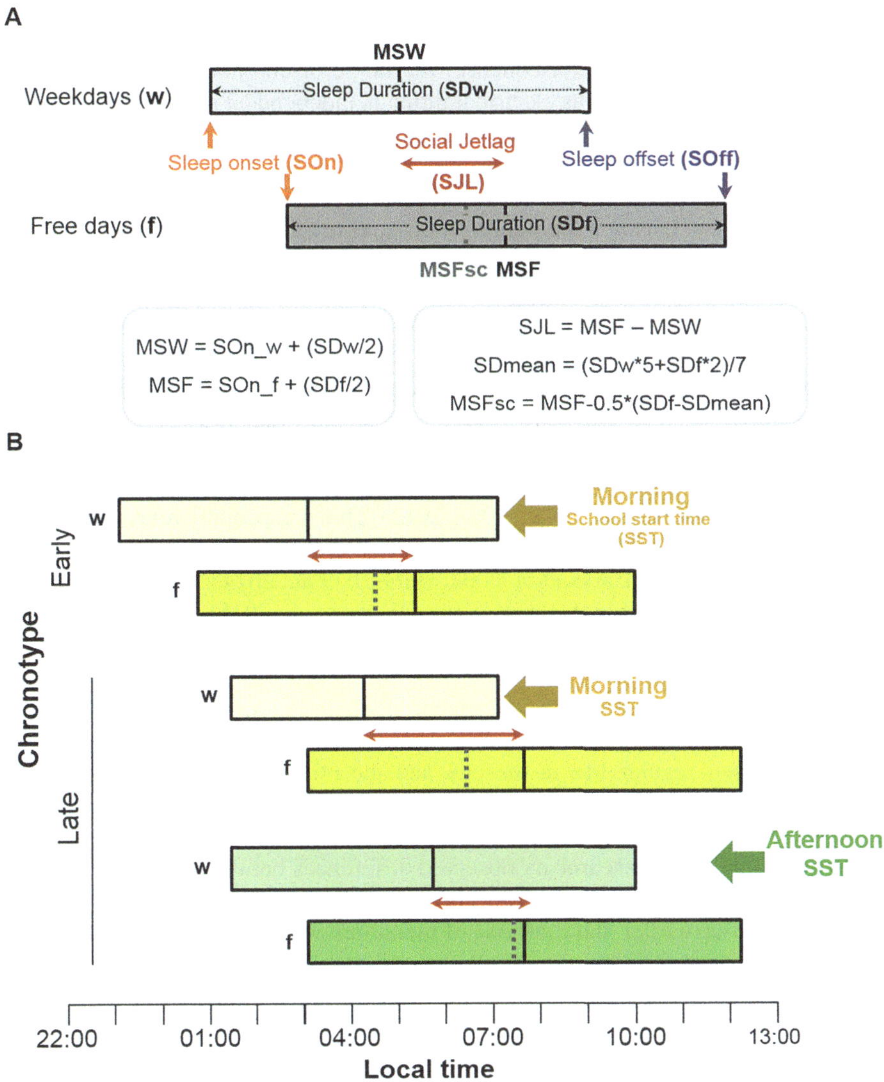

Fig. 1 Chronotype, sleep duration, social jetlag, and school timing. (**a**) Chronotype and sleep-associated variable calculations. The Midpoint of Sleep on Free Days—sleep-corrected—or MSFsc is a chronotype index based on sleep timing. Sleep duration (SD) is the amount of hours of sleep. Social Jetlag (SJL) is the discrepancy on sleep timing between free and weekdays. SOn_w: sleep onset on weekdays; SOn_f: sleep onset on free days; SDw: sleep duration on weekdays; SDf: sleep duration on free days. (**b**) Relations between chronotype sleep duration on weekdays (w, lighter bar), sleep duration on free days (f, darker bar), and SJL. From top to bottom: a student with an early chronotype attending school in the morning, a student with a late chronotype also attending school in the morning, and, finally, a student with a late chronotype attending school in the afternoon

Wright et al., 2013). However, in modern societies, especially since the global access to electric lighting, the strength of sunlight as a synchronizer has weakened (Wright et al., 2013). As a consequence, the range of chronotypes in the human population has widened and, as electric lighting is independent of the actual time of the day, there is a tendency towards later chronotypes (Roenneberg, Pilz, et al., 2019; Wright et al., 2013). Consistently, most people exhibit short sleep duration on weekdays (SDw) (Ford et al., 2015; Keyes et al., 2015; Matricciani et al., 2012) and also a misalignment between their endogenous and social timings, as people tend to go to sleep later but still have to wake up early due to their school or work schedule (Komada et al., 2019; Roenneberg, Pilz, et al., 2019; Wittmann et al., 2006). This misalignment is known as social jetlag (SJL), and it is calculated as the difference between the MSF (midpoint of sleep on free days, where individuals can, theoretically, freely choose their sleep timing because they have not social obligations) and the midpoint of sleep in weekdays (days with social obligations, e.g., school or work) (Fig. 1) (Roenneberg, Pilz, et al., 2019; Wittmann et al., 2006). Both SJL and SDw correlate with chronotype, with late chronotypes associated with higher SJL and shorter SDw (Roenneberg et al., 2003, 2012). This is especially relevant because both high SJL and short SDw have been recurrently associated with physical and mental health problems (Carskadon, 2002; Haraszti et al., 2014; Levandovski et al., 2011; Patel & Hu, 2008; Steptoe et al., 2006; Wong et al., 2015).

Specific examples are useful to fully illustrate the concept of chronotype and its relationship with sleep duration and SJL, together with their modulation by light exposure. First, there is an iconic experiment in which the rest-activity cycle (measured by wrist actigraphy) and melatonin were evaluated on people for 2 weeks: one while living their regular life in the city and the other while camping outdoors (Wright et al., 2013). In the latter, participants were exposed to more intense light during the day (compared with typical indoor lightning) and almost no light during the night. Consistently, researchers observed differences between both weeks, both at the behavioral and physiological levels: during their outdoor days, they went to sleep and woke-up earlier and their rise of melatonin was earlier too (Wright et al., 2013). Other studies also obtained similar results at more naturalistic conditions. A rural population exhibited higher light exposure, earlier chronotypes and lower SJL than an urban comparable population of Brazil (Carvalho et al., 2014). Similar results were also obtained from the comparison of two Toba/Qom communities (with and without access to electric lighting): the community with access to electric lighting showed later sleep timing, shorter sleep duration (de la Iglesia et al., 2015) and a later onset of melatonin secretion (Casiraghi et al., 2020). Independently of the experimental setting, results show that when individuals are exposed to sunlight during the morning and/or reduce their light exposure at night, their chronotype becomes earlier, SJL levels decrease, and sleep duration increases.

The variety of available methods to quantify chronotype can sometimes make difficult the comparison between results in the literature. Altogether, it is clear that chronotype is a complex concept that can be studied from different perspectives, and which is highly modulated by environmental cues, including society imposed schedules.

The Perfect Storm: Adolescents' Chronotype and School Start Time

Chronotype has been repeatedly and consistently reported to change during development in several animal species (Hummer & Lee, 2016). In humans, specifically, chronotype becomes progressively delayed throughout adolescence, reaching a peak of lateness at the end of this period (Randler et al., 2017; Roenneberg et al., 2004). Consistently with this developmental chronotype delay, the sleep onset (both during week days and on free days) gets later during adolescence, but sleep offset (i.e., wake-up time) on weekdays usually remains unmodified because of the very early school start times (Fig. 1). This situation creates a clash between the internal biological timing and the social timing (i.e., school timing) of adolescents, clearly illustrated by M. A. Carskadon as "the perfect storm" (Carskadon, 2011). Importantly, this metaphor refers not only to the misalignment between the students' social and biological clocks, but also to its consequences. A growing body of literature reports that a high proportion of adolescents around the world do not reach the recommended minimum 8 hours of sleep (Hirshkowitz et al., 2015; Shalini et al., 2021) in school days (Carskadon, 2011; Carskadon et al., 1998; Estevan et al., 2020; Goldin et al., 2020). Importantly, adolescents also exhibit considerably high levels of SJL (Goldin et al., 2020; Roenneberg et al., 2012). These consequences are particularly worrying because both SJL and short sleep duration have been associated with several undesirable outcomes, such as depressive feelings (Levandovski et al., 2011; Talbot et al., 2010), obesity (Carskadon, 2002; Parsons et al., 2015; Roenneberg et al., 2012), substance abuse (McKnight-Eily et al., 2011; Nguyen-Louie et al., 2018) as well as lower cognitive and academic performance (Díaz-Morales & Escribano, 2015; Mak et al., 2012; Ng et al., 2009), among others. Altogether, the scenario illustrated here suggests that early school start times leave adolescents in a vulnerable position regarding their functioning, behavior and health.

Although early school start times are challenging for most high school students, those exhibiting later chronotypes are the most affected. Specifically, later chronotypes during high school has been associated with lower SDw, higher SJL and lower academic performance (Arrona-Palacios & Díaz-Morales, 2018; Estevan et al., 2018; Goldin et al., 2020; Rahafar et al., 2017; Scherrer & Preckel, 2021; Tonetti et al., 2015; Zerbini & Merrow, 2017). Regarding academic performance, this effect appears to be generally stronger for math and science-related subjects than for languages, where the effect is usually smaller (Goldin et al., 2020; Preckel et al., 2013) or even absent (Zerbini et al., 2017). In this last study, an association between late chronotypes and lower academic performance was found for math, chemistry, biology and geography but not for Dutch, English, history and physics (Zerbini et al., 2017). One possible explanation of these results is that scientific subjects mainly rely on fluid intelligence while linguistics and humanistic subjects rely more on crystallized intelligence (Goldstein et al., 2007; Hasher et al., 2005). Thus, the differential effect of school subject on chronotype could be explained by the fact that, when evaluated in the morning, fluid but not crystallized intelligence shows a positive association with earlier chronotypes (Goldstein et al., 2007; Hasher et al., 2005). Importantly, the global association between later chronotypes and lower performance during the morning could be explained by the fact that people usually perform better when the activity timing is aligned with their chronotype (i.e., synchrony effect) (May, 1999; May et al., 1993). Consistently, some studies have shown that performance differences between early and late chronotypes disappear when adolescents are evaluated in the afternoon (Arrona-Palacios & Díaz-Morales, 2018; Estevan et al., 2018; Goldin et al., 2020; Itzek-Greulich et al., 2016; van der Vinne et al., 2015). Overall, later chronotypes attending school early in the morning are at cognitive disadvantage, which might lead to negative effects in their future academic and professional life (French et al., 2015; Geiser & Santelices, 2007).

Three main types of interventions have been proposed to mitigate this "perfect storm" and to improve adolescents' sleep health and academic performance. First, educational interventions aimed at promoting sleep hygiene awareness seem to succeed at improving students' knowledge about healthy sleep habits. However, their effectiveness in changing adolescents' sleep habits has not been consistent between reports (Blunden et al., 2012; Cain et al., 2011; John et al., 2016; Kira et al., 2014; Moseley & Gradisar, 2009; Sousa et al., 2013). Second, a few studies suggest that adolescents benefit from interventions that either increase light exposure during the morning (Bonnar et al., 2015; Gradisar et al., 2011)or which decrease it at night, mainly by limiting the use of electronic devices (Perrault et al., 2019; van der Lely et al., 2015). In both types of light-associated interventions, the main result is that the adolescents' chronotype becomes earlier and morning-attending students aligns better with their school timing, thus reducing their SJL. Third, interventions delaying school start times have become relevant in the last two decades. Overall, these interventions show consistent improvements regarding students' sleep duration, SJL and well-being (Boergers et al., 2014; Dunster et al., 2018; Owens et al., 2017; Widome et al., 2020). However, the impact on academic performance is not conclusive: some studies found improvements (Dunster et al., 2018; Kelley et al., 2017), but others did not (Boergers et al., 2014; Owens et al., 2010; Thacher & Onyper, 2016). Surprisingly, most studies assessing the effect of delays in school start time on academic achievement do not report whether the performance gap between early and late chronotypes finally closes, or not. One possible explanation is that the magnitude of the tested delays, which ranges from 15 min to 90 min, might not be always sufficient to observe changes neither in academic achievement nor in the performance gap between chronotypes. Relevantly, some school systems include morning, afternoon and even evening school timings, offering naturalistic setups to assess the impact of very different school start times on adolescents' sleep habits and chronotype. Precisely, this is the case of Latin America.

The Perfect Hurricane: Chronotype and School Start Times in Latin America

When compared with their peers of other continents, many Latin American adolescents show extremely late chronotypes but attend school very early in the morning. Thus, the "storm" becomes a "hurricane." In Argentina, for example, students in their first year of secondary school (12–13 y.o.) present a mean MSFsc of 05:49, while in older students (17–18 y.o.), the mean MSFsc is 06:10 (Goldin et al., 2020). These chronotypes values are extremely late, and even the earliest are more than 1 h 30 min later than the ones reported for adolescents of the same ages at central Europe (Roenneberg et al., 2004) (Fig. 2). However, school start times are similar, or even earlier, in Argentina. In sum, the combination of extremely late chronotypes and very early school start times leads to extraordinarily high levels of SJL (around 3.8 h) and worryingly sleep-deprived students (more than 90% sleep less than 8 hours on weekdays nights) (Goldin et al., 2020). This is especially alarming for older students, who sleep less than 6 h on each school night, on average. These extreme levels of SJL and sleep deprivation in adolescents have not been previously reported neither in North America or in Europe (Carskadon, 2011; Matricciani et al., 2012; Roenneberg et al., 2012) (Fig. 2). It is worth noting that even though studies from some Asian countries reported chronotype and SJL levels similar to their European peers, levels of sleep deprivation in China, Korea and Japan are closer to those presented by Argentinian adolescents (Chung & Cheung, 2008; Han & Chung, 2020; Komada et al., 2019; Steptoe et al., 2006; Yang et al., 2005). Conversely, adolescents from Uruguay exhibited chronotypes, SJL and sleep duration on weekdays similar to their Argentinian peers (Estevan et al., 2020; Goldin et al., 2020). Importantly, during free days, Argentinian and Uruguayan adolescents do not sleep more than Europeans do (Fig. 2), which seems to indicate that the average sleep duration of the students from these countries is indeed considerably shorter. Of course, an alternative possibility is that Argentinian and Uruguayan students compensate their short nocturnal sleep duration on weekdays with more and/or longer diurnal naps. However, napping was insufficient to reach the 8 h of daily sleep on weekdays for Argentinian adolescents (Goldin et al., 2020) and it is difficult to compare this result with other countries because naptimes are rarely reported. Overall, adolescents from Argentina and Uruguay seem to be even more negatively affected by early school start times due to their later chronotypes and, consistently, they show even more unhealthier sleep habits (higher social jetlag and shorter sleep duration) than teenagers from many other countries.

It is not clear yet if such extremely late chronotypes are mostly found in the region of "Río de la Plata" (Argentina and Uruguay) or whether they are widespread throughout Latin America. However, several studies including Brazilian adolescents seem to indicate that extremely late chronotypes are not necessarily the norm in Latin America (Brandalize et al., 2011; Carvalho-Mendes et al., 2020; de Souza et al., 2012). Although most of the studies conducted in Brazil do not inform MSFsc, they do report sleep habits of adolescents and then it is possible to compare these

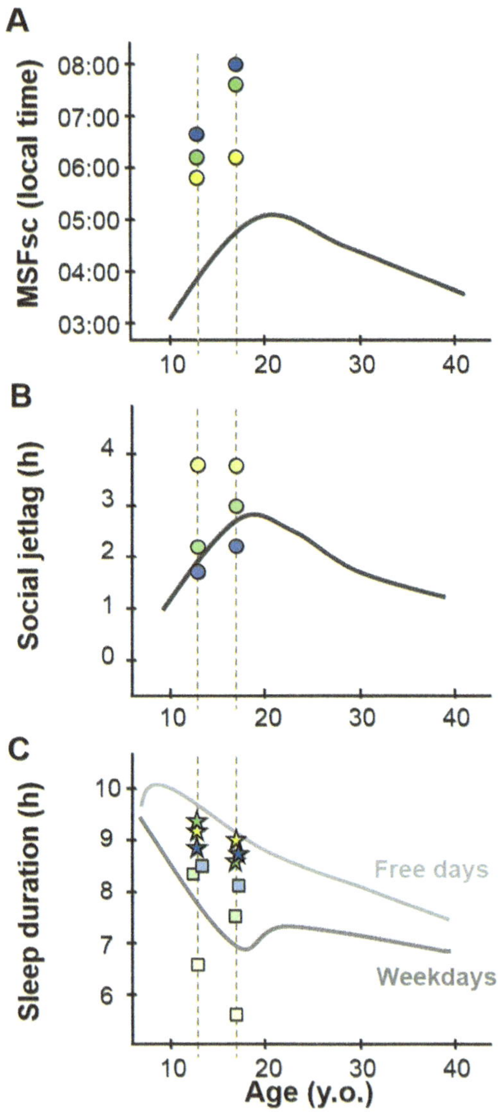

Fig. 2 Comparison between Argentinian and central European adolescents. (**a**) Chronotype, (**b**) social jetlag, and (**c**) sleep duration. Lines represent results obtained in central Europe for chronotype (Modified from Roenneberg et al. (2004), Copyright (2004), with permission from Elsevier), SJL, and sleep duration (Modified from Roenneberg et al. (2012), Copyright (2012), with permission from Elsevier) (from top to bottom). Dots, squares (weekdays), and stars (free days) represent the results obtained for Argentinian students in their first (13–14 y.o.) and last (17–18 y.o.) year of high school (Goldin et al., 2020). Color indicates the school timing. Yellow: morning school timing; green: afternoon school timing; blue: evening school timing

results with those from Argentina and Uruguay. Brazilian morning-attending students present considerably earlier sleep onset (bedtime time) and offset (wake-up time) on free days than their Argentinian and Uruguayan peers. In particular, the mean sleep onset on free days was between 23:30 and 00:30 for Brazilian students, while for Argentinian and Uruguayan students, it was between 02:00 and 02:46 (Brandalize et al., 2011; Carvalho-Mendes et al., 2020; de Souza et al., 2012; Estevan et al., 2020; Goldin et al., 2020). A similar relation is observed in the sleep offset on free days: from 08:30 to 10:00 for Brazilians and from 11:00 to 11:46 for students from the "Río de la Plata." Thus, the mean MSF of Brazilian students results around 2.5 h earlier, and they probably present chronotypes more similar to those of European adolescents (Brandalize et al., 2011; Carvalho-Mendes et al., 2020; de Souza et al., 2012; Roenneberg et al., 2004). Consistently, Brazilian students show lower levels of SJL and longer sleep duration than their Argentinian or Uruguayan peers do.

Interestingly, although adolescents from Mexico (Arrona-Palacios et al., 2015) show sleep patterns similar to their Brazilian peers, little is known about chronotype and sleep habits of adolescents of other Latin American countries. A study comparing adolescents attending German schools around the world showed similar results to the ones described above, including some other Latin American countries such as Guatemala and Costa Rica (Randler, 2008a). Students from those countries exhibit similar or even earlier chronotypes than Brazilian and Mexican students. The authors propose that the difference observed in chronotype could be explained by differences on the climate of these countries (Randler, 2008a). Nonetheless, the available data is not yet sufficient to reach robust conclusions. Results are not consistent between different studies and that is not surprising because chronotype is modulated by a great number of factors (which also include geographical latitude and longitude, and seasonality) (Bekinschtein et al., 2004; Bjorvatn et al., 2021; Leocadio-Miguel et al., 2017; Masal et al., 2015; Roenneberg, Winnebeck, & Klerman, 2019). Thus, it is crucial to seek local knowledge regarding adolescents' chronotype and sleep: if we improve our understanding of Latin American students' sleep habits and which factors negatively affects them, we will be able to think and implement interventions to improve their sleep health. This is important not only for countries where studies are scarce or inexistent, such as Guatemala, Chile, Bolivia, or Peru, but also for countries where some studies already have taken place, such as in Argentina, Uruguay, Brazil, and Mexico.

Consequences of the Perfect Hurricane

"The perfect hurricane" also affects academic performance. Morning-attending students with later chronotypes are more susceptible to have lower academic performance than their peers with earlier chronotypes (Arrona-Palacios & Díaz-Morales, 2018; Estevan et al., 2018; Goldin et al., 2020) (Fig. 3). Unfortunately, the magnitude of the effect cannot be easily compared between countries because chronotype was assessed with different questionnaires. While the study from Argentina (Goldin et al., 2020) obtained chronotypes using the MCTQ (Roenneberg et al., 2003) (a questionnaire based on sleep timing), studies in Uruguay (Estevan et al., 2018) and Mexico (Arrona-Palacios & Díaz-Morales, 2018) used the MEQ (Horne & Östberg, 1976) and MESC (Carskadon et al., 1993), respectively, which are both based on diurnal preferences. Thus, here we can schematically compare the results obtained for Argentina with those of other regions, but we cannot quantitatively compare results of different Latin American countries.

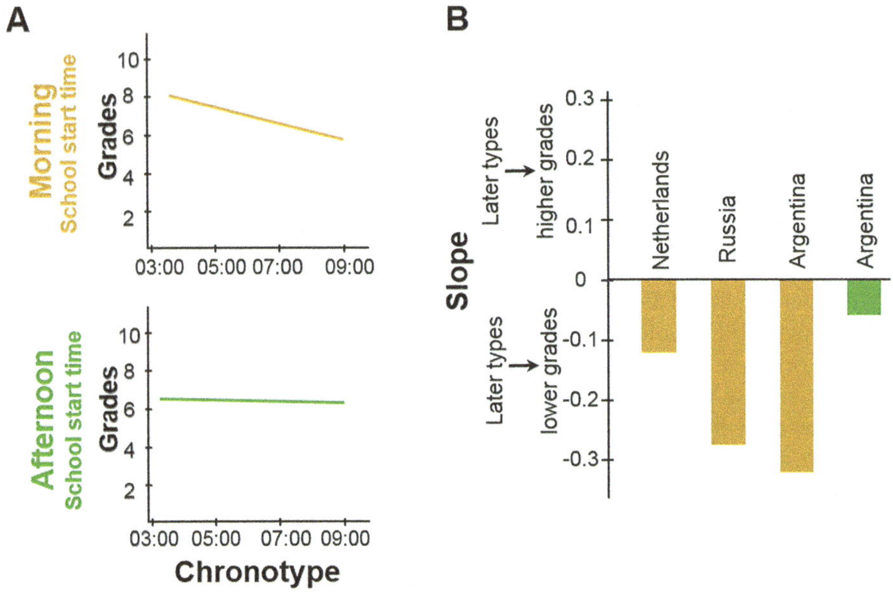

Fig. 3 Chronotype and school timing effects on academic performance in Latin America. (**a**) Correlation between grades and chronotype for morning and afternoon school timings. The effect of chronotype on academic performance depends on school timing, with a negative effect found in the morning school timing (top) and no effect in the afternoon school timing (bottom). (**b**) Comparison of the effect of chronotype on academic performance in different countries. The slopes illustrated here are the beta coefficients associated to MSFsc in linear models with academic performance as dependent variable (Goldin et al., 2020; Kolomeichuk et al., 2016; Zerbini et al., 2017). The models also include several control variables, which vary between studies: 11–16 y.o. ages and only math grades (the Netherlands), 13–14 y.o. students and only math grades (Argentina), and 10–18 y.o. ages considering the average of all school subjects' grades (Russia). Color indicates school timing. Yellow: morning school timing; green: afternoon school timing

Figure 3 shows a particularly relevant comparison including data from three different populations with gradually later mean chronotypes to compare the effect on academic performance. As we previously explained, Argentinian students present extremely late chronotypes (mean MSFsc for 13–14 y.o. students: 05:49; mean MSFsc for 17–18 y.o. students: 06:10) (Goldin et al., 2020), especially when compared with Dutch adolescents (MSFsc mean for 11–16 y.o.: 04:16) (Zerbini et al., 2017), but also with the Russian cohort (MSFsc mean for 10–18 y.o.: 05:04) (Kolomeichuk et al., 2016). Later chronotypes are associated with lower grades on the three studies, but the magnitude of the effect differs among countries: 1 h later, chronotype is associated with a grade reduction of 0.315 (0.338), 0.282 and 0.124 in younger (older) students from Argentina, Russia and the Netherlands, respectively (grades scaled from 1 to 10). Although interesting and illustrative, this magnitude comparison should be cautiously analyzed because the three studies have important differences between them. For example, linear models including grades and chronotype includes different covariables between studies, and, in the Russian study, the grades were self-reported and ranged from 1 to 5 (the coefficient reported above has been corrected to allow the comparison). In addition, the grades included in these three studies correspond to different school subjects: for Argentina and the Netherlands, the values presented here are only for math while for Russia it includes many different school subjects (Goldin et al., 2020; Kolomeichuk et al., 2016; Zerbini et al., 2017). Thus, considering that math is one of the school subjects where the effect of chronotype has been reported to be stronger (Goldin et al., 2020; Preckel et al., 2013; Zerbini et al., 2017), the result reported on the Russian sample might be underestimated. Consequently, more evidence is needed to clarify whether the relation between chronotype and academic performance is stronger for students' populations exhibiting later chronotypes for students attending school in the morning.

Importantly, the effect of different school timings can be assessed in Latin America, including not only different morning school start times. Some Latin American countries have schools with two or three different school timings: morning, approximately from 07:30 to 12:30; afternoon, approximately from 12:30 to 17:30; and only in some exceptional cases, evening, approximately from 17:30 to 22:30. This particular organization of the educational system allows studying whether, and how, school timing affects students' chronotypes, sleep habits, and the relation between chronotype and academic performance.

First, as a social cue, school timing could modulate adolescents' chronotype. If this effect exists, afternoon-attending students would exhibit later chronotypes than morning-attending students. Consistently with this idea, studies from Argentina (Goldin et al., 2020), Uruguay (Estevan et al., 2020) and Mexico (Arrona-Palacios et al., 2015) have shown that adolescents attending school in the afternoon present considerably later chronotypes than their peers attending school in the morning. Interestingly, in the cases of Uruguayan and Argentinian studies, the assignation of students to one of the two (or three) school schedules was random and then it is possible to assume that the differences observed are not due to, for example, previous diurnal preferences (Estevan et al., 2020; Goldin et al., 2020). Additionally, the Argentinian study also includes evening attending-students, who exhibit even later chronotypes than their afternoon-attending peers (Goldin et al., 2020): for example, for older students (17–18 y.o.), the mean MSFsc were 06:10, 07:34, and 08:01 in morning, afternoon, and evening school timings, respectively (Fig. 2). It is important to notice these differences on MSFsc indicate that school schedules (on weekdays) are affecting sleep timing on free days. Thus, school schedules are modulating adolescents' internal timing and not only affecting weekdays sleep timing: even though on free days all adolescents can theoretically freely choose their sleep times, earlier school schedules are associated with earlier chronotypes. As expected because of the late chronotypes and the early morning school timing, the chronotype modulation is only partial: social jetlag levels increases as school timing get earlier, showing that chronotype is not earlier enough to equal sleep timing along the whole week. In contrast, two studies from Brazil reported no differences on students' sleep timing on free days (and, consistently, chronotype) between afternoon and morning school timings (Brandalize et al., 2011; Carvalho-Mendes et al., 2020). Importantly, one of these studies have an additional strength: the same students were evaluated at both morning and afternoon school timings, allowing to compare school timing effects discarding interindividual differences associated with cross-sectional studies (Brandalize et al., 2011). In sum, the lack of chronotype differences between morning and afternoon school timings in Brazilian students might be explained by the fact that these adolescents exhibited earlier chronotypes than their Argentinian and Uruguayan peers. Consistently, morning school timing might be not early enough to change the earlier internal timing of Brazilian students.

Second, both sleep duration and social jetlag differed among school timings. Afternoon-attending students present longer sleep duration on weekdays, without differences in sleep duration on free days. This result was found on different studies, including those Brazilian where chronotype did not differ between school timings (Arrona-Palacios et al., 2015; Brandalize et al., 2011; Carvalho-Mendes et al., 2020; Estevan et al., 2020; Goldin et al., 2020). To exemplify this effect, older Argentinian students sleep 5.95 h, 7.43 h, and 8.1 8 h, on weekdays when attending morning, afternoon, or evening school timing, respectively (Fig. 2). Similar results were observed in SJL levels: afternoon-attending students exhibit lower levels of SJL than their morning-attending peers (Arrona-Palacios et al., 2015; Brandalize et al., 2011; Carvalho-Mendes et al., 2020; Estevan et al., 2020; Goldin et al., 2020). In particular, older Argentinian students exhibited 3.80 h, 2.93 h, or 2.24 h of SJL when attending morning, afternoon, or evening school timings, respectively (Goldin et al., 2020) (Fig. 2). It is worth noticing that SJL levels of evening-attending students are still considerably high, which is not the case for Brazilian students attending school in the afternoon, who present low (Brandalize et al., 2011), and even close to zero (Carvalho-Mendes et al., 2020), SJL levels. The absence of SJL is a particularly rare event in urban modern societies and, especially, in the adolescent population and might be explained by cultural and/or geographical characteristics of these samples. Overall, afternoon and evening school timings seem to benefit students, who present lower levels of SJL and longer sleep duration on weekdays than their morning-attending peers. This is particularly important for extremely late populations such as Argentinian adolescents,

Third, the existence of afternoon and evening school timings is an excellent naturalistic setup to study whether the relation between early chronotypes and better academic performance, is modulated by school timing or not. This association has been repeatedly reported on students attending school in the morning and then it might depend on school timing. If the negative association between chronotype and grades is independent from school timing, earlier chronotypes would perform better in both morning and afternoon (evening) school timings. On the contrary, if the relation is modulated by school timing, and particularly, if the negative association disappears or reverts in the afternoon (evening) school timing, it would indicate the existence of a synchrony effect (Fig. 3). That is, students would perform better when the school schedule (i.e., the learning and/or the evaluation times) is aligned with their internal timing or chronotype. Studies in Mexico and Uruguay showed that the negative association between chronotypes and academic performance (late chronotypes, low grades) disappears in afternoon school timing (Arrona-Palacios &

Díaz-Morales, 2018; Estevan et al., 2018). In the case of the Argentinian study, results are more complex as it considers two students' ages (13–14 and 17–18 y.o.) and three different categories of school subjects (math, language and other subjects) (Goldin et al., 2020). In order to simplify, only results for older students in math and language are considered here. Regarding maths, the negative effect disappeared in both afternoon and evening school timing (Fig. 3). Interestingly, for language the negative association persisted in the afternoon school timing but it even reverted (the slope became positive) in the evening school timing (Goldin et al., 2020). This last result suggests an advantage of later chronotypes over earlier chronotypes when attending school in the evening. It should be note that even though the evening slope was significantly higher than the afternoon slope and positive, it was not statistically different from zero. As explained, this reversion was observed in language but not in math. Globally, early chronotypes appear to be having an advantage over late types, but this advantage is lower or absent in later school timings. Altogether, afternoon and evening school timings seem to be more fair educational schedules, where early chronotypes at least do not show an advantage over later chronotypes.

Conclusion

Adolescents'chronotypes and sleep habits showed significant differences not only between Latin America and geographically and/or culturally distant regions, but also among Latin American countries. These differences might be explained by a combination of factors. For example, geographical longitude and, in particular, the distance from the center of the adopted time zone affect the phase relationship between the social and the solar clock, which in turn could affect individuals' chronotype (Masal et al., 2015; Roenneberg, Winnebeck, & Klerman, 2019). Furthermore, culture is one of the most relevant factors explaining the variability of chronotype between countries: regular activities occur at different timings and might help to differentially synchronize circadian rhythms (Randler et al., 2015; Short et al., 2013). Consistently, in order to further understand the chronotype differences all over the world, it is important to know which is the relevance of sleep health and which are the regular activities or social cues that modulate chronotype on different regions/cultures. For instance, "Río de la Plata" population have dinner later than the United States or even Brazil, which might be associated with the observed delayed chronotype. Interestingly, differences exist between cities of the same country (Carvalho et al., 2014) and might even exist between areas with different socioeconomic status within the same city. Thus, it is extremely important to generate local knowledge to plan and implement public policies designed to improve adolescents' sleep health and performance. The existence of different school timings on the educational systems in Latin America constitutes a useful setup that allows us to ask and find answers to questions that are extremely difficult to address in other regions of the world. Therefore, these type of studies from Latin America are extremely relevant not only for its inhabitants but also for the rest of the globe:

the evaluation of these otherwise ignored scenarios allows not only to improve the field knowledge but also to have a global and complete overview of the effects of chronotype on academic performance. In sum, understanding the relation between chronotype and school timing and its consequences on sleep and academic achievement will allow us to improve adolescents' health and performance, not only in Latin America but also all over the world.

References

Adan, A., Archer, S. N., Hidalgo, M. P., Milia, L. D., Natale, V., & Randler, C. (2012). Circadian typology: A comprehensive review. *Chronobiology International, 29*(9), 1153–1175. https://doi.org/10.3109/07420528.2012.719971

Arrona-Palacios, A., & Díaz-Morales, J. F. (2018). Morningness–eveningness is not associated with academic performance in the afternoon school shift: Preliminary findings. *British Journal of Educational Psychology, 88*(3), 480–498. https://doi.org/10.1111/bjep.12196

Arrona-Palacios, A., García, A., & Valdez, P. (2015). Sleep–wake habits and circadian preference in Mexican secondary school. *Sleep Medicine, 16*(10), 1259–1264. https://doi.org/10.1016/j.sleep.2015.05.026

Baehr, E. K., Revelle, W., & Eastman, C. I. (2000). Individual differences in the phase and amplitude of the human circadian temperature rhythm: With an emphasis on morningness–eveningness. *Journal of Sleep Research, 9*(2), 117–127. https://doi.org/10.1046/j.1365-2869.2000.00196.x

Bekinschtein, T. A., Negro, A., Goldin, A. P., Fernández, M. P., Rosenbaum, S., & Golombek, D. A. (2004). Seasonality in a Mapuche native population. *Biological Rhythm Research, 35*(1–2), 145–152. https://doi.org/10.1080/09291010412331313296

Bjorvatn, B., Saxvig, I. W., Waage, S., & Pallesen, S. (2021). Self-reported seasonality is strongly associated with chronotype and weakly associated with latitude. *Chronobiology International, 38*(2), 278–285. https://doi.org/10.1080/07420528.2020.1844725

Blunden, S. L., Chapman, J., & Rigney, G. A. (2012). Are sleep education programs successful? The case for improved and consistent research efforts. *Sleep Medicine Reviews, 16*(4), 355–370. https://doi.org/10.1016/j.smrv.2011.08.002

Boergers, J., Gable, C. J., & Owens, J. A. (2014). Later school start time is associated with improved sleep and daytime functioning in adolescents. *Journal of Developmental & Behavioral Pediatrics, 35*(1), 11–17. https://doi.org/10.1097/DBP.0000000000000018

Bonnar, D., Gradisar, M., Moseley, L., Coughlin, A.-M., Cain, N., & Short, M. A. (2015). Evaluation of novel school-based interventions for adolescent sleep problems: Does parental involvement and bright light improve outcomes? *Sleep Health, 1*(1), 66–74. https://doi.org/10.1016/j.sleh.2014.11.002

Brandalize, M., Pereira, É. F., Leite, N., Filho, G. L., & Louzada, F. M. (2011). Effect of morning school schedule on sleep and anthropometric variables in adolescents: A follow-up study. *Chronobiology International, 28*(9), 779–785. https://doi.org/10.3109/07420528.2011.603452

Cain, N., Gradisar, M., & Moseley, L. (2011). A motivational school-based intervention for adolescent sleep problems. *Sleep Medicine, 12*(3), 246–251. https://doi.org/10.1016/j.sleep.2010.06.008

Carrier, J., Monk, T. H., Buysse, D. J., & Kupfer, D. J. (1997). Sleep and morningness-eveningness in the 'middle' years of life (20–59y). *Journal of Sleep Research, 6*(4), 230–237. https://doi.org/10.1111/j.1365-2869.1997.00230.x

Carskadon, M. A. (2002). *Adolescent sleep patterns: Biological, social, and psychological influences*. Cambridge University Press.

Carskadon, M. A. (2011). Sleep in adolescents: The perfect storm. *Pediatric Clinics of North America, 58*(3), 637–647. https://doi.org/10.1016/j.pcl.2011.03.003

Carskadon, M. A., Vieira, C., & Acebo, C. (1993). Association between puberty and delayed phase preference. *Sleep, 16*(3), 258–262. https://doi.org/10.1093/sleep/16.3.258
Carskadon, M. A., Wolfson, A. R., Acebo, C., Tzischinsky, O., & Seifer, R. (1998). Adolescent sleep patterns, circadian timing, and sleepiness at a transition to early school days. *Sleep, 21*(8), 871–881. https://doi.org/10.1093/sleep/21.8.871
Carvalho, F. G., Hidalgo, M. P., & Levandovski, R. (2014). Differences in circadian patterns between rural and urban populations: An epidemiological study in countryside. *Chronobiology International, 31*(3), 442–449. https://doi.org/10.3109/07420528.2013.846350
Carvalho-Mendes, R. P., Dunster, G. P., de la Iglesia, H. O., & Menna-Barreto, L. (2020). Afternoon school start times are associated with a Lack of both social jetlag and sleep deprivation in adolescents. *Journal of Biological Rhythms, 35*(4), 377–390. https://doi.org/10.1177/0748730420927603
Casiraghi, L. P., Martino, D., Marengo, E., Igoa, A., Ais, E., Strejilevich, S., André, D., & Golombek, S. (2010). Human period-3 gene involvement in diurnal preference among Argentinean bipolar disorders patients. *Sleep Science, 3*(1), 22–26.
Casiraghi, L. P., Plano, S. A., Fernández-Duque, E., Valeggia, C., Golombek, D. A., & de la Iglesia, H. O. (2020). Access to electric light is associated with delays of the dim-light melatonin onset in a traditionally hunter-gatherer Toba/Qom community. *Journal of Pineal Research, 69*(4), e12689. https://doi.org/10.1111/jpi.12689
Chung, K.-F., & Cheung, M.-M. (2008). Sleep-wake patterns and sleep disturbance among Hong Kong Chinese adolescents. *Sleep, 31*(2), 185–194. https://doi.org/10.1093/sleep/31.2.185
de la Iglesia, H. O., Fernández-Duque, E., Golombek, D. A., Lanza, N., Duffy, J. F., Czeisler, C. A., & Valeggia, C. R. (2015). Access to electric light is associated with shorter sleep duration in a traditionally hunter-gatherer community. *Journal of Biological Rhythms, 30*(4), 342–350. https://doi.org/10.1177/0748730415590702
de Souza, J. C., de Sousa, I. C., Maia, A. P. L., & de Azevedo, C. V. M. (2012). Sleep patterns of teachers and adolescents who attend school in the morning. *Biological Rhythm Research, 43*(1), 65–72. https://doi.org/10.1080/09291016.2011.638156
Díaz-Morales, J. F., & Escribano, C. (2015). Social jetlag, academic achievement and cognitive performance: Understanding gender/sex differences. *Chronobiology International, 32*(6), 822–831. https://doi.org/10.3109/07420528.2015.1041599
Duffy, J. F., Dijk, D. J., Hall, E. F., & Czeisler, C. A. (1999). Relationship of endogenous circadian melatonin and temperature rhythms to self-reported preference for morning or evening activity in young and older people. *Journal of Investigative Medicine, 47*(3), 141–150.
Dunster, G. P., de la Iglesia, L., Ben-Hamo, M., Nave, C., Fleischer, J. G., Panda, S., & de la Iglesia, H. O. (2018). Sleepmore in Seattle: Later school start times are associated with more sleep and better performance in high school students. *Science Advances, 4*(12), eaau6200. https://doi.org/10.1126/sciadv.aau6200
Estevan, I. (2020). Psychometric properties of the Morningness/Eveningness scale for children among Uruguayan adolescents: The role of school start times. *Biological Rhythm Research, 0*(0), 1–11. https://doi.org/10.1080/09291016.2020.1846284
Estevan, I., Silva, A., & Tassino, B. (2018). School start times matter, eveningness does not. *Chronobiology International, 35*(12), 1753–1757. https://doi.org/10.1080/07420528.2018.1504785
Estevan, I., Silva, A., Vetter, C., & Tassino, B. (2020). Short sleep duration and extremely delayed chronotypes in Uruguayan youth: The role of school start times and social constraints. *Journal of Biological Rhythms, 35*(4), 391–404. https://doi.org/10.1177/0748730420927601
Ford, E. S., Cunningham, T. J., & Croft, J. B. (2015). Trends in self-reported sleep duration among US adults from 1985 to 2012. *Sleep, 38*(5), 829–832. https://doi.org/10.5665/sleep.4684
French, M. T., Homer, J. F., Popovici, I., & Robins, P. K. (2015). What you do in high school matters: High school GPA, educational attainment, and labor market earnings as a young adult. *Eastern Economic Journal, 41*(3), 370–386. https://doi.org/10.1057/eej.2014.22
Geiser, S., & Santelices, M. V. (2007). Validity of high-school grades in predicting student success beyond the freshman year: High-school record vs. standardized tests as indicators of four-year

college outcomes. Research & Occasional Paper Series: CSHE.6.07. In *Center for studies in higher education*. Center for Studies in Higher Education. https://eric.ed.gov/?id=ED502858

Goldin, A. P., Sigman, M., Braier, G., Golombek, D. A., & Leone, M. J. (2020). Interplay of chronotype and school timing predicts school performance. *Nature Human Behaviour*, 1–10. https://doi.org/10.1038/s41562-020-0820-2

Goldstein, D., Hahn, C. S., Hasher, L., Wiprzycka, U. J., & Zelazo, P. D. (2007). Time of day, intellectual performance, and behavioral problems in Morning versus Evening type adolescents: Is there a synchrony effect? *Personality and Individual Differences, 42*(3), 431–440. https://doi.org/10.1016/j.paid.2006.07.008

Gradisar, M., Dohnt, H., Gardner, G., Paine, S., Starkey, K., Menne, A., Slater, A., Wright, H., Hudson, J. L., Weaver, E., & Trenowden, S. (2011). A randomized controlled trial of cognitive-behavior therapy plus bright light therapy for adolescent delayed sleep phase disorder. *Sleep, 34*(12), 1671–1680. https://doi.org/10.5665/sleep.1432

Han, C. H., & Chung, J. (2020). Late Chronotype is associated with adolescent asthma: Assessment using the Korean-version MCTQ. *International Journal of Environmental Research and Public Health, 17*(9), 3000. https://doi.org/10.3390/ijerph17093000

Haraszti, R. Á., Ella, K., Gyöngyösi, N., Roenneberg, T., & Káldi, K. (2014). Social jetlag negatively correlates with academic performance in undergraduates. *Chronobiology International, 31*(5), 603–612. https://doi.org/10.3109/07420528.2013.879164

Hasher, L., Goldstein, D., & May, C. P. (2005). It's about time: Circadian rhythms, memory, and aging. In *Human learning and memory: Advances in theory and application: The 4th Tsukuba International Conference on Memory* (pp. 199–217). Lawrence Erlbaum Associates Publishers.

Hirano, A., Shi, G., Jones, C. R., Lipzen, A., Pennacchio, L. A., Xu, Y., Hallows, W. C., McMahon, T., Yamazaki, M., Ptáček, L. J., & Fu, Y.-H. (2016). A Cryptochrome 2 mutation yields advanced sleep phase in humans. *eLife, 5*, e16695. https://doi.org/10.7554/eLife.16695

Hirshkowitz, M., Whiton, K., Albert, S. M., Alessi, C., Bruni, O., DonCarlos, L., Hazen, N., Herman, J., Adams Hillard, P. J., Katz, E. S., Kheirandish-Gozal, L., Neubauer, D. N., O'Donnell, A. E., Ohayon, M., Peever, J., Rawding, R., Sachdeva, R. C., Setters, B., Vitiello, M. V., & Ware, J. C. (2015). National Sleep Foundation's updated sleep duration recommendations: Final report. *Sleep Health, 1*(4), 233–243. https://doi.org/10.1016/j.sleh.2015.10.004

Horne, J. A., & Östberg, O. (1976). A self-assessment questionnaire to determine morningness-eveningness in human circadian rhythms. *International Journal of Chronobiology, 4*, 97–110.

Hummer, D. L., & Lee, T. M. (2016). Daily timing of the adolescent sleep phase: Insights from a cross-species comparison. *Neuroscience & Biobehavioral Reviews, 70*, 171–181. https://doi.org/10.1016/j.neubiorev.2016.07.023

Itzek-Greulich, H., Randler, C., & Vollmer, C. (2016). The interaction of chronotype and time of day in a science course: Adolescent evening types learn more and are more motivated in the afternoon. *Learning and Individual Differences, 51*, 189–198. https://doi.org/10.1016/j.lindif.2016.09.013

John, B., Bellipady, S. S., & Bhat, S. U. (2016). Sleep promotion program for improving sleep behaviors in adolescents: A randomized controlled pilot study. *Scientifica, 2016*, e8013431. https://doi.org/10.1155/2016/8013431

Kandeger, A., Selvi, Y., & Tanyer, D. K. (2019). The effects of individual circadian rhythm differences on insomnia, impulsivity, and food addiction. *Eating and Weight Disorders – Studies on Anorexia, Bulimia and Obesity, 24*(1), 47–55. https://doi.org/10.1007/s40519-018-0518-x

Kantermann, T., Sung, H., & Burgess, H. J. (2015). Comparing the morningness-eveningness questionnaire and Munich ChronoType Questionnaire to the dim light melatonin onset. *Journal of Biological Rhythms, 30*(5), 449–453. https://doi.org/10.1177/0748730415597520

Kelley, P., Lockley, S. W., Kelley, J., & Evans, M. D. R. (2017). Is 8:30 a.m. still too early to start school? A 10:00 a.m. school start time improves health and performance of students aged 13–16. *Frontiers in Human Neuroscience, 11*. https://doi.org/10.3389/fnhum.2017.00588

Keyes, K. M., Maslowsky, J., Hamilton, A., & Schulenberg, J. (2015). The great sleep recession: Changes in sleep duration among US adolescents, 1991–2012. *Pediatrics, 135*(3), 460–468. https://doi.org/10.1542/peds.2014-2707

Kira, G., Maddison, R., Hull, M., Blunden, S., & Olds, T. (2014). Sleep education improves the sleep duration of adolescents: A randomized controlled pilot study. *Journal of Clinical Sleep Medicine, 10*(07), 787–792. https://doi.org/10.5664/jcsm.3874

Kitamura, S., Hida, A., Aritake, S., Higuchi, S., Enomoto, M., Kato, M., Vetter, C., Roenneberg, T., & Mishima, K. (2014). Validity of the Japanese version of the Munich ChronoType Questionnaire. *Chronobiology International, 31*(7), 845–850. https://doi.org/10.3109/07420528.2014.914035

Kolomeichuk, S. N., Randler, C., Shabalina, I., Fradkova, L., & Borisenkov, M. (2016). The influence of chronotype on the academic achievement of children and adolescents – Evidence from Russian Karelia. *Biological Rhythm Research, 47*(6), 873–883. https://doi.org/10.1080/09291016.2016.1207352

Komada, Y., Okajima, I., Kitamura, S., & Inoue, Y. (2019). A survey on social jetlag in Japan: A nationwide, cross-sectional internet survey. *Sleep and Biological Rhythms, 17*(4), 417–422. https://doi.org/10.1007/s41105-019-00229-w

Kuula, L., Pesonen, A.-K., Merikanto, I., Gradisar, M., Lahti, J., Heinonen, K., Kajantie, E., & Räikkönen, K. (2018). Development of late circadian preference: Sleep timing from childhood to late adolescence. *The Journal of Pediatrics, 194*, 182–189.e1. https://doi.org/10.1016/j.jpeds.2017.10.068

Leocadio-Miguel, M. A., Louzada, F. M., Duarte, L. L., Areas, R. P., Alam, M., Freire, M. V., Fontenele-Araujo, J., Menna-Barreto, L., & Pedrazzoli, M. (2017). Latitudinal cline of chronotype. *Scientific Reports, 7*(1), 5437. https://doi.org/10.1038/s41598-017-05797-w

Leone, M. J., Sigman, M., & Golombek, D. (2020). *Effects of social isolation on human sleep and Chronotype during the COVID-19 pandemic* (SSRN scholarly paper ID 3624469). Social Science Research Network. https://doi.org/10.2139/ssrn.3624469.

Levandovski, R., Dantas, G., Fernandes, L. C., Caumo, W., Torres, I., Roenneberg, T., Hidalgo, M. P. L., & Allebrandt, K. V. (2011). Depression scores associate with chronotype and social jetlag in a rural population. *Chronobiology International, 28*(9), 771–778. https://doi.org/10.3109/07420528.2011.602445

Mak, K.-K., Lee, S.-L., Ho, S.-Y., Lo, W.-S., & Lam, T.-H. (2012). Sleep and academic performance in Hong Kong adolescents. *Journal of School Health, 82*(11), 522–527. https://doi.org/10.1111/j.1746-1561.2012.00732.x

Masal, E., Randler, C., Beşoluk, Ş., Önder, İ., Horzum, M. B., & Vollmer, C. (2015). Effects of longitude, latitude and social factors on chronotype in Turkish students. *Personality and Individual Differences, 86*, 73–81. https://doi.org/10.1016/j.paid.2015.05.019

Matricciani, L., Olds, T., & Petkov, J. (2012). In search of lost sleep: Secular trends in the sleep time of school-aged children and adolescents. *Sleep Medicine Reviews, 16*(3), 203–211. https://doi.org/10.1016/j.smrv.2011.03.005

May, C. P. (1999). Synchrony effects in cognition: The costs and a benefit. *Psychonomic Bulletin & Review, 6*(1), 142–147. https://doi.org/10.3758/BF03210822

May, C. P., Hasher, L., & Stoltzfus, E. R. (1993). Optimal time of day and the magnitude of age differences in memory. *Psychological Science, 4*(5), 326–330. https://doi.org/10.1111/j.1467-9280.1993.tb00573.x

McKnight-Eily, L. R., Eaton, D. K., Lowry, R., Croft, J. B., Presley-Cantrell, L., & Perry, G. S. (2011). Relationships between hours of sleep and health-risk behaviors in US adolescent students. *Preventive Medicine, 53*(4–5), 271–273. https://doi.org/10.1016/j.ypmed.2011.06.020

Mecacci, L., & Zani, A. (1983). Morningness-eveningness preferences and sleep-waking diary data of morning and evening types in student and worker samples. *Ergonomics, 26*(12), 1147–1153. https://doi.org/10.1080/00140138308963450

Moseley, L., & Gradisar, M. (2009). Evaluation of a school-based intervention for adolescent sleep problems. *Sleep, 32*(3), 334–341. https://doi.org/10.1093/sleep/32.3.334

Ng, E. P., Ng, D. K., & Chan, C. H. (2009). Sleep duration, wake/sleep symptoms, and academic performance in Hong Kong Secondary School Children. *Sleep and Breathing, 13*(4), 357–367. https://doi.org/10.1007/s11325-009-0255-5

Nguyen-Louie, T. T., Brumback, T., Worley, M. J., Colrain, I. M., Matt, G. E., Squeglia, L. M., & Tapert, S. F. (2018). Effects of sleep on substance use in adolescents: A longitudinal perspective. *Addiction Biology, 23*(2), 750–760. https://doi.org/10.1111/adb.12519

Owens, J. A., Belon, K., & Moss, P. (2010). Impact of delaying school start time on adolescent sleep, mood, and behavior. *Archives of Pediatrics & Adolescent Medicine, 164*(7), 608–614. https://doi.org/10.1001/archpediatrics.2010.96

Owens, J. A., Dearth-Wesley, T., Herman, A. N., Oakes, J. M., & Whitaker, R. C. (2017). A quasi-experimental study of the impact of school start time changes on adolescent sleep. *Sleep Health, 3*(6), 437–443. https://doi.org/10.1016/j.sleh.2017.09.001

Pandi-Perumal, S. R., Smits, M., Spence, W., Srinivasan, V., Cardinali, D. P., Lowe, A. D., & Kayumov, L. (2007). Dim light melatonin onset (DLMO): A tool for the analysis of circadian phase in human sleep and chronobiological disorders. *Progress in Neuro-Psychopharmacology and Biological Psychiatry, 31*(1), 1–11. https://doi.org/10.1016/j.pnpbp.2006.06.020

Parsons, M. J., Moffitt, T. E., Gregory, A. M., Goldman-Mellor, S., Nolan, P. M., Poulton, R., & Caspi, A. (2015). Social jetlag, obesity and metabolic disorder: Investigation in a cohort study. *International Journal of Obesity, 39*(5), 842–848. https://doi.org/10.1038/ijo.2014.201

Shalini, P., Brooks, L. J., Carolyn, D.'. A., Hall, W. A., Suresh, K., Lloyd, R. M., Malow, B. A., Kiran, M., Cynthia, N., Quan, S. F., Rosen, C. L., Troester, M. M., & Wise, M. S. (2021). Recommended amount of sleep for pediatric populations: A consensus statement of the American academy of sleep medicine. *Journal of Clinical Sleep Medicine, 12*(06), 785–786. https://doi.org/10.5664/jcsm.5866

Patel, S. R., & Hu, F. B. (2008). Short sleep duration and weight gain: A systematic review. *Obesity, 16*(3), 643–653. https://doi.org/10.1038/oby.2007.118

Patke, A., Murphy, P. J., Onat, O. E., Krieger, A. C., Özçelik, T., Campbell, S. S., & Young, M. W. (2017). Mutation of the human circadian clock gene CRY1 in familial delayed sleep phase disorder. *Cell, 169*(2), 203–215.e13. https://doi.org/10.1016/j.cell.2017.03.027

Perrault, A. A., Bayer, L., Peuvrier, M., Afyouni, A., Ghisletta, P., Brockmann, C., Spiridon, M., Hulo Vesely, S., Haller, D. M., Pichon, S., Perrig, S., Schwartz, S., & Sterpenich, V. (2019). Reducing the use of screen electronic devices in the evening is associated with improved sleep and daytime vigilance in adolescents. *Sleep, 42*(zsz125)., Article zsz125. https://doi.org/10.1093/sleep/zsz125

Piggins, H. D. (2002). Human clock genes. *Annals of Medicine, 34*(5), 394–400. https://doi.org/10.1080/078538902320772142

Preckel, F., Lipnevich, A. A., Boehme, K., Brandner, L., Georgi, K., Könen, T., Mursin, K., & Roberts, R. D. (2013). Morningness-eveningness and educational outcomes: The lark has an advantage over the owl at high school. *British Journal of Educational Psychology, 83*(1), 114–134. https://doi.org/10.1111/j.2044-8279.2011.02059.x

Rahafar, A., Randler, C., Vollmer, C., & Kasaeian, A. (2017). Prediction of school achievement through a multi-factorial approach – The unique role of chronotype. *Learning and Individual Differences, 55*, 69–74. https://doi.org/10.1016/j.lindif.2017.03.008

Randler, C. (2008a). Morningness-Eveningness comparison in adolescents from different countries around the world. *Chronobiology International, 25*(6), 1017–1028. https://doi.org/10.1080/07420520802551519

Randler, C. (2008b). Morningness–Eveningness and satisfaction with life. *Social Indicators Research, 86*(2), 297–302. https://doi.org/10.1007/s11205-007-9139-x

Randler, C., Faßl, C., & Kalb, N. (2017). From Lark to Owl: Developmental changes in morningness-eveningness from new-borns to early adulthood. *Scientific Reports, 7*. https://doi.org/10.1038/srep45874

Randler, C., Prokop, P., Sahu, S., & Haldar, P. (2015). Cross-cultural comparison of seven morningness and sleep–wake measures from Germany, India and Slovakia. *International Journal of Psychology, 50*(4), 279–287. https://doi.org/10.1002/ijop.12098

Refinetti, R., & Menaker, M. (1992). The circadian rhythm of body temperature. *Physiology & Behavior, 51*(3), 613–637. https://doi.org/10.1016/0031-9384(92)90188-8

Roenneberg, T. (2015). Having trouble typing? What on earth is chronotype? *Journal of Biological Rhythms, 30*(6), 487–491. https://doi.org/10.1177/0748730415603835

Roenneberg, T., Allebrandt, K. V., Merrow, M., & Vetter, C. (2012). Social jetlag and obesity. *Current Biology, 22*(10), 939–943. https://doi.org/10.1016/j.cub.2012.03.038

Roenneberg, T., Keller, L. K., Fischer, D., Matera, J. L., Vetter, C., & Winnebeck, E. C. (2015). Chapter twelve—Human activity and rest in situ. In A. Sehgal (Ed.), *Methods in enzymology* (Vol. 552, pp. 257–283). Academic Press. https://doi.org/10.1016/bs.mie.2014.11.028

Roenneberg, T., Kuehnle, T., Juda, M., Kantermann, T., Allebrandt, K., Gordijn, M., & Merrow, M. (2007). Epidemiology of the human circadian clock. *Sleep Medicine Reviews, 11*(6), 429–438. https://doi.org/10.1016/j.smrv.2007.07.005

Roenneberg, T., Kuehnle, T., Pramstaller, P. P., Ricken, J., Havel, M., Guth, A., & Merrow, M. (2004). A marker for the end of adolescence. *Current Biology, 14*(24), R1038–R1039. https://doi.org/10.1016/j.cub.2004.11.039

Roenneberg, T., Kumar, C., & Merrow, M. (2007). The human circadian clock entrains to sun time. *Current Biology, 17*, R44–R45. https://doi.org/10.1016/j.cub.2006.12.011

Roenneberg, T., Pilz, L. K., Zerbini, G., & Winnebeck, E. C. (2019). Chronotype and social jetlag: A (self-) critical review. *Biology, 8*(3), 54. https://doi.org/10.3390/biology8030054

Roenneberg, T., Winnebeck, E. C., & Klerman, E. B. (2019). Daylight saving time and artificial time zones – A Battle between biological and social times. *Frontiers in Physiology, 10*. https://doi.org/10.3389/fphys.2019.00944

Roenneberg, T., Wirz-Justice, A., & Merrow, M. (2003). Life between clocks: Daily temporal patterns of human chronotypes. *Journal of Biological Rhythms, 18*(1), 80–90. https://doi.org/10.1177/0748730402239679

Roepke, S. E., & Duffy, J. F. (2010). Differential impact of chronotype on weekday and weekend sleep timing and duration. *Nature and Science of Sleep, 2*, 213–220. https://doi.org/10.2147/NSS.S12572

Ryu, H., Joo, E. Y., Choi, S. J., & Suh, S. (2018). Validation of the Munich ChronoType Questionnaire in Korean older adults*. *Psychiatry Investigation, 15*(8), 775–782. https://doi.org/10.30773/pi.2018.04.09

Santisteban, J. A., Brown, T. G., & Gruber, R. (2018, May 9). *Association between the Munich Chronotype Questionnaire and Wrist Actigraphy* [Research Article]. Sleep Disorders; Hindawi. https://doi.org/10.1155/2018/5646848.

Scherrer, V., & Preckel, F. (2021). Circadian preference and academic achievement in school-aged students: A systematic review and a longitudinal investigation of reciprocal relations. *Chronobiology International, 38*(8), 1195–1214. https://doi.org/10.1080/07420528.2021.1921788

Short, M. A., Gradisar, M., Lack, L. C., Wright, H. R., Dewald, J. F., Wolfson, A. R., & Carskadon, M. A. (2013). A cross-cultural comparison of sleep duration between U.S. and Australian adolescents: The effect of school start time, parent-set bedtimes, and extracurricular load. *Health Education & Behavior, 40*(3), 323–330. https://doi.org/10.1177/1090198112451266

Skeldon, A. C., Phillips, A. J. K., & Dijk, D.-J. (2017). The effects of self-selected light-dark cycles and social constraints on human sleep and circadian timing: A modeling approach. *Scientific Reports, 7*(1), 45158. https://doi.org/10.1038/srep45158

Sousa, I. C., Souza, J. C., Louzada, F. M., & Azevedo, C. V. M. (2013). Changes in sleep habits and knowledge after an educational sleep program in 12th grade students. *Sleep and Biological Rhythms, 11*(3), 144–153. https://doi.org/10.1111/sbr.12016

Steptoe, A., Peacey, V., & Wardle, J. (2006). Sleep duration and health in young adults. *Archives of Internal Medicine, 166*(16), 1689–1692. https://doi.org/10.1001/archinte.166.16.1689

Stothard, E. R., McHill, A. W., Depner, C. M., Birks, B. R., Moehlman, T. M., Ritchie, H. K., Guzzetti, J. R., Chinoy, E. D., LeBourgeois, M. K., Axelsson, J., & Wright, K. P. (2017). Circadian entrainment to the natural light-dark cycle across seasons and the weekend. *Current Biology, 27*(4), 508–513. https://doi.org/10.1016/j.cub.2016.12.041

Talbot, L. S., McGlinchey, E. L., Kaplan, K. A., Dahl, R. E., & Harvey, A. G. (2010). Sleep deprivation in adolescents and adults: Changes in affect. *Emotion, 10*(6), 831–841. https://doi.org/10.1037/a0020138

Thacher, P. V., & Onyper, S. V. (2016). Longitudinal outcomes of start time delay on sleep, behavior, and achievement in high school. *Sleep, 39*(2), 271–281. https://doi.org/10.5665/sleep.5426

Tonetti, L., Natale, V., & Randler, C. (2015). Association between circadian preference and academic achievement: A systematic review and meta-analysis. *Chronobiology International, 32*(6), 792–801. https://doi.org/10.3109/07420528.2015.1049271

van der Lely, S., Frey, S., Garbazza, C., Wirz-Justice, A., Jenni, O. G., Steiner, R., Wolf, S., Cajochen, C., Bromundt, V., & Schmidt, C. (2015). Blue blocker glasses as a countermeasure for alerting effects of evening light-emitting diode screen exposure in male teenagers. *Journal of Adolescent Health, 56*(1), 113–119. https://doi.org/10.1016/j.jadohealth.2014.08.002

van der Vinne, V., Zerbini, G., Siersema, A., Pieper, A., Merrow, M., Hut, R. A., Roenneberg, T., & Kantermann, T. (2015). Timing of examinations affects school performance differently in early and late chronotypes. *Journal of Biological Rhythms, 30*(1), 53–60. https://doi.org/10.1177/0748730414564786

Waterhouse, J., Drust, B., Weinert, D., Edwards, B., Gregson, W., Atkinson, G., Kao, S., Aizawa, S., & Reilly, T. (2005). The circadian rhythm of core temperature: Origin and some implications for exercise performance. *Chronobiology International, 22*(2), 207–225. https://doi.org/10.1081/CBI-200053477

Widome, R., Berger, A. T., Iber, C., Wahlstrom, K., Laska, M. N., Kilian, G., Redline, S., & Erickson, D. J. (2020). Association of delaying school start time with sleep duration, timing, and quality among adolescents. *JAMA Pediatrics, 174*(7), 697–704. https://doi.org/10.1001/jamapediatrics.2020.0344

Wittmann, M., Dinich, J., Merrow, M., & Roenneberg, T. (2006). Social jetlag: Misalignment of biological and social time. *Chronobiology International, 23*(1–2), 497–509. https://doi.org/10.1080/07420520500545979

Wong, P. M., Hasler, B. P., Kamarck, T. W., Muldoon, M. F., & Manuck, S. B. (2015). Social jetlag, chronotype, and cardiometabolic risk. *The Journal of Clinical Endocrinology & Metabolism, 100*(12), 4612–4620. https://doi.org/10.1210/jc.2015-2923

Wright, K. P., McHill, A. W., Birks, B. R., Griffin, B. R., Rusterholz, T., & Chinoy, E. D. (2013). Entrainment of the human circadian clock to the natural light-dark cycle. *Current Biology, 23*(16), 1554–1558. https://doi.org/10.1016/j.cub.2013.06.039

Yang, C.-K., Kim, J. K., Patel, S. R., & Lee, J.-H. (2005). Age-related changes in sleep/wake patterns among Korean teenagers. *Pediatrics, 115*(Supplement 1), 250–256. https://doi.org/10.1542/peds.2004-0815G

Zanello, S. B., Jackson, D. M., & Holick, M. F. (2000). Expression of the circadian clock genes clock and period1 in human skin. *Journal of Investigative Dermatology, 115*(4), 757–760. https://doi.org/10.1046/j.1523-1747.2000.00121.x

Zavada, A., Gordijn, M. C. M., Beersma, D. G. M., Daan, S., & Roenneberg, T. (2005). Comparison of the Munich Chronotype Questionnaire with the Horne-Östberg's Morningness-Eveningness score. *Chronobiology International, 22*(2), 267–278. https://doi.org/10.1081/CBI-200053536

Zerbini, G., & Merrow, M. (2017). Time to learn: How chronotype impacts education. *PsyCh Journal, 6*(4), 263–276. https://doi.org/10.1002/pchj.178

Zerbini, G., van der Vinne, V., Otto, L. K. M., Kantermann, T., Krijnen, W. P., Roenneberg, T., & Merrow, M. (2017). Lower school performance in late chronotypes: Underlying factors and mechanisms. *Scientific Reports, 7*(1), 4385. https://doi.org/10.1038/s41598-017-04076-y

Stereotype Threat and Professional Motivation: Assessing Career Expectations of Undergraduate and Graduate Female Students

Juliana Reina and Marcos Emanoel Pereira

To stigmatized individuals, stereotypes convey negative views of their category or group's social identity. Women (belonging to societies that hold sexist beliefs) in general know that stereotypes define them as emotional, empathetic, kind beings, but with reduced mathematical ability, no leadership skills, and with academic and professional skills only in specific areas such as the Humanities (Deemer et al., 2015; von Hippel et al., 2011).

Gender-based stereotypes and attitudes toward the socially disadvantaged category of women often have favorable content but harmful consequences. Eagly and Mladinic (1989) have shown that while most people have favorable stereotypes about women as being helpful, kind, warm, and empathetic, these beliefs can actually invisibilize their competence (Langford & MacKinnon, 2000). Glick and Fiske (1996, 2001) have argued that the perceived friendliness and competence of stigmatized social groups are often inversely related and that forms of benevolent sexism, in which women are seen as warm (but incompetent), endorse a system of gender inequality (Glick et al., 2000, 2004).

Regarding the category of Black women, stereotypes are associated to depersonalization, infrahumanization, low social, and intellectual prestige (or lack of intellectual capacity). In a study on the content of stereotypes surrounding Black characters in Brazilian educational publications, Silva (2001) verified that these stereotypes pointed toward a depersonalization and animalization. Black characters were often unnamed and generally associated to animals. They were also associated to slavery and occupations such as house maids, singers, and dancers.

The relation between perceived skin color and social status and their effects on the attribution of typically human characteristics were assessed by Lima and Vala (2004). The results showed that socially successful Black people were perceived as

J. Reina (✉) · M. E. Pereira
Universidade Federal da Bahia (UFBA), Salvador, Bahia, Brazil
e-mail: juliana.reina@ufba.br

M. V. Alves et al. (eds.), *Cognitive Sciences and Education in Non-WEIRD Populations*, https://doi.org/10.1007/978-3-031-06908-6_14

whiter than the socially unsuccessful. A mediation analysis showed that as much successful blacks are perceived as white, more typically human characteristics are attributed to them. The inverse happened with the Black people perceived as Black. The findings of Wade and Bielitz (2005) demonstrate that Afro-American women with lighter skin color and Afro-American men with darker skin color were perceived as being more socially successful by White participants. Light-skinned women were considered more intelligent than dark-skinned men and women.

Tajfel (1982) argues that stereotypes are based on three central processes: social categorization, social identity, and social comparison. Social categorizations are the cognitive tools that segment, classify, and order the social environment, thus enabling individuals to engage in different forms of social activity. Beyond systematizing the social world, the categorizations also provide a system of orientation for self-reference, since they create and define the individual's place in society. Social groups, in this sense, provide its members with a form of self-identifying in social terms. These identifications are largely relational and comparative, defining individuals as similar or different, "better" or "worse" than the members of other groups (Tajfel & Turner, 1986).

In this perspective, social identity is understood as the individual's self-concept that originates from consciously belonging to a social group or groups. This belonging has affective-emotional meaning and value. The main assumption of the theory of social identity (Tajfel, 1982) is the notion that individuals strive to achieve or maintain a positive type of social identity, that is, they strive to maintain a high self-esteem through a positive self-image. This image is obtained when individuals positively differentiate themselves from members of other social groups in a process of social comparison (Tajfel & Turner, 1986). When the social identity is not satisfactory, individuals strive to leave their group and join another more valued one and/or make their group more positively distinctive.

Based on research on social categorization and on the concept of social identity, the self-categorization theory is characterized by a set of hypotheses and premises about the functioning of the social self-concept. That is, the self-concept based on comparisons with other people that are important for the social interaction. This theory proposes that individuals who are more identified with their social group will have their group identity easily activated in situations that are relevant for their identity (Turner et al., 1987).

Building on this notion, the theory of stereotype threat (Steele & Aronson, 1995) assumes that each person has multiple group identities (e.g. gender, race/ethnicity, age, socioeconomic status, etc.). When situational cues signal the value or importance of an identity in a given environment, the association with a specific group becomes more prominent and a process of vigilance begins (Murphy & Taylor, 2012).

During the stage of vigilance, the individual's attention is directed to situational cues to determine whether the group identity is a variable to be assessed. Two assessments are possible. If the cues in the social environment fail to confirm that the social identity may be a source of stigma, devaluing or mistreatment, vigilance decreases. Performance and functioning will therefore be contingent only on the task at hand. However, if the situational cues confirm the possibility that the social identity is being negatively evaluated, vigilance increases, leading to impaired performance and functioning (Cohen & Garcia, 2008; Cohen et al., 2012).

Even apparently harmless situational cues such as filling in race or gender on a form may be meaningful, once people are always trying to investigate if they are possibly being devalued in a given context (Wout et al., 2009). Individuals differ as to the probability and intensity with which they engage in the process of vigilance, that is, people have different thresholds for evaluating whether there is threat to identity.

Some people put almost every environment under scrutiny, in search of situational cues that may signal the value of their identities. This may happen if they are, for instance, particularly sensitive to identity-based rejection (Mendoza-Denton et al., 2002) or highly aware of the stigma associated to their identities (Pinel, 2004). Some will require a strong situational cue, such as a sexist comment from a fellow student or professor, while others may feel threatened when several more subtle suggestions converge (Murphy & Taylor, 2012).

Besides, certain contextual signs are less threatening for individuals who did not personally invest in specific domains, for example, women who avoid careers in the hard sciences as a strategy for protecting their self-concept and personal self-esteem. According to this idea, stereotype threat is more related to maladaptive psychological reactions to the stressor than to the stressor itself (i.e., the stereotype) (Liu et al., 2021).

Negative stereotypes on social identities provide social blueprints for interpreting individual behavior in certain domains, generating concerns, for the individuals, regarding whether they are being judged or treated in terms of these stereotypes. Experienced as a threatening state, this phenomenon, known as stereotype threat, can hinder the performance of members of devalued social groups in different domains (Kaye et al., 2018; Steele, 1997; Steele & Aronson, 1995; Steele et al., 2002).

Stereotype threat weakens performance because when the person tries to refute a negative stereotype, anxiety is automatically activated. Anxiety distracts the individual from the task at hand due to the efforts to reduce it. Additionally, concerns regarding stereotypes are also a distraction and require regulation. The attempt to regulate the anxiety and concerns steals the necessary resources from the working memory for the cognitive demand(s) of the daily tasks related to stereotypes (Schmader et al., 2008).

These experiences and the consequences of the process of stereotype threat can influence people's desire to identify and persist in academic and professional domains (Nussbaum & Steele, 2007). These vigilance processes can also shape people's experience in the future, directing their attention to similar situational cues in new environments (Murphy & Taylor, 2012).

Any individual may be confronted with the psychological threat of being reduced to a negative stereotype. This concern results in impaired performance in tasks that are considered to have a high level of difficulty and/or a high level of social importance (Steele et al., 2002). The harmful effects of stereotype threat on performance have been found in a variety of domains and social groups: in the academic domain in general for Black people (Steele & Aronson, 1995); women in the hard sciences, in traffic, and at work (Deemer et al., 2015; Kinias & Sim, 2016; Yeung & von Hippel, 2008); elderly people in relation to memory (Lamont et al., 2015); and White men in athletics (Stone et al., 2012), among others. Therefore, we understand that stereotype threat is situational, because it arises when people become aware that they may be the target of negative stereotypes for belonging to devalued social groups, during their daily activities (Steele, 1997).

In this study, we investigated the social groups of undergraduate and graduate female students. We believe the threat of being stereotyped interferes in the intellectual functioning and academic engagement of these students, for belonging to a stigmatized group or groups. This sets the ground for possible differences in later educational performance and professional career expectations and choices (Murphy & Taylor, 2012; Ryan & Sackett, 2013; Walton et al., 2015).

The results of the study by von Hippel et al. (2015) with a sample of 512 women from the financial sector showed that they perceived stereotype threat in their workplace, reported a decrease in their well-being in work and were less prone to recommend their field to other women. These findings were mediated by an identity separation, that is, the women perceived their gender identity as incompatible with their work identity.

Despite the presence of women in various spheres of society nowadays, gender stereotypes reduce women's ability to achieve certain goals or aspirations, especially in areas dominated by men (Roberson & Kulik, 2007). Studies show that women are less seen as a hiring option in typically male domains, have less opportunities for promotions and still earn less than their male coworkers (Lyness & Heilman, 2006). Furthermore, they are evaluated as inferior employees for being perceived as less committed to their career and more focused on their families (Correll et al., 2007).

We may consider that gender inequalities and sexist stereotypes have highly damaging impacts on women's career aspirations and on their beliefs about their professional future, to the extent that most of these stereotypes are incongruous with professional success (Powell et al., 2002; Schein, 2007). Relationships with organizations marked by gender (and race) inequality produce intense stress in women when they plan their professional career and perceive they have more barriers than men (Lyness & Heilman, 2006; Lyness & Thompson, 2000). In this study, we considered stereotype threat as a specific barrier that may affect the career expectations of female university students in different study fields.

From the perspective of the target, we are interested in the expectations that undergraduate and graduate female students have for their professional future and the professional future of other women, as a social category. Studies on stereotype threat usually infer its presence based on a pattern of outcomes where the dependent variable is the performance of members of socially devalued groups. However, we sought to assess stereotype threat through a different variable, that is, expectation, in order to investigate whether this phenomenon may affect other dimensions of the lives of female university students, in this case, their beliefs about their own professional future and that of their social category.

Research Problem and Hypotheses

Our study assumes that in our society stereotypes, as legitimators of a sexist system, devalue the social category of women and, even when positive, hold beliefs that classify them as incompetent (Santos & Amâncio, 2014; Jost & Kay, 2005). We also take into consideration women individual psychological mechanism of striving to maintain a positive personal self-esteem, once they're members of a social devalued group, especially in academic and professional contexts, in which competence is constantly being evaluated (Oswald & Lindstedt, 2006). One of the main consequences of having a devalued social identity is the attempt to make oneself more positively distinctive (Tajfel & Turner, 1986). With these elements, we place the hypothesis that the participants will have a lower group expectation regarding professional career in comparison to their personal expectation; the students will expect more for themselves than for their social category (H1).

Understanding how these threat processes can produce different outcomes for stigmatized groups is particularly important for educators and employers who interact with diverse populations on a daily basis. One of the initial goals of research on stereotype threat was not only to reveal the processes that negatively impact members of stereotyped groups but also to develop strategies for removing these threats from socially valued environments such as schools, universities, and workplaces (Steele et al., 2002).

According to Murphy and Taylor (2012), cues of a secure identity can reverse this process by signaling to people that their social identity is valued. The literature has shown that highlighting the achievements and positive attributes of the target group of stereotypes and prejudice seems to produce a significant effect in reducing the harmful impacts of stereotype threat (McIntyre et al., 2003; da Silva, 2007).

The theory of stereotype threat has shed light on the harmful effects of this process, specifically on performance, due to the fear that members of socially stigmatized groups have of being judged or treated in terms of negative stereotypes (Deemer et al., 2015; Steele & Aronson, 1995). This has led to our second hypothesis, which postulates that the participants exposed to stereotype threat will have lower career expectations compared to participants exposed to the control condition and participants exposed to the positive evaluation of attributes of their social category, who will have higher career expectations (H2).

It is essential to point out that most research on stereotype threat has been conducted in experimental laboratories, and while this offers experimental control and strong evidence of the power of situational factors in reducing performance, the question of generalizability to real-world contexts is important. Studies assume that while members of devalued groups, including women, are unlikely to experience the kinds of stereotype threats that are induced in the laboratory in real-life contexts (Liu et al., 2021), subtle experiences in school, university, and work may cause them to worry about being evaluated based on their gender and/or race (von Hippel et al., 2011).

In this research, we are interested in the social comparisons that women engage in to evaluate their achievements, and consequently their career expectations. Given that men generally earn more, are promoted more quickly, and are given jobs of greater value to organizations (Agars, 2004; Heilman, 2001), it can be problematic for women to compare themselves to their male counterparts. Through social comparisons with men, women may become aware that they are paid less, are climbing the corporate ladder more slowly, and are working on less visible projects. Social comparisons with those who perform "better" (Schwinghammer et al., 2006) or are more valued negatively affect self-assessment.

Studies suggest that social comparisons with men are associated with feelings of stereotype threat in the workplace while social comparisons with other women are not (von Hippel et al., 2011; Marx et al., 2005). Stereotype threat has been associated with conflicts around women's identity as employees, less perceived likelihood of achieving career goals, negative attitudes toward the workplace, and intentions of changing jobs (higher turnover rates) (von Hippel et al., 2011).

From this perspective, the present research is interested in the activation of the social identity of female university students through two types of social comparisons, one that devalues and another that values the social category. Thus, the third hypothesis postulates a differential effect produced by identity activation through social comparisons. In this scenario, participants subjected to stereotype threat with social comparison will have even lower career expectations, while those exposed to positive evaluation of the category with social comparison will have higher expectations than the control, stereotype threat and positive evaluation (with no social comparison) conditions (H3).

Method

This study was developed with a 3 × 2 design. The first factor refers to the experimental conditions (stereotype threat, positive evaluation, control), and the second factor refers to the social comparisons (with identity activation, without identity activation). Professional career expectations for oneself and for the social category were evaluated in the different conditions as well as the impact of the following mediators of the effects of stereotype threat and positive evaluation of attributes on career expectations: domain identification, racial identity, domain disengagement, personal self-esteem, collective self-esteem, academic, and professional goals.

Participants

The study was carried out with undergraduate and graduate university students. A total of 331 women participated in the study, randomly allocated in the three experimental conditions according to the distribution exposed in Table 1, which shows there was no discrepancy of the values in each condition ($\chi^2 = 331.000$, $p < 0.001$).

Table 1 Sample distribution by experimental condition

Experimental condition	Without comparison	With comparison	Total
Threat	56	60	116
Positive evaluation	53	45	98
Control	52	65	117
Total	161	170	331

Procedure

The call for participants was done through social networks. Female university students were invited to voluntarily participate in an experimental online survey about career expectations. After agreeing to participate in the research through an informed consent form, ensuring the anonymity of their information, they answered a sociodemographic questionnaire and chose the band that most resembled their skin tone on a skin color scale. Upon starting the task, participants were randomly allocated to one of the three experimental conditions. Students in the stereotype threat condition, for example, were exposed to a priming screen containing a text with characteristics that threatened the social category of women. After reading the text, they would answer a questionnaire about their expectations regarding their own professional career and that of their social category. Afterward, they were asked to answer the instruments that evaluated the mediators. Participants in the control group were exposed to a priming screen with a text with no stereotype threat and that did not value the attributes of their social group.

Experimental Manipulation
Participants were randomly assigned to one of three experimental conditions. The experimental conditions were activated by reading a short text at the initial stage of the experiment. See Table 2.

Instruments

Assessment of Expectations

Two self-report measures with 10 questions each assessed the students‘expectations regarding their personal professional career and that of their social category. The scales were an adaptation of the Scale of Career Expectations in Organizations by Vasconcellos and Neira (2016). On two 5-point Likert scales ranging from strongly disagree to strongly agree, students rated items such as "I will achieve relevant professional achievements in my career"/"Women will achieve relevant professional achievements in their careers," "I will be well paid in relation to my professional responsibilities"/"Women will be well paid in relation to their professional responsibilities," "I will be recognized by the organization for my contributions at work"/"Women will be recognized by the organization for their contributions at work."

Table 2 Texts referring to the three experimental conditions

Threat	Positive evaluation	Control
In recent years, due to the expansion of higher education in Brazil, numerous studies have been conducted on the experiences of students in the academic environment. Research evaluating the academic performance of university students at the undergraduate and graduate levels indicates that women present poor academic performance in examinations (tests and exams) in hard science courses, such as mathematics, engineering, statistics, and technology.	In recent years, due to the expansion of higher education in Brazil, numerous studies have been conducted on the experiences of students in the academic environment. Research evaluating the academic performance of university students at the undergraduate and graduate levels indicates that women present satisfactory academic performance in examinations (tests and exams) in hard science courses, such as mathematics, engineering, statistics, and technology.	In recent years, due to the expansion of higher education in Brazil, numerous studies have been conducted on the experiences of students in the academic environment. Data released by the Ministry of Education indicated that more than seven million students enrolled in higher education institutions in the years 2011 and 2012 in Brazil. According to the School Census of Higher Education, there were more than one million enrollments in federal institutions during the years 2011 and 2012.
Comparison Therefore, the data show that in areas predominantly occupied by men, women tend to perform significantly worse than their male colleagues.	*Comparison* Therefore, the data show that in areas predominantly occupied by men, women tend to perform significantly better than their male colleagues.	*Comparison* Recent data indicate that the private school system already has more than 6.2 million students, which represents more than 75% of the higher education system. More than 1.1 million students are enrolled in the federal universities.

Participant Variables Assessment

The following self-report measures were used to assess the impact of psychosocial and sociodemographic variables:

Racial Identity Two scales were used to assess racial identity: the Skin Color Scale and an adaptation of Helms' Racial Identity Scale (1990). In the Skin Color Scale, the participant was asked to indicate the color that most closely matched her skin tone out of 20 options. This scale was included given the complexity of racial self-identification in Brazil, where miscegenation resulted in a "gradient" of skin tones. In this sense, racial appearance, found in elements such as skin color, intervenes as an identity marker that may lead to different interpretations (Cunin, 2003). We considered it important to assess whether the students'skin color affected their professional expectations. The adaptation of the Racial Identity Scale was used to assess how far individual self-definition is related to belonging to a social group in which race is the marker and whether this identification would produce differences in expectations.

Identification with the Academic Domain An adaptation of the Identification with School Questionnaire (Voelkl, 1996) was used to assess the extent to which the student's self-esteem is connected to and dependent on academic performance. Studies on stereotype threat show that individuals who are most identified with this domain will be the most affected.

Domain Engagement An adaptation of Stone's Student Engagement Questionnaire (2002) assessed the students' skills in relation to the academic domain.

Personal Self-esteem The Rosenberg Self-Esteem Scale (1965, cited by Vargas et al., 2005) was used. According to this scale, having high self-esteem means feeling "good enough," respecting oneself, considering oneself capable and, at the same time, not feeling superior or inferior to others. On the other hand, low self-esteem suggests self-rejection, dissatisfaction with and contempt for oneself. We believe that there is a close relationship between personal self-esteem and professional expectations.

Collective Self-esteem The Collective Self-Esteem Scale was developed by Luhtanen and Crocker (1992) and has a version adapted for the Brazilian population by de Souza and Ferreira (2004).

Academic and Professional Achievement Goals An adaptation of the 3 × 2 Achievement Goal Questionnaire (AGQ; Elliot et al., 2011) was used. Theorists posit that achievement goals reflect purpose and anticipated outcomes that guide behavior.

Sociodemographic Questionnaire The instrument was applied with the objective of characterizing the participants in sociodemographic terms and consisted of questions about age, race/color, education, income, marital status, number of children, professional status, and academic field.

Analysis and Discussion of the Results

To test hypothesis 1, in which we expected to find that participants would have higher career expectations for themselves (personal expectation) than for the social category of women (group expectation), we conducted a Bayesian analysis. Comparison of the differences showed an expectation score 0.42 higher for their own future than for the category of women (personal expectations: 3.42 ± 0.60; group expectations: 3.07 ± 0.69). A two-tailed analysis using Student's t-test for repeated measures revealed a Bayesian factor suggesting that the substantive hypothesis of difference between means was 6.58e + 23 times better than the null hypothesis. After respecification and subsequent application of a one-tailed test to verify whether personal expectations were higher than group expectations, the Bayesian test indicated an extremely robust model, 1.32e + 24 better than the null hypothesis, with a median effect size of 0.67 (0.51–0.84). This result confirmed the first hypothesis, demonstrating that female university students have more positive beliefs about their personal professional future than about the future of their social group.

Self-categorization theory, linked to social identity theory, argues that people strive to maintain a positive social identity and that self-stereotyping can be found related to all relevant attributes or dimensions of the group (Cohen et al., 2012; Oswald & Lindstedt, 2006; Tajfel, 1982; Turner et al., 1987). Thus, positive gender stereotypes are likely to be easily integrated into the self-concept and promote positive personal and collective self-esteem. Negative gender stereotypes, on the other hand, present a problem, since complete denial of these stereotypes tends to be unrealistic, given the prevalence of stereotypes in society.

Internalizing negative stereotypes about social groups can threaten social identity and result in low personal and collective self-esteem (Katz et al., 2002). Gender self-stereotyping poses a major problem for maintaining a positive self-concept. In a study by Oswald and Lindstedt (2006), participants readily accepted positive gender stereotypes to describe themselves and negative stereotypes were used more often to describe the group than themselves, demonstrating an attempt to maintain positive personal self-esteem.

The results derived from the first hypothesis point to the impact of the internalization of sexist stereotypes that devalue women. Since stereotypes shape social identities, in spaces where this identity is devalued, women need to constantly demonstrate competence and personal effort to maintain a positive self-image and personal self-esteem. Thus, we understand why female university students have more positive expectations about themselves than about their stigmatized social group.

Hypothesis 2 postulated that participants exposed to the stereotype threat condition would have lower expectations relative to participants in the control and positive evaluation condition, who in turn would have higher expectations. Analyses conducted using a Bayesian ANOVA for repeated measures (specifying a Cauchy prior for the effects) pointed to a Bayes factor consistent with the understanding that the data is 6.33e + 24 times better for a model that includes time and expectations compared to the null model. As noted in the graph found in Fig. 1, and taking into consideration the post hoc comparisons conducted regarding the different experimental conditions, it is clear that both threat priming (posterior odds = 3.494) and positive evaluation priming (posterior odds = 274.439) created higher personal and group expectations than the control condition, while no differences in posterior odds scores were identified between the positive evaluation and threat conditions (0.155).

Thus, hypothesis 2 was partially confirmed, since in the positive evaluation condition the participants' expectations increased, but for the stereotype threat condition, which should have reduced the score, it did not prove effective.

In this study, we threatened the group of women with stereotypes that signaled a devaluation of their social category within the hard sciences (a male-dominated field of knowledge) and expected that this condition would reduce their professional expectations. A number of studies show that the mere existence of social stereotypes highlighting a supposed intellectual inferiority of specific social groups creates threatening intellectual environments for its members, where anything they say or do can be interpreted through the lens of low expectations. This corrosive effect of stereotyping can distort individuals' interpretation of critical assessments, produce gender and race disparities in career ambitions, and undermine efforts aimed at improving diversity and inclusion in organizations (Roberson & Kulik, 2007; Shapiro & Neuberg, 2007; Steele & Aronson, 1995).

We believe that stereotype threat priming signaled to the participants that they should differentiate themselves personally from their social category, i.e., show they are competent and "can have a promising future" even though they belong to a devalued social category. Therefore, by using a self-report measure, the female students, in an attempt to protect their personal self-image and self-esteem, reported having high personal career expectations. It is noteworthy that this effect would probably not occur with a performance measure, since performance is assessed directly and not indirectly as with a self-report measure.

The last hypothesis was that participants in the stereotype threat condition with social comparison (devalued identity activation) would have lower career expectations, while in the positive evaluation condition with social comparison (valued identity activation), expectations would be higher than for participants exposed to the control, stereotype threat, and positive evaluation conditions without social comparison. To test hypothesis 3, we conducted a Bayesian repeated measures ANOVA including as comparison factors between conditions (control, threat and positive evaluation) and identity activation (yes or no). The data pointed to a model 2.06e + 24 better than the null model, with BF_{10} values of 25.325 for the effect of

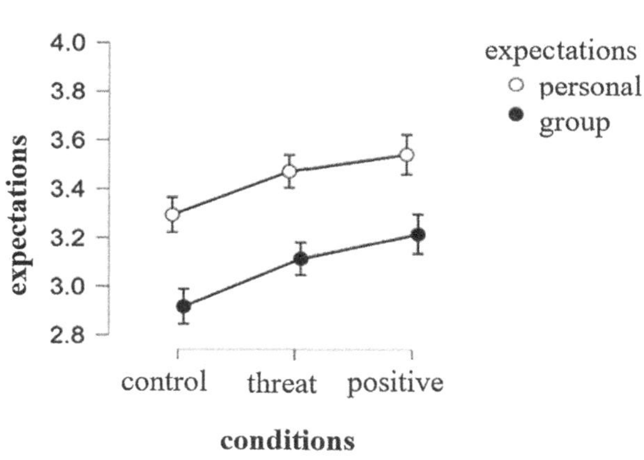

Fig. 1 Participants personal and group expectations by experimental conditions

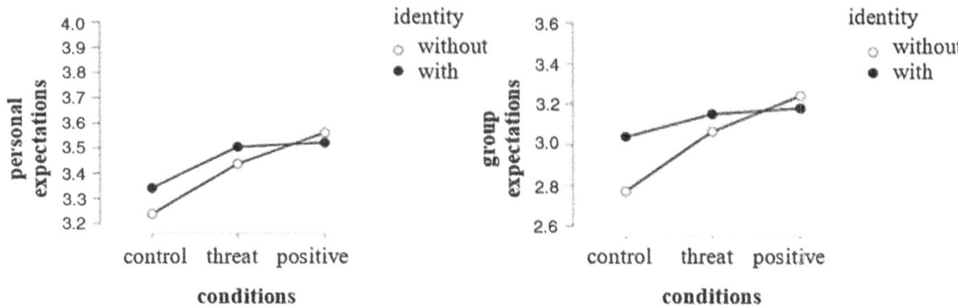

Fig. 2 Participants personal and group expectations with and without identity activation by experimental conditions

condition and only 4.276 for identity activation, indicating only a modest effect of the latter variable.

The analysis of the graph found in Fig. 2 and the post hoc test values confirm that the effect of identity occurred only in relation to group expectations and only in the control condition, in which an increase in expectation is identified as a function of identity activation, a result that opposes the postulate in hypothesis 3.

We believe this result can be explained by considering the profile of participants in the sample. The female students in the present study may not have been sensitive to the type of experimental manipulation, i.e., comparing the performance of women with that of men did not activate the social identity of these women. Studies on the stereotype threat effect on the category of women are mostly conducted with students or professionals in the hard sciences field, in which they are threatened with logic tests. However, in this study, the profile was of humanities students' field. This is an effect of reality itself, since many courses in the hard sciences are predominantly taken by men. This is an important indicator of how sexist stereotypes interfere in the distribution of students by area. Perhaps for students in the humanities, a field predominantly composed of women, comparing them to men is not enough to activate their social identity.

A marking aspect of our discussion refers to a predominance of participants from the humanities. To analyze this effect, we conducted a new analysis considering the participants' psychosocial and sociodemographic variables in the result pattern of the expectations. To formulate predictive models, we ran Bayesian multiple linear regressions, with an a priori value [P(M)] of 0.5 per model for each expectation, using the sociodemographic and psychosocial variables as predictors. We found evidence supporting a regression model for personal expectations ($BF_{10} = 3.499e + 08$), in which the following predictors were included, in order of inclusion factor: personal self-esteem ($BF_{Inc} = 3.22e + 07$), positive evaluation priming ($BF_{Inc} = 96.08$), education in the humanities ($BF_{Inc} = 7.66$), and skin color ($BF_{Inc} = 3.03$).

Analysis of the model indicates an r2 value = 0.208 and indicate an intercept of *3.42* and values of 0.256 for personal self-esteem, 0.248 for positive evaluation priming, −0.140 for education in the humanities, and −0.012 for skin color. Since the inclusion of the skin color variable contributed little to the overall model, as observed by the BF_{Inc}, the model generated for personal expectations indicates that the higher the personal self-esteem and the activation of the positive evaluation priming, the higher the level of personal expectations, although this effect is clearly reduced among participants in the humanities.

We verified a discrepancy in the intensity of personal expectation between students from the humanities and other areas, that is, students from the humanities have lower personal expectation. Moreover, the relationship between personal self-esteem (main predictor variable) and personal expectation is higher for nonhumanities students and this effect is independent of positive evaluation. On the other hand, students from the humanities only show a marked relation between self-esteem and personal expectation when they are subjected to the condition of positive evaluation of the social category (Fig. 3).

Conclusion

The results together indicate that female university students demonstrated higher expectations regarding their own professional future than that of the category of women. This effect was intensified by priming activation, regardless of whether it

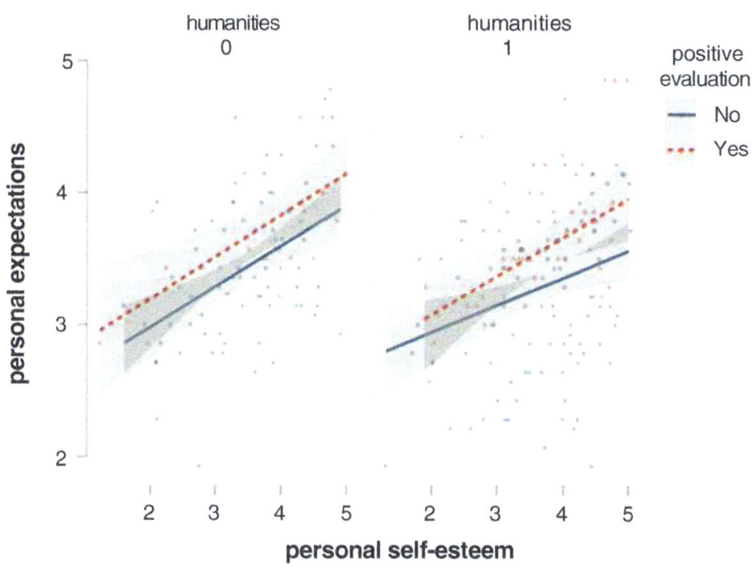

Fig. 3 Relationship between personal self-esteem and personal expectations of humanities and nonhumanities field students exposed or not to the positive evaluation condition

was stereotype threat or positive evaluation of the category, although expectations were not affected by social identity activation. The evidence presented in this study points to the negative impacts of sexist beliefs on expectations regarding the professional future of the social category of women. Individually, women strive to disconfirm the negative stereotypes related to their category, demonstrating positive expectations for themselves. However, with regards to their social category, they show they have internalized socially shared negative stereotypes about women, since they reported having low expectations for the professional future of their social group. This study contributed by measuring the effect of stereotype threat in a dimension other than performance on expectations about the future of a socially devalued category. The threat condition produced an unexpected pattern with respect to women's career expectations. However, social identity theory and self-categorization theory are able to explain this effect. The experimental manipulation of the identity activation priming was not effective. This leads us to consider the possibility of problems in the profile of the participants (mostly from the field of humanities). More research is needed to evaluate the relationship between these constructs and with a larger number of women from other fields of knowledge.

References

Agars, M. D. (2004). Reconsidering the impact of gender stereotypes on the advancement of women in organizations. *Psychology of Women Quarterly., 28*(2). https://doi.org/10.1111/j.1471-6402.2004.00127.x

Cohen, G. L., & Garcia, J. (2008). Identity, belonging, and achievement: A model, interventions, implications. *Current Directions in Psychological Science, 17*(6), 365–369. https://doi.org/10.1111/j.1467-8721.2008.00607.x

Cunin, E. (2003). *Identidades a flor de piel. Lo "negro" entre apariencias y pertenencias: categorías raciales y mestizaje en Cartagena*. Arfo Editores.

Cohen, G. L., Purdie-Vaughns, V., & Garcia, J. (2012). An identity threat perspective on intervention. In M. Inzlicht & T. Schmader (Eds.), *Stereotype threat: Theory, process, and application* (pp. 280–296). Oxford University Press.

Correll, S. J., Benard, S., & Paik, I. (2007). Getting a job: Is there a motherhood penalty? *American Journal of Sociology, 112*(5), 1297–1338. https://doi.org/10.1086/511799

Deemer, E. D., Lin, C., & Soto, C. (2015). Stereotype threat and women's science motivation: Examining the disidentification effect. *Journal of Career Assessment, 1-14*. https://doi.org/10.1177/1069072715616064

Eagly, A. H., & Mladinic, A. (1989). Gender stereotypes and attitudes toward women and men. *Personality and Social Psychology Bulletin, 15*(543), 558. https://doi.org/10.1177/0146167289154008

Elliot, A. J., Murayama, K., & Pekrun, R. (2011). A 3 × 2 achievement goal model. *Journal of Educational Psychology, 103*(3), 632–648. https://doi.org/10.1037/a0023952

Glick, P., & Fiske, S. T. (1996). The Ambivalent Sexism Inventory: Differentiating hostile and benevolent sexism. *Journal of Personality and Social Psychology, 70*(3), 491–512. https://doi.org/10.1037/0022-3514.70.3.491

Glick, P., & Fiske, S. T. (2001). An ambivalent alliance: Hostile and benevolent sexism as complementary justifications for gender inequality. *American Psychologist, 56*(2), 109–118. https://doi.org/10.1037/0003-066X.56.2.109

Glick, P., Fiske, S. T., Mladinic, A., Saiz, J. L., Abrams, D., Masser, B., & López López, W. (2000). Beyond prejudice as simple antipathy: Hostile and benevolent sexism across cultures. *Journal of Personality and Social Psychology, 79*, 763–775. https://doi.org/10.1037/0022-3514.79.5.763

Glick, P., Lameiras, M., Fiske, S. T., Eckes, T., Masser, B., Volpato, C., & Wells, R. (2004). Bad but bold: Ambivalent attitudes toward men predict gender inequality in 16 nations. *Journal of Personality and Social Psychology, 86*, 713–728. https://doi.org/10.1037/0022-3514.86.5.713

Heilman, M. E. (2001). Description and prescription: How gender stereotypes prevent women's ascent up the organizational ladder. *Journal of Social Issues, 57*(4), 657–674. https://doi.org/10.1111/0022-4537.00234

Helms, J. E. (1990). *Black and white racial identity: Theory, research, and practice*. Westport.

Jost, J. T., & Kay, A. C. (2005). Exposure to benevolent sexism and complementary gender stereotypes: Consequences for specific and diffuse forms of system justification. *Journal of Personality and Social Psychology, 88*(3), 498–509. https://doi.org/10.1037/0022-3514.88.3.498

Katz, J., Joiner, T. E., Jr., & Kwon, P. (2002). Membership in a devalued social group and emotional well-being: Developing a model of personal self-esteem, collective self-esteem, and group socialization. *Sex Roles: A Journal of Research, 47*(9–10), 419–431. https://doi.org/10.1023/A:1021644225878

Kaye, L. K., Pennington, C. R., & McCann, J. J. (2018). Do casual gaming environments evoque stereotype threat? Examining the effects of explicit priming and avatar gender. *Computers in Human Behavior, 78*, 142–150. https://doi.org/10.1016/j.chb.2017.09.031

Kinias, Z., & Sim, J. (2016). Facilitating women's success in business: Interrupting the process of stereotype threat through affirmation of personal values. *Journal of Applied Psychology, 101*(11), 1585–1597. https://doi.org/10.1037/apl0000139

Lamont, R. A., Swift, H. J., & Abrams, D. (2015). A review and meta-analysis of age-based stereotype threat: Negative stereotypes, not facts, do the damage. *Psychology and Aging, 30*(1), 180–193. https://doi.org/10.1037/a0038586

Langford, T., & Mackinnon, N. J. (2000). The affective bases for the gendering of traits: Comparing the United States and Canada. *Social Psychology Quarterly, 63*(1), 34–48. https://doi.org/10.2307/2695879

Lima, M. E. O., & Vala, J. (2004). As novas formas de expressão do preconceito e do racismo. *Estudos de Psicologia, 9*(3), 4001–4411. https://doi.org/10.1590/S1413-294X2004000300002

Liu, S., Liu, P., Wang, M., & Zhang, B. (2021). Effectiveness of stereotype threat interventions: A meta-analytic review. *Journal of Applied Psychology, 106*(6), 921–949. https://doi.org/10.1037/apl0000770

Luhtanen, R., & Crocker, J. (1992). A collective self-esteem scale: Self-evaluation of one's social identity. *Personality and Social Psychology Bulletin, 18*(3), 302–318. https://doi.org/10.1177/0146167292183006

Lyness, K. S., & Heilman, M. E. (2006). When fit is fundamental: Performance evaluations and promotions of upper-level female and male managers. *Journal of Applied Psychology, 91*(4), 777–785. https://doi.org/10.1037/0021-9010.91.4.777

Lyness, K. S., & Thompson, D. E. (2000). Climbing the corporate ladder: Do female and male executives follow the same route? *Journal of Applied Psychology, 85*(1), 86–101. https://doi.org/10.1037/0021-9010.85.1.86

Marx, D. M., Stapel, D. A., & Muller, D. (2005). We can do it: The interplay of construal orientation and social comparisons under threat. *Journal of Personality and Social Psychology, 88*(3), 432–446. https://doi.org/10.1037/0022-3514.88.3.432

McIntyre, R. B., Paulson, R. M., & Lord, C. G. (2003). Alleviating women's mathematics stereotype threat through salience of group achievements. *Journal of Experimental Social Psychology, 39*, 83–90. https://doi.org/10.1016/S0022-1031(02)00513-9

Mendoza-Denton, R., Downey, G., Purdie, V. J., Davis, A., & Pietrzak, J. (2002). Sensitivity to status-based rejection: Implications for African American students' college experience. *Journal of Personality and Social Psychology, 83*(4), 896–918. https://doi.org/10.1037/0022-3514.83.4.896

Murphy, M. C., & Taylor, V. J. (2012). The role of situational cues in signaling and maintaining stereotype threat. In M. Inzlicht & T. Schmader (Eds.), *Stereotype threat: Theory, process, and application* (pp. 17–33). Oxford University Press.

Nussbaum, A. D., & Steele, C. M. (2007). Situational disengagement and persistence in the face of adversity. *Journal of Experimental Social Psychology, 43*(1), 127–134. https://doi.org/10.1016/j.jesp.2005.12.007

Oswald, D., & Lindstedt, K. (2006). The content function of gender self-stereotypes: Na exploratory investigation. *Sex Roles, 54*, 447–458. https://doi.org/10.1007/s11199-006-9026-y

Pinel, E. C. (2004). You're just saying that because I'm a woman: Stigma consciousness and attributions to discrimination. *Self and Identity, 3*, 39–51. https://doi.org/10.1080/13576500342000031

Powell, G. N., Butterfield, D. A., & Parent, J. D. (2002). Gender and managerial stereotypes: Have the times changed? *Journal of Management, 28*(2), 177–193. https://doi.org/10.1177/014920630202800203

Roberson, L., & Kulik, C. T. (2007). Stereotype threat at work. *The Academy of Management Perspectives, 21*(2), 24–40. https://doi.org/10.5465/AMP.2007.25356510

Ryan, A. M., & Sackett, P. R. (2013). Stereotype threat in workplace assessments. In K. F. Geisinger, B. A. Bracken, J. F. Carlson, J.-I. C. Hansen, N. R. Kuncel, S. P. Reise, & M. C. Rodriguez (Eds.), *APA handbook of testing and assessment in psychology* (Test theory and testing and assessment in industrial and organizational psychology) (Vol. 1, pp. 661–673). American Psychological Association. https://doi.org/10.1037/14047-036

Schein, V. (2007). Women in management: Reflections and projections. *Women in Management Review, 22*(1), 6–18. https://doi.org/10.1108/09649420710726193

Schmader, T., Johns, M., & Forbes, C. (2008). An integrated process model of stereotype threat effects on performance. *Psychological Review, 115*(2), 336–356. https://doi.org/10.1037/0033-295X.115.2.336

Schwinghammer, S. A., Stapel, D. A., & Blanton, H. (2006). Different selves have different effects: Self-activation and defensive social comparison. *Personality and Social Psychology Bulletin, 32*(1), 27–39. https://doi.org/10.1177/0146167205277096

Santos, M. H., & Amâncio, L. (2014). Percepção de justiça, discriminação e sexismo. *Revista PSICOLOGIA, 28*(1), 67–81. https://doi.org/10.17575/rpsicol.v28i1.623

Shapiro, J. R., & Neuberg, S. L. (2007). From stereotype threat to stereotype threats: Implications of a multi-threat framework for causes, moderators, mediators, consequences, and interventions. *Personality and Social Psychology Review, 11*(2), 107–130. https://doi.org/10.1177/1088868306294790

Silva, A. C. (2001). Desconstruindo a discriminação do negro no livro didático. *EDUFBA*.

Silva, J. F. da (2007). *Ameaça dos Estereótipos na performance intelectual de Estudantes Universitários Ingressos pelo Sistema de Cotas*. [Dissertação de Mestrado, Universidade Federal da Bahia].

Steele, C. (1997). A threat in the air: How stereotypes shape intellectual identity and performance. *American Psychologist, 52*(6), 613–629. https://doi.org/10.1037/0003-066X.52.6.613

Steele, C. M., & Aronson, J. (1995). Stereotype threat and the intellectual test performance of African Americans. *Journal of Personality and Social Psychology, 69*(5), 797–811. https://doi.org/10.1037/0022-3514.69.5.797

Steele, C. M., Spencer, S. J., & Aronson, J. (2002). Contending with group image: The psychology of stereotype and social identity threat. In M. P. Zanna (Ed.), *Advances in experimental social psychology* (Vol. 34, pp. 379–440). Academic Press. https://doi.org/10.1016/S0065-2601(02)80009-0

Stone, F. (2002). Battling doubt by avoiding practice: The effects of stereotype threat on self-handicapping in white athletes. *Personality and Social Psychology Bulletin, 77*, 1213–1227. https://doi.org/10.1177/014616702237648

Stone, J., Chalabaev, A., & Harrison, C. K. (2012). The impact of stereotype threat on performance in sports. In M. Inzlicht & T. Schmader (Eds.), *Stereotype threat: Theory, process, and application* (pp. 217–230). Oxford University Press.

de Souza, D. B. L. S., & Ferreira, M. C. (2004). Propriedades Psicométricas da Escala de Auto-Estima Coletiva. In *Anais da XXXIV Reunião Anual da Sociedade Brasileira de Psicologia.*
Tajfel, H. (1982). *Social identity and intergroup relations*. Cambridge University Press.
Turner, J. C., Hogg, M. A., Oakes, P. J., Reicher, S. D., & Wetherell, M. S. (1987). *Rediscovering the social group: A self-categorization theory*. Basil Blackwell.
Tajfel, H., & Turner, J. C. (1986). The social identity theory of intergroup behavior. *Psychology of Intergroup Relations, 5*, 7–24.
Vargas, T. V., Dantas, R. A. S., & Gois, C. F. L. (2005). A autoestima de indivíduos que foram submetidos à cirurgia de revascularização do miocárdio. *Revista da Escola de Enfermagem da U.S.P., 39*(1). https://doi.org/10.1590/S0080-62342005000100003
Vasconcellos, V. C., & Neira, E. R. (2016). Escala de Expectativas de Carreira na Organização: Desenvolvimento e Evidências de Validade. *Psico-USF, 21*(2), 245–257. https://doi.org/10.1590/1413-82712016210203
Voelkl, K. E. (1996). Measuring students' identification with school. *Educational and Psychological Measurement, 56*(5), 760–770. https://doi.org/10.1177/0013164496056005003
von Hippel, C., Issa, M., Ma, R., & Stokes, A. (2011). Stereotype threat: Antecedents and consequences for working women. *European Journal of Social Psychology, 41*(2), 151–161. https://doi.org/10.1002/ejsp.749
von Hippel, C., Sekaquaptewa, D., & McFarlane, M. (2015). Stereotype threat among women in finance: Negative effects on identity, workplace well-being, and recruiting. *Psychology of Women Quarterly, 39*(3), 405–414. https://doi.org/10.1177/0361684315574501
Wade, T. J., & Bielitz, S. (2005). The differential effect of skin color on attractiveness, personality evaluations, and perceived life success of African Americans. *Journal of Black Psychology, 31*(3), 215–236. https://doi.org/10.1177/0095798405278341
Walton, G. M., Murphy, M. C., & Ryan, A. M. (2015). Stereotype threat in organizations: Implications for equity and performance. *Annual Review of Organizational Psychology and Organizational Behavior, 2*, 523–550. https://doi.org/10.1146/annurev-orgpsych-032414-111322
Wout, D., Shih, M. J., Jackson, J. S., & Sellers, R. M. (2009). Targets as perceivers: How people determine when they will be negatively stereotyped. *Journal of Personality and Social Psychology, 96*, 349–362. https://doi.org/10.1037/a0012880
Yeung, N. C. J., & von Hippel, C. (2008). Stereotype threat increases the likelihood that female drivers in a simulator run over jaywalkers. *Accident Analysis and Prevention, 40*(2), 667–674. https://doi.org/10.1016/j.aap.2007.09.003

Retrieval Practice as a Learning Strategy for Diverse Populations

Roberta Ekuni and Antônio Jaeger

Introduction

The 2018 PISA (Programme for International Student Assessment) showed that there are no Latin American countries in the top positions regarding reading, mathematics, and science (OECD, 2019). Although the reasons for this fact are undoubtedly complex and require solutions in multiple levels, current research on human learning shows that evidence-based education may have an important role toward the improvement of education in these countries (Slavin, 2020). As demonstrated by hundreds of studies conducted in the last two decades (for a review, see Dunlosky et al., 2013), one of the best strategies to promote long-lasting learning is called "retrieval practice," and consists in trying to remember materials that were previously studied. That is, when performing retrieval practice, students engage in attempts to remember the studied materials (i.e., to practice retrieval) instead of restudying the materials. Putting it simply, retrieval practice focuses on "taking information out of students' heads" instead of "inserting information into the students' heads" (Agarwal et al., 2013).

Recent studies demonstrate that this approach facilitates learning through a variety of processes, such as reducing forgetting (McDermott, 2021), promoting transfer to different test formats and inference questions (Pan & Rickard, 2018), improving reconsolidation of nonretrieved contents (Zaromb & Roediger, 2010), and minimizing retroactive interference (for a review, see Roediger et al., 2011). Interestingly, the benefits of retrieval practice are found even when individuals fail

R. Ekuni (✉)
Universidade Estadual do Norte do Paraná, Jacarezinho, Paraná, Brazil
e-mail: robertaekuni@uenp.edu.br

A. Jaeger
Universidade Federal de Minas Gerais, Belo Horizonte, Minas Gerais, Brazil

M. V. Alves et al. (eds.), *Cognitive Sciences and Education in Non-WEIRD Populations*, https://doi.org/10.1007/978-3-031-06908-6_15

to produce accurate memory responses during the practice of retrieval (Kornell et al., 2009).

Even though empirical evidence supporting memory retrieval as a learning strategy have been documented for over a century (for a meta-analysis, see Rowland, 2014), the great majority of the research on this topic was conducted on WEIRD (Western, Educated, Industrialized, Rich, and Democratic) individuals, which are not representative of most of the world's population (Henrich et al., 2010). In fact, a recent systematic review which included 37 studies on the effects of retrieval practice in classroom settings reported only three studies conducted on non-WEIRD populations (Agarwal et al., 2021), and just a few with children and nontypical populations. Considering the need to make psychological science more inclusive (Buchanan et al., 2020), it is essential that future research approach the question of whether retrieval practice is beneficial for diverse populations (Klein et al., 2018).

Retrieval Practice: An Overview

Retrieval practice can take many formats and can be conducted by both teachers and students. When students practice retrieval on their own, they might do this through activities such as self-testing or responding to quizzes, for example. Teachers, on the other hand, can administer a variety of practices that engage retrieval in their classes. For example, after giving a short power-point presentation, teachers can administer a short-answer quiz about the presented subject, which can be followed by corrective feedback (i.e., exposing students to the correct responses after the quiz is completed). Alternatively, teachers may ask students to write down all they remember from the presentation, and then expose the main contents of the presentation again, allowing students to verify whether they responded correctly and to add any missing information.

Even though retrieval practice may have different formats, all these formats have in common an active memory search for previously studied information (i.e., activities engaging retrieval). Thus, a comprehension of the theoretical basis of retrieval practice is essential for its effective application. Furthermore, retrieval practice is an evidence-based learning strategy, a perspective that naturally assumes that its theoretical foundations are based on empirical data. Thus, we briefly discuss here the empirical approach typically adopted to study retrieval practice, as well as two current important theoretical accounts.

In a typical retrieval practice experiment (see Rowland, 2014), participants first study the to-be-learned materials, which can be done by reading, watching lectures, watching videos, and so on. After this study phase is finished, participants are asked to engage in retrieval practice. That is, they are asked to retrieve the studied information through some kind of test, such as multiple-choice tests, short-answer questions, or fill-in-the-blank tasks. This condition is normally compared to a control condition in which participants most often restudy the material through rereading or a similar task (see Moreira, Pinto, et al., 2019; van den Broek et al., 2016). Finally, to verify whether retrieval practice produced greater memory retention than the "restudy" condition, a final criterion test is conducted. The final test also can take many formats, such as multiple choice or cued recall, for example. Also, it may be conducted after various intervals, and greater intervals (over a couple of days) typically produce greater benefits for retrieval practice over restudy.

A growing body of research has been shown that retrieving previously studied information is more beneficial for memory retention not only in comparison to rereading (Rowland, 2014), but also in comparison to note taking (Heitmann et al., 2018), to engaging in self-explanations (i.e., try to self-explain the learning material) (Larsen et al., 2013), to copying notes (Gingerich et al., 2014), and even in comparison to other learning strategies, such as conceptual mapping (Karpicke & Blunt, 2011). It is also worth noting that a range of different materials has been used to show its benefits, such as statistical content (Lyle & Crawford, 2011), biology textbook (Uner & Roediger, 2017), medical topics (Larsen et al., 2009), educational psychology course (Ekuni & Pompeia, 2020a), second language vocabulary (Candry et al., 2020), and orthography learning (Jones et al., 2015).

Thus, due to the compelling benefits of retrieval practice for long-term memory, a few theoretical accounts have been proposed to explain this phenomenon (see Rowland, 2014). A prominent account assumes that memory retrieval promotes the semantic elaboration of the retrieved content (elaborative retrieval hypothesis). That is, every time a given information is retrieved, concepts that are semantically related to that information are activated, which eventually produce alternative pathways through which that information can be later retrieved (Carpenter, 2009; McKenzie & Eichenbaum, 2011; Wing et al., 2013).

Another prominent hypothesis, the episodic context account (Karpicke, Lehman, & Aue, 2014), proposes that because the studied item and its respective episodic context are retrieved in a new context during retrieval practice, both the initial context and the new context become associated to the item. At the final test, the memory search will be facilitated, since only items associated with both contexts will be considered. It is worth noting that these accounts are not mutually exclusive, and that further research will be necessary to reach any consensus regarding the theoretical account for the retrieval practice effect (for further accounts, see Van Den Broek et al., 2016).

In sum, retrieval practice is an empirically well-defined and replicable phenomenon. Also, it has already generated well specified and falsifiable theoretical accounts. The question arises then as to whether the benefits for retrieval practice found in laboratory experiments are also found in real classroom environments, in a manner that can be adopted as a teaching or learning strategy.

Evidence from Classroom Studies

There are several differences between the application of retrieval practice in laboratory versus in real classroom settings. Such differences include, for instance, simpler materials in laboratory than in classroom studies, or different motivation to perform retrieval practice in the laboratory than in the classroom. In this sense, besides showing retrieval practice effects in laboratory settings, it is vital to show whether the benefits of retrieval practice are also found in real classroom environments (see Wooldridge et al., 2012). Below, we discuss some studies on retrieval practice in classroom environments, as well as the main issues surrounding such application.

In order to use retrieval practice at class, there are a variety of test formats that can be adopted, such as tests requiring students to choose the correct among a few alternative answers (i.e., multiple-choice tests; McDaniel et al., 2007), tests in which students judge whether a given statement is "true" or "false" (Uner et al., 2021), tests requiring that students produce a short-answer (McDaniel et al., 2007), and exercises requiring that students fill in blank spaces in a phrase or text (Hinze & Wiley, 2011).

As the research on retrieval practice applied to classroom environments advances, it becomes clearer that all these types of tests may be beneficial for students. For example, early studies suggested that short-answer questions lead to greater retention than multiple-choice tests (Kang et al., 2007; McDaniel et al., 2007). More recent meta-analyses reviewing both laboratory and classroom studies, however, suggested that this is not always the case (Adesope et al., 2017; Yang et al., 2020). In reality, both types of tests seem to be similarly beneficial for learning, suggesting that several factors should be considered when determining the type of retrieval practice to be adopted in class. These factors can be as general as the familiarity of the student with the to-be-learned materials, to as specific as the quality of the alternatives made on the multiple-choice questions (Little et al., 2012).

Another variable that is important to consider for the application of retrieval practice in classroom environments is the presence or absence of feedback. It should be noted that researches that aim to investigate the isolated benefit of retrieval should avoid including feedback in their research protocol (Karpicke et al., 2016). However, when it comes to practical applications, it is well demonstrated that the benefits of retrieval are boosted by feedback (Ekuni & Pompeia, 2020b; Yang et al., 2020). For instance, when students are exposed to incorrect information when answering multiple-choice questions, feedback is important to dismiss such information (Roediger & Butler, 2011). Furthermore, when a teacher offers elaborated feedback, the addition of more detailed information may favor the transfer of knowledge (Pan & Rickard, 2018). Thus, whenever possible, students should receive feedback regarding the accuracy of their remembrances during retrieval practice.

An additional issue that should be considered for the application of retrieval practice in classroom settings is whether retrieval practice should be conducted with open- versus closed book (Agarwal et al., 2008). That is, whether students will be allowed to search for the answers in their textbook during retrieval practice or not. Recent research suggests that retrieval practice is more advantageous when administered with closed book (Tauber et al., 2018). It is important to note, however, that the expectation of an open-book final test may hinder performance on the final test (Agarwal & Roediger, 2011), probably because students exert less effort when they know that a certain amount of information will be available at the final test. In any case, retrieval practice is beneficial for learning in classroom contexts with both open- or closed book, especially when there is an expectation for later closed-book tests.

In sum, a growing body of research has been showing that retrieval practice can be successfully applied as a learning strategy in real classroom environments. We discussed above some issues that should be considered for such application. A remaining question, however, is whether retrieval practice can be successfully applied to more diverse populations. We discuss this possibility below.

Retrieval Practice in Diverse Populations

As discussed above, retrieval practice is well demonstrated to benefit learning for college students from WEIRD countries (see Agarwal et al., 2021; Moreira, Pinto, et al., 2019). The question arises then as to whether retrieval practice is replicated in more diverse populations, such as children, students with special educational needs, individuals from non-WEIRD populations, and thus, can be used as an effective educational technique in these contexts (Klein et al., 2018). Therefore, this section has the goal of discussing the potential use of retrieval practice to improve learning in these populations.

Although research on retrieval practice in children is scarce, the evidence shows that it can improve children's learning both from preschool and elementary school children (Fazio & Marsh, 2019). That is, in children, retrieving information can improve the learning of a wide range of materials, from learning how to write new vocabulary words (Jones et al., 2015) to increasing the retention of book contents (Cornell et al., 1988).

Evidence that retrieval can boost learning for elementary school children was found in a study wherein children responded to multiple-choice questions after reading a short story (Brojde & Wise, 2008). In the final criterion test, children were asked to summarize and to answer to short open-ended questions about the story. The results were favorable for retrieval practice, since responding to multiple-choice questions after reading the story increased recall in both final tests. Unfortunately, however, there was no control condition in this study (i.e., multiple-choice questions were compared to "no activity"), a limitation that should be considered in future replications of these findings.

Interestingly, another study shows that although retrieval practice can be beneficial for elementary school children, they might need some guidance to properly perform retrieval practice (Karpicke et al., 2012). More specifically, in that study, children tried to freely recall and to make conceptual maps of contents read in texts from their school curricula. When they were left alone to perform these tasks their performances were extremely low, whereas when they received guidance (e.g., a conceptual map partially completed), they were successful in the use of retrieval practice. Thus, further than showing that retrieval practice can be beneficial for elementary school children, this study highlights the importance of adapting the practices tested with college students to elementary school children.

In a recent review on retrieval practice in children (Fazio & Marsh, 2019), only one study made with a non-WEIRD sample is cited. This study was conducted in Brazil and used a short text as learning material (Jaeger et al., 2014). After children read the text, they either received the full text again and reread it (i.e., restudy), or received the text with missing words, which they should remember from their first reading (i.e., retrieval practice). The final criterion test also consisted in remembering missing words from the text and was administered after an interval of seven days. The group that practiced retrieval showed a significantly greater performance on the final test than the group that just reread the materials. That is, the retrieval practice effect typically found in WEIRD populations was replicated on non-WEIRD children.

The studies above are examples of the benefits of retrieval practice for elementary school children. Because this population is younger, it is important to adapt the activities involving retrieval practice according to their needs. For example, it might be particularly beneficial to use cued-recall tests and to provide immediate feedback for children in this age range (Kliegl et al., 2018). Also, it is important to note that retrieval practice is not as beneficial when students remember less than 50% of the studied contents when performing the practice (Rowland, 2014). Thus, retrieval practice protocols that are beneficial for college students may not be beneficial for children, unless they are properly adapted.

Retrieval practice typically involves responding to questions after a study phase. A possibility that has been increasingly investigated, however, is whether questions made before the study phase are also beneficial for learning. This possibility was investigated in a study involving elementary school children in Brazil (de Lima & Jaeger, 2020). In this study, children first responded to "prequestions," then read a text and responded to "postquestions." Both pre- and postquestions consisted of completing the text with missing words, whereas the control condition involved only rereading the text again. Criterion final tests were conducted after an interval of seven days, and showed that while postquestions (i.e., retrieval practice) produced greater memory retention than prequestions, both pre- and postquestions produced greater memory retention than rereading. These findings strongly suggest that retrieval practice, as well as the administration of prequestions, enhances long-term learning in non-WEIRD elementary school children.

Studies involving middle and high school WEIRD children also showed benefits of retrieval practice. For instance, responding to a cued-recall test resulted in greater long-term retention of a second language vocabulary than just copying or reading the vocabulary (Candry et al., 2020). Also, middle and high school students learned more Science and History contents when multiple opportunities to retrieve those contents were administered (via multiple-choice or short-answer questions) than when they were asked to reread the contents (McDermott et al., 2014). Unfortunately, however, although studies on WEIRD populations indicate that retrieval practice benefits learning of middle and high school students, we did not find studies in a real classroom environment for non-WEIRD populations within this age range.

Regarding non-WEIRD undergraduate students, a study conducted in Brazil investigated whether the placement of retrieval practice, relative to study, affected memory retention (Ekuni & Pompeia, 2020a). Retrieval practice consisted of short-answer questions, and were placed either at the end of the class in which the content was taught, or at the beginning of the following class. Memory retention was measured through a multiple-choice test, which showed that placing retrieval practice at the end of the class was more beneficial for memory than placing retrieval practice at the beginning of the next class. Interestingly, retrieval practice at the end of the class was also more beneficial than a teacher-led review conducted during the same class or at the beginning of the next class.

Another study involving non-WEIRD college students was conducted in Pakistan. The study focused on medical school students, and investigated whether retrieval practice in the form of multiple-choice quizzes was beneficial for the learning of curricula contents (Ayyub & Mahboob, 2017). This study showed that students had a strikingly greater performance when quizzed after each class, during a period of 4 weeks, in comparison to when no quiz was administered, a result that support the view that retrieval practice is consistently beneficial for students from different cultures and populations.

In the same line, during a review nursing course conducted in Taiwan, students either participated of didactic instructions (i.e., teacher reviewed the studied contents) or performed retrieval practice in the form of multiple-choice questions followed by feedback (Tu et al., 2017). Importantly, based on a prior exam, the students were divided into three groups: low, intermediate, and high performance. Although low performance individuals were especially benefited by retrieval practice, intermediate performance individuals were particularly benefited by didactic instructions. This is an important finding, since it suggests that individual differences can influence the effects of retrieval practice.

Another study involving non-WEIRD college students was conducted in Turkey (Yiğit et al., 2014) and showed equivalent memory retention for chemistry contents learned through multiple-choice questions or through rereading. Furthermore, no differences between these learning strategies were found after intervals of one day or one week. Intriguingly, however, research involving WEIRD students enrolled in pharmacology or medicinal chemistry courses (Hernick, 2015), or just studying chemistry contents (Marsh et al., 2009) found significant benefits of retrieval practice for learning such contents.

Although the studies reviewed so far focused on neurotypical populations, below we review the few studies that focused on verifying the efficiency of retrieval practice for neurodivergent populations. In one of these studies, which was conducted in Brazil, children with diverse reading skills were equally benefited by remembering words previously studied in a text (i.e., retrieval practice) in comparison to rereading these same words (Moreira, Pinto, et al., 2019). That is, retrieval practice was equally beneficial for children with low and high reading skills.

Regarding special education students, another study conducted in Brazil found that individuals with Down syndrome had a numerical advantage for the retention of the contents read in a text when they engaged in a fill-in-the-blank test (i.e., retrieval practice), compared to when they simply reread the text (Starling et al., 2019). The limitation of this study, however, was that the sample was small. Thus, although the effect did not reach statistical significance, this study has pioneered the application of retrieval practice on special education students.

We found two studies focusing on assessing the benefits of retrieval practice for individuals with attention-deficit hyperactivity disorder (ADHD; Knouse et al., 2020; Knouse et al., 2016). These studies were conducted in the United States and involved exclusively college students. Both studies found that individuals with or without ADHD were equally benefited by retrieval practice to learn key-terms or words from texts or lists of categorized words. Thus, they show that retrieval practice is an efficient learning strategy for ADHD individuals, as it is for individuals without ADHD.

In sum, it is important that future studies consider whether the efficiency of retrieval practice is affected by cultural as well as individual differences. A growing awareness about this need is noticeable in the recent literature, as revealed by studies focusing on whether working memory capacity affects the benefits of retrieval practice, for example (Agarwal et al., 2016). Furthermore, once research begin focusing more intensely on individual differences and diversity, attention should also be paid to socioeconomic factors, considering for example, whether potential educational gaps cause by socioeconomic inequality can be reduced by efficient and low-cost learning strategies, such as retrieval practice.

Practical Implications for Students and Educators

As discussed above, evidence-based strategies such as retrieval practice can be useful to improve learning. However, research on the benefits of retrieval practice has focused mostly on WEIRD populations. Above, we reviewed the literature on non-WEIRD and other populations and showed that retrieval practices are also beneficial for those individuals. Therefore, below we provide some practical guidance for the use of retrieval practice by teachers and students from the populations discussed above.

As we stated before, one variable that influences the benefits of retrieval is the success of the first retrieval attempt (should be above 50%, Rowland, 2014). Considering children and students with special needs, it might be important to provide scaffolding strategies to have such high rate of initial retrieval success. For instance, fill-in-the-blank question with word stem for the target word (e.g., Jaeger et al., 2014; Moreira, da Silva Pinto, et al., 2019; Starling et al., 2019) or whatever guidance that allows students to retrieve more information during retrieval practice (Karpicke et al., 2012). Also, it is important to provide as much feedback as possible after retrieval, a factor consistently shown to enhance the long-term benefit of retrieval practice (Kliegl et al., 2018).

Regarding older students (from high school students to undergraduates), it is important to consider that students from both WEIRD and non-WEIRD populations prefer to study by rereading, highlighting, and summarizing, instead of engaging in retrieval practice (Karpicke et al., 2009; Persky & Hudson, 2016; Ekuni et al., 2020). Even though most teachers report giving tips about how students should study (Morehead et al., 2015), students report that they were not informed about which study strategies are more effective to promote long-lasting learning (Hartwig & Dunlosky, 2012; Karpicke et al., 2009; Hartwig & Dunlosky, 2012; Morehead et al., 2015). In any case, when students prefer strategies such as highlighting, rereading, or outlining, further than instructing them about more efficient learning strategies, they can be instructed on how to use those habitual strategies in a more effective way. For instance, if a given student prefers to highlight, she or he can be instructed to first read the section one time and only highlight the important parts (see Miyatsu et al., 2018). It is important to note, however, that strategies as highlighting and outlining are more beneficial for students that are able to identify the important contents of the materials (Dunlosky et al., 2013).

As for children, another recommendation to maximize the benefits of retrieval practice in older students is to provide tips or cues to increase retrieval performance during retrieval practice. For example, if a given question is too hard for a student, educators can provide specific guidance to enhance the chances that the student will respond correctly to the question. Such strategy can also increase the motivation of students to practice retrieval more often (Vaughn & Kornell, 2019). In the same line, enhancing the awareness of students about the efficiency of the available learning strategies, may lead them to choose more efficient strategies to study (Brown-Kramer, 2021).

Interestingly, practicing retrieval through high-stakes quizzes or tests may intuitively seem to enhance the motivation of student to learn, and therefore to produce greater long-term learning than practicing retrieval through low-stakes quizzes and tests. Research shows, however, that high- and low-stakes quizzes produce similar learning performances (Hinze & Rapp, 2014). In fact, low-stakes quizzes may produce even greater memory retention than high-stakes quizzes when longer intervals between the quizzes and the final criterion tests are adopted. In the light of the studies discussed above, to use effective learning techniques such as retrieval practice does not require only organization and preparation from teachers, but also student engagement and often more effort to study, factors that if not carefully considered may discourage the use of those study strategies in real learning contexts (Karpicke, 2009). Also, considering students from non-WEIRD countries that may be poverty-stricken or highly unequal, it is important to note that retrieval practice does not require large financial investments (Carpenter, 2012; Roediger & Pyc, 2012), not even more time spent in the classroom (Agarwal et al., 2012), so it is a promising strategy to implement even in schools with low financial resources.

Final Considerations

Considering the growing demand for evidence-based education strategies, we reviewed here the empirical basis and the applications of one of the most effective study strategies, namely, retrieval practice. We reviewed evidence for this practice from elementary school children to college students, and in more diverse populations, including the few non-WEIRD studies available in the literature. We also pointed practical ways that both students and educators could use this practice. As a final note, we highlight the need for future research to increase its focus on more diverse populations, and to verify whether retrieval practice is equally effective for a wider range of cultures, considering the specific needs of such distinct populations.

References

Adesope, O. O., Trevisan, D. A., & Sundararajan, N. (2017). Rethinking the use of tests: A meta-analysis of practice testing. *Review of Educational Research, 87*(3), 659–701. https://doi.org/10.3102/0034654316689306

Agarwal, P. K., Bain, P. M., & Chamberlain, R. W. (2012). The value of applied research: Retrieval practice improves classroom learning and recommendations from a teacher, a principal, and a scientist (I. of E. Sciences & W. U. in St. Louis, Eds.). *Educational Psychology Review, 24*, 437–448. https://doi.org/10.1007/s10648-012-9210-2

Agarwal, P. K., Finley, J. R., Rose, N. S., & Roediger, H. L. (2016). Benefits from retrieval practice are greater for students with lower working memory capacity. *Memory, 8211*(August), 1–8. https://doi.org/10.1080/09658211.2016.1220579

Agarwal, P. K., Karpicke, J. D., Kang, S. H. K., Roediger, H. L., & McDermott, K. B. (2008). Examining the testing effect with open- and closed-book tests. *Applied Cognitive Psychology, 22*(7), 861–876. https://doi.org/10.1002/acp.1391

Agarwal, P. K., Nunes, L. D., & Blunt, J. R. (2021). Retrieval practice consistently benefits student learning: a systematic review of applied research in schools and classrooms. *Educational Psychology Review*. https://doi.org/10.1007/s10648-021-09595-9

Agarwal, P. K., & Roediger, H. L. (2011). Expectancy of an open-book test decreases performance on a delayed closed-book test. *Memory, 19*(8), 836–852. https://doi.org/10.1080/09658211.2011.613840

Agarwal, P. K., Roediger, H. L., McDaniel, M. A., & McDermott, K. B. (2013). *How to use retrieval practice to improve learning*. Retrieved from Washington University in St. Louis website: www.retrievalpractice.org

Ayyub, A., & Mahboob, U. (2017). Effectiveness of Test-Enhanced Learning (TEL) in lectures for undergraduate medical students. *Pakistan Journal of Medical Sciences, 33*(6), 1339–1343. https://doi.org/10.12669/pjms.336.13358

Brojde, C. L., & Wise, B. W. (2008). An evaluation of the testing effect with third grade students. *Proceedings of the 3-Th Annual Meeting of the Cognitive Science Society*, 1362–1367.

Brown-Kramer, C. R. (2021). Improving students' study habits and course performance with a "learning how to learn" assignment. *Teaching of Psychology, 48*(1), 48–54. https://doi.org/10.1177/0098628320959926

Buchanan, N. T., Perez, M., Prinstein, M. J., & Thurston, I. B. (2020). *Upending racism in psychological science: Strategies to change how our science is conducted, reported, reviewed & disseminated.* Accepted AAM. https://doi.org/10.31234/osf.io/6nk4x

Candry, S., Decloedt, J., & Eyckmans, J. (2020). Comparing the merits of word writing and retrieval practice for L2 vocabulary learning. *System, 89*, 102206. https://doi.org/10.1016/j.system.2020.102206

Carpenter, S. K. (2009). Cue strength as a moderator of the testing effect: The benefits of elaborative retrieval. *Journal of Experimental Psychology: Learning, Memory, and Cognition, 35*(6), 1563–1569. https://doi.org/10.1037/a0017021

Carpenter, S. K. (2012). Improving student learning in low-maintenance and cost-effective ways. *Journal of Applied Research in Memory and Cognition, 3*(3), 121–123. https://doi.org/10.1016/j.jarmac.2014.07.004

Cornell, E. H., Sénéchll, M., & Broda, L. S. (1988). Recall of picture books by 3-year-old children: Testing and repetition effects in joint reading activities. *Journal of Educational Psychology, 80*(4), 537–542. https://doi.org/10.1037/0022-0663.80.4.537

Dunlosky, J., Rawson, K. A., Marsh, E. J., Nathan, M. J., & Willingham, D. T. (2013). Improving students' learning with effective learning techniques: Promising directions from cognitive and educational psychology. *Psychological Science in the Public Interest, Supplement, 14*(1), 4–58. https://doi.org/10.1177/1529100612453266

Ekuni, R., & Pompeia, S. (2020a). Improving retention by placing retrieval practice at the end of class: A naturalistic study. *Revista Latinoamericana de Psicología, 52*(3), 22–32. https://doi.org/10.14349/rlp.2020.v52.3

Ekuni, R., & Pompeia, S. (2020b). Retrieval practice: Which factors should educators pay attention to? *Psicologia Escolar e Educacional, 24*, 1–10. https://doi.org/10.1590/2175-35392020220284

Ekuni, R., Souza, B. M. N., Agarwal, P. K., & Pompeia, S. (2020). A conceptual replication of survey research on study strategies in a diverse, non-WEIRD student population. *Scholarship of Teaching and Learning in Psychology*, (online first publication),, 1–14. https://doi.org/10.1037/stl0000191

Fazio, L. K., & Marsh, E. J. (2019). Retrieval-based learning in children. *Current Directions in Psychological Science, 28*(2), 111–116. https://doi.org/10.1177/0963721418806673

Gingerich, K. J., Bugg, J. M., Doe, S. R., Rowland, C. A., Richards, T. L., Tompkins, S. A., & McDaniel, M. A. (2014). Active processing via write-to-learn assignments: Learning and retention benefits in introductory psychology. *Teaching of Psychology, 41*(4), 303–308. https://doi.org/10.1177/0098628314549701

Hartwig, M. K., & Dunlosky, J. (2012). Study strategies of college students: Are self-testing and scheduling related to achievement? *Psychonomic Bulletin & Review, 19*(1), 126–134. https://doi.org/10.3758/s13423-011-0181-y

Heitmann, S., Grund, A., Berthold, K., Fries, S., & Roelle, J. (2018). Testing is more desirable when it is adaptive and still desirable when compared to note-taking. *Frontiers in Psychology, 9*(DEC), 1–13. https://doi.org/10.3389/fpsyg.2018.02596

Henrich, J., Heine, S. J., & Norenzayan, A. (2010). The weirdest people in the world? *Behavioral and Brain Sciences, 33*(2), 1–23. https://doi.org/10.1017/S0140525X0999152X

Hernick, M. (2015). Test-enhanced learning in an immunology and infectious disease medicinal chemistry/pharmacology course. *American Journal of Pharmaceutical Education, 79*(7). https://doi.org/10.5688/ajpe79797

Hinze, S. R., & Rapp, D. N. (2014). Retrieval (sometimes) enhances learning: Performance pressure reduces the benefits of retrieval practice. *Applied Cognitive Psychology, 28*(4), 597–606. https://doi.org/10.1002/acp.3032

Hinze, S. R., & Wiley, J. (2011). Testing the limits of testing effects using completion tests. *Memory, 19*(3), 290–304. https://doi.org/10.1080/09658211.2011.560121

Jaeger, A., Eisenkraemer, R. E., & Stein, L. M. (2014). Test-enhanced learning in third-grade children. *Educational Psychology, 3410*(December), 1–9. https://doi.org/10.1080/01443410.2014.963030

Jones, A. C., Wardlow, L., Pan, S. C., Zepeda, C., Heyman, G. D., Dunlosky, J., & Rickard, T. C. (2015). Beyond the rainbow: Retrieval practice leads to better spelling than does rainbow writing. *Educational Psychology Review, 28*(2), 385–400. https://doi.org/10.1007/s10648-015-9330-6

Kang, S. H. K., McDermott, K. B., & Roediger, H. L. (2007). Test format and corrective feedback modify the effect of testing on long-term retention. *European Journal of Cognitive Psychology, 19*(4–5), 528–558. https://doi.org/10.1080/09541440601056620

Karpicke, J. D. (2009). Metacognitive control and strategy selection: Deciding to practice retrieval during learning. *Journal of Experimental Psychology: General, 138*(4), 469–486. https://doi.org/10.1037/a0017341

Karpicke, J. D., & Blunt, J. (2011). Retrieval practice produces more learning than elaborative studying with concept mapping. *Science, 331*(February), 772–775. https://doi.org/10.1126/science.1199327

Karpicke, J. D., Blunt, J. R., & Smith, M. A. (2016). Retrieval-based learning: Positive effects of retrieval practice in elementary school children. *Frontiers in Psychology, 7*(MAR), 1–9. https://doi.org/10.3389/fpsyg.2016.00350

Karpicke, J. D., Blunt, J. R., Smith, M. A., & Karpicke, S. S. (2012). Retrieval-based learning: The need for guided retrieval in elementary school children. *Journal of Applied Research in Memory and Cognition, 3*(3), 198–206. https://doi.org/10.1016/j.jarmac.2014.07.008

Karpicke, J. D., Lehman, M., & Aue, W. R. (2014). Retrieval-based learning: An episodic context account. In *Psychology of learning and motivation* (Vol. 61, pp. 237–284). Academic Press.

Karpicke, J. D., Butler, A. C., & Roediger, H. L. (2009). Metacognitive strategies in student learning: Do students practise retrieval when they study on their own? *Memory, 17*(4), 471–479. https://doi.org/10.1080/09658210802647009

Klein, R. A., Vianello, M., Hasselman, F., Adams, B. G., Adams, R. B., Alper, S., … Nosek, B. A. (2018). Many labs 2: Investigating variation in replicability across samples and settings. *Advances in Methods and Practices in Psychological Science, 1*(4), 443–490. https://doi.org/10.1177/2515245918810225

Kliegl, O., Abel, M., & Bäuml, K. H. T. (2018). A (preliminary) recipe for obtaining a testing effect in preschool children: Two critical ingredients. *Frontiers in Psychology, 9*(AUG), 1–8. https://doi.org/10.3389/fpsyg.2018.01446

Knouse, L. E., Rawson, K. A., & Dunlosky, J. (2020). How much do college students with ADHD benefit from retrieval practice when learning key-term definitions? *Learning and Instruction, 68*(July 2019), 101330. https://doi.org/10.1016/j.learninstruc.2020.101330

Knouse, L. E., Rawson, K. A., Vaughn, K. E., & Dunlosky, J. (2016). Does testing improve learning for college students with attention-deficit/hyperactivity disorder? *Clinical Psychological Science, 4*(1), 136–143. https://doi.org/10.1177/2167702614565175

Kornell, N., Hays, M. J., & Bjork, R. A. (2009). Unsuccessful retrieval attempts enhance subsequent learning. *Journal of Experimental Psychology: Learning, Memory, and Cognition, 35*(4), 989–998. https://doi.org/10.1037/a0015729

Larsen, D. P., Butler, A. C., & Roediger, H. L. (2009). Repeated testing improves long-term retention relative to repeated study: A randomised controlled trial. *Medical Education, 43*(12), 1174–1181. https://doi.org/10.1111/j.1365-2923.2009.03518.x

Larsen, D. P., Butler, A. C., & Roediger, H. L. (2013). Comparative effects of test-enhanced learning and self-explanation on long-term retention. *Medical Education, 47*(7), 674–682. https://doi.org/10.1111/medu.12141

de Lima, N. K., & Jaeger, A. (2020). The effects of prequestions versus postquestions on memory retention in children. *Journal of Applied Research in Memory and Cognition, 9*(4), 555–563. https://doi.org/10.1016/j.jarmac.2020.08.005

Little, J. L., Bjork, E. L., Bjork, R. A., & Angello, G. (2012). Multiple-choice tests exonerated, at least of some charges: Fostering test-induced learning and avoiding test-induced forgetting. *Psychological Science, 23*(11), 1337–1344. https://doi.org/10.1177/0956797612443370

Lyle, K. B., & Crawford, N. A. (2011). Retrieving essential material at the end of lectures improves performance on statistics exams. *Teaching of Psychology, 38*(2), 94–97. https://doi.org/10.1177/0098628311401587

Marsh, E. J., Agarwal, P. K., & Roediger, H. L. (2009). Memorial consequences of answering SAT II questions. *Journal of Experimental Psychology. Applied, 15*(1), 1–11. https://doi.org/10.1037/a0014721

McDaniel, M. A., Roediger, H. L., & McDermott, K. B. (2007). Generalizing test-enhanced learning from the laboratory to the classroom. *Psychonomic Bulletin & Review, 14*(2), 200–206. https://doi.org/10.3758/BF03194052

McDermott, K. B. (2021). Practicing retrieval facilitates learning. *Annual Review of Psychology, 72*, 609–633. https://doi.org/10.1146/annurev-psych-010419-051019

McDermott, K. B., Agarwal, P. K., D'Antonio, L., Roediger, H. L., & McDaniel, M. A. (2014). Both multiple-choice and short-answer quizzes enhance later exam performance in middle and high school classes. *Journal of Experimental Psychology: Applied, 20*(1), 3–21. https://doi.org/10.1037/xap0000004

McKenzie, S., & Eichenbaum, H. (2011). Consolidation and reconsolidation: Two lives of memories? *Neuron, 71*(2), 224–233. https://doi.org/10.1016/j.neuron.2011.06.037

Miyatsu, T., Nguyen, K., & McDaniel, M. A. (2018). Five popular study strategies: Their pitfalls and optimal implementations. *Perspectives on Psychological Science, 13*(3), 390–407. https://doi.org/10.1177/1745691617710510

Morehead, K., Rhodes, M. G., & DeLozier, S. (2015). Instructor and student knowledge of study strategies. *Memory, 24*(2), 257–271. https://doi.org/10.1080/09658211.2014.1001992

Moreira, B. F. T., da Silva Pinto, T. S., Justi, F. R. R., & Jaeger, A. (2019). Retrieval practice improves learning in children with diverse visual word recognition skills. *Memory, 27*(10), 1423–1437. https://doi.org/10.1080/09658211.2019.1668017

Moreira, B. F. T., Pinto, T. S. S., Starling, D. S. V., & Jaeger, A. (2019). Retrieval practice in classroom settings: A review of applied research. *Frontiers in Education, 4*(February). https://doi.org/10.3389/feduc.2019.00005

OECD. (2019). PISA 2018 results (volume II): Where all students can succeed. In *OECD Publishing*. Retrieved from https://www.oecd.org/pisa/publications/PISA2018_CN_IDN.pdf

Pan, S. C., & Rickard, T. C. (2018). Transfer of test-enhanced learning: Meta-analytic review and synthesis. *Psychological Bulletin, 144*(7), 710–756. https://doi.org/10.1037/bul0000151

Persky, A. M., & Hudson, S. L. (2016). A snapshot of student study strategies across a professional pharmacy curriculum: Are students using evidence-based practice? *Currents in Pharmacy Teaching and Learning, 8*(2), 141–147. https://doi.org/10.1016/j.cptl.2015.12.010

Roediger, H. L., & Butler, A. C. (2011). The critical role of retrieval practice in long-term retention. *Trends in Cognitive Sciences, 15*(1), 20–27. https://doi.org/10.1016/j.tics.2010.09.003

Roediger, H. L., Putnam, A. L., & Smith, M. A. (2011). Ten benefits of testing and their applications to educational practice. In J. Mestre & B. Ross (Eds.), *Psychology of learning and motivation: Cognition in education* (Vol. 55, pp. 1–36). https://doi.org/10.1016/B978-0-12-387691-1.00001-6

Roediger, H. L., & Pyc, M. A. (2012). Inexpensive techniques to improve education: Applying cognitive psychology to enhance educational practice. *Journal of Applied Research in Memory and Cognition, 1*(4), 242–248.

Rowland, C. A. (2014). The effect of testing versus restudy on retention: A meta-analytic review of the testing effect. *Psychological Bulletin, 140*(6), 1–32. https://doi.org/10.1037/a0037559

Slavin, R. E. (2020). How evidence-based reform will transform research and practice in education. *Educational Psychologist, 55*(1), 21–31. https://doi.org/10.1080/00461520.2019.1611432

Starling, D. S. V., Moreira, B. F. T., & Jaeger, A. (2019). Retrieval practice as a learning strategy for individuals with Down syndrome: A preliminary study. *Dementia e Neuropsychologia, 13*(1), 104–110. https://doi.org/10.1590/1980-57642018dn13-010012

Tauber, S. K., Witherby, A. E., Dunlosky, J., Rawson, K. A., Putnam, A. L., & Roediger, H. L. (2018). Does covert retrieval benefit learning of key-term definitions? *Journal of Applied Research in Memory and Cognition, 7*(1), 106–115. https://doi.org/10.1016/j.jarmac.2016.10.004

Tu, Y. C., Lin, Y. J., Lee, J. W., & Fan, L. W. (2017). Effects of didactic instruction and test-enhanced learning in a nursing review course. *Journal of Nursing Education, 56*(11), 683–687. https://doi.org/10.3928/01484834-20171020-09

Uner, O., & Roediger, H. L. (2017). The effect of question placement on learning from textbook chapters. *Journal of Applied Research in Memory and Cognition*. https://doi.org/10.1016/j.jarmac.2017.09.002

Uner, O., Tekin, E., & Roediger, H. L. (2021). True–false tests enhance retention relative to rereading. *Journal of Experimental Psychology. Applied*. https://doi.org/10.1037/xap0000363

Van Den Broek, G., Takashima, A., Wiklund-hörnqvist, C., Wirebring, L. K., Segers, E., Verhoeven, L., & Nyberg, L. (2016). Neurocognitive mechanisms of the " testing effect ": A review. *Trends in Neuroscience and Education, 5*(2), 52–66. https://doi.org/10.1016/j.tine.2016.05.001

Vaughn, K. E., & Kornell, N. (2019). How to activate students' natural desire to test themselves. *Cognitive Research: Principles and Implications, 4*(1). https://doi.org/10.1186/s41235-019-0187-y

Wing, E. A., Marsh, E. J., & Cabeza, R. (2013). Neural correlates of retrieval-based memory enhancement: An fMRI study of the testing effect. *Neuropsychologia, 51*(12), 2360–2370. https://doi.org/10.1016/j.neuropsychologia.2013.04.004

Wooldridge, C. L., Bugg, J. M., McDaniel, M. A., & Liu, Y. (2012). The testing effect with authentic educational materials: A cautionary note. *Journal of Applied Research in Memory and Cognition, 3*(3), 214–221. https://doi.org/10.1016/j.jarmac.2014.07.001

Yang, C., Luo, L., Vadillo, M. A., & Shanks, D. R. (2020). Testing (quizzing) boosts classroom learning : A systematic and meta-analytic review. In *Psychological bulletin* (Vol. 2).

Yiğit, E. A., Kıyıcı, F. B., & Çetinkaya, G. (2014). Evaluating the testing effect in the classroom: An effective way to retrieve learned information. *Eurasian Journal of Educational Research, 14*(54), 99–116. https://doi.org/10.14689/ejer.2014.54.6

Zaromb, F. M., & Roediger, H. L. (2010). The testing effect in free recall is associated with enhanced organizational processes. *Memory & Cognition, 38*(8), 995–1008. https://doi.org/10.3758/MC.38.8.995

Part IV
Educational Practices and Interventions

Consideration of Individual Differences in Cognitive Interventions for Children at Risk for Poverty

Carolina Soledad Fracchia, Federico Giovannetti, and Marcos Luis Pietto

Introduction

Development of Executive Processes

The study of child development is fundamental for creating and implementing actions that contribute to the design of interventions and policy sensitive to the particulars of such development (Farah, 2018; Lerner, 2018; Lipina & Segretin, 2015). In this context, it is relevant to consider that cognitive and socio-affective processes experience a high amount of integration during the first 5 years of life. In this period, the development of executive functions (EF) tends to accelerate, showing performance increases in tasks with EF demands. In turn, those functions have been identified as fundamental mechanisms for human cognition, associated with a variety of daily life aspects such as school readiness, academic achievement, and long-term health and educational outcomes (Doebel, 2020). They refer to high-level cognitive processes that direct, coordinate, or engage lower-level processes, enabling individuals to flexibly control and regulate their thoughts and actions during goal-directed behavior. However, the definition of EFs leads to important debates within developmental sciences. Garon et al. (2008) proposed an integrative model of EFs by which the development of such processes is based on the development and coordination of simpler processes (e.g., sustained attention and simple representation memory) (Diamond, 2013; Garon et al. 2008; Munakata et al., 2012; Posner & Rothbart, 2007). On the other hand, Friedman and Miyake consider EF as high-level cognitive processes that allow individuals to regulate their thoughts and actions during goal-oriented behavior (Banich, 2009; Diamond, 2013; Friedman & Miyake, 2017;

C. S. Fracchia (✉) · F. Giovannetti · M. L. Pietto
Unidad de Neurobiología Aplicada (UNA, CEMIC-CONICET), Buenos Aires, Argentina
e-mail: carolinafracchia@gmail.com

M. V. Alves et al. (eds.), *Cognitive Sciences and Education in Non-WEIRD Populations*, https://doi.org/10.1007/978-3-031-06908-6_16

Jurado & Rosselli, 2007). In turn, Diamond and Ling consider working memory, inhibitory control, and cognitive flexibility to be the three core processes implied in all EFs (Collins & Koechlin, 2012; Diamond & Ling, 2016; Miyake et al., 2000).

Cognitive control processes, executive attention, working memory, flexibility, and planning are within the most studied EF. In particular, attention is associated with a wide range of cognitive processes with common characteristics. Posner et al. (2013) consider attention as a system consisting of three specialized networks: alerting, orienting, and executive networks. In the present work, the latter will be considered of special interest given its implication in monitoring and conflict resolution mechanisms (Posner et al., 2013). As an example, in the Flanker task, the participant is told to control the interference produced by distracting stimuli placed next to the target stimulus.

Another process of interest for this work is working memory. This process refers to the ability for online manipulation of goal-relevant information (Diamond, 2013; Schelble et al., 2012). Working memory allows individuals to sustain a limited amount of information that could be required for solving tasks oriented to a particular goal (Bergman Nutley et al., 2011). This involves performing mental operations with information that is no longer perceptually available by, for example, remembering rules, reorganizing elements, and making plans, among others (Baddeley & Hitch, 1994; Smith & Jonides, 1999; D'Esposito & Postle, 2015).

On the other hand, cognitive flexibility could be defined as the ability to adjust behavior according to the challenges imposed by the context (Dajani & Uddin, 2015). Diamond (2013) defines flexibility as the ability to change perspectives and being as flexible as possible for adjusting to changing demands, priorities, and objectives. It should be mentioned that this is considered to be a late-emerging process because the ability to switch adequately between tasks involves specific working memory performance levels in order to maintain different rules in mind and specific inhibitory control performance levels in order to inhibit previous rules or interferences (Diamond, 2013; Müller & Kerns, 2015).

Finally, planning consists of the elaboration of strategies and execution of plans for solving a given problem (Shallice, 1982; Unterrainer & Owen, 2006). Programming actions involve assessing different possibilities, predicting consequences, and, by means of that, organizing and executing the desired plan (Kaller et al., 2011; McCormack & Atance, 2011).

Development of Children From Poor Contexts

In the last decades, there have flourished multiple studies showing associations between early experiences and cognitive, emotional, and social development in children. Such experiences influence development by means of a series of individual and contextual modulators. Among them, the literature covers factors related to income, educational levels, parental occupational levels, and dwelling conditions (Bradley & Corwyn, 2002; Gassman-Pines et al., 2015; Li et al., 2017;

Lipina, 2016; Roy & Raver, 2014; Spencer & Swanson, 2013; Walker et al., 2011; Yoshikawa et al., 2012). Depending on their characteristics, these factors can work as protective (e.g., absence of neurotoxic elements in the parenting environment, an available social network in the community) or risk factors (e.g., presence of neurotoxic elements in the parenting environment, absence of any social network in the community). Related to risk factors, several researchers from psychology and cognitive neurosciences have been dedicated to identifying the cognitive, behavioral, and emotional effects of living in poor contexts as well as the mediating factors (Lipina et al., 2017). Recent reviews (Johnson et al., 2016; Noble & Giebler, 2020) reported the vast range of studies showing such associations through different levels of analysis (i.e., neural and cognitive). In particular, EF development appears as a mediator between socioeconomic level and different measures such as academic performance in math and reading (Lawson et al., 2018).

In summary, poverty involves multiple factors and it has specific effects during infancy. Evidence shows that poverty-related factors such as overcrowding, sanitation, and parental occupation modulate the emergence and development of cognitive and executive processes. With regard to Latin-American evidence, a systematic revision (Segretin et al., 2016b) showed a significant variability in the type of studies, ages, processes, and criteria considered to measure poverty. In addition, only few studies assessed mediating factors.

Cognitive Training Interventions with Children From Poor Homes

Several researchers have been dedicated to the design of cognitive training interventions for promoting child development. Recently, different types of behavioral and cognitive interventions showed effectiveness in changing performance, structure, and neural activity in EF demanding tasks with children from poor homes. In Argentina, although there are still few studies on the issue, different intervention efforts showed changes in executive functions and self-regulatory processes (Goldin et al., 2014; Korzeniowski et al., 2017; Segretin et al., 2014).

However, evidence shows different patterns of results, which would be related to different aspects such as the size of the effects, the sustaining of those effects in the time, and the lack of far transfer to untrained processes (Diamond & Ling, 2016; Scionti et al., 2020). A significant aspect linked with such limitations could be related to the inter- and intra-individual differences in the studied samples. Several authors highlight the need to consider the differential effects of diverse samples in distinct groups of children (Diamond & Ling, 2016; Karbach et al., 2017; Katz et al., 2021; Scionti et al., 2020). In fact, cognitive processes usually involve qualitative and quantitative mechanisms that vary significantly between individuals (Espinet et al., 2012; Fracchia et al., 2016). This means that, potentially, not all children would be equally susceptible to their developmental contexts.

In consequence, performance trajectories within cognitive tasks could be modulated by such individual differences (Dunlosky et al., 2013; Raver et al., 2013).

The Relational Developmental Systems questions whether the interest of developmental sciences is the study of common characteristics shared by all humans or the study of what makes us unique. This involves considering developmental trajectories both as *multifinal* (processes that, even sharing a similar starting point, may show diverse trajectories through time) and *equifinal* (processes that end in the same point even after showing different trajectories) (McClelland et al., 2015). Such individual trajectories could be linked with factors such as age, initial performance, socioeconomic status, personality, and motivation among others (Katz et al., 2021). In addition, cognitive training intervention approaches could benefit from the consideration of other factors such as children's values, beliefs, experiences, and knowledge (Doebel, 2020).

In general, intervention designs in the field offer a particular program to all participants without considering individual variability among them (Katz et al., 2021). However, incorporating the study of individual factors to the design of cognitive training interventions could contribute to boosting efficacy levels by generating targeted approaches. This could imply considering the design of person-oriented approaches in order to better understand the influence of individual differences in child development. This perspective defies the idea of homogeneity among samples (von Eye et al., 2015) since this could mask intra- and inter-individual differences.

On behalf of what has been told, the present chapter aims to describe two examples related to different ways of analyzing individual differences in the performance of EF demanding tasks with preschool children from different socioeconomic contexts. In the first study, analysis of different population samples in Argentina were described. We aimed to identify performance profiles in tasks with EF demands by analyzing performance trajectories within them, with children coming from different sociodemographic contexts. In a second study, we describe a cognitive training intervention in which initial performance differences were considered for the design of specific individualized and adaptive intervention menus. This work included, in addition, the analysis of the impact of the intervention both at the behavioral and neural levels. For the sake of manuscript length and clarity, we omitted informing significance and effect size values in both studies. However, such data is available in the respective published articles.

Individual Profiles of Cognitive Task Resolution

A cross-sectional study was implemented to evaluate the association between poverty, environmental factors, and individual aspects of cognitive performance in 3–5-year-old Argentinian children. In particular, this section will focus on the analysis of specific aspects of four samples considered between 2005 and 2013 by a team of researchers from the UNA-CEMIC.[1] These programs were carried out in

Table 1 Sociodemographic characteristics of the samples from different Argentine regions

	M	SD	M	SD	M	SD
SES variables						
Parental education	7.00	2.70	5.80	2.70	8.30	2.70
Parental occupation	3.70	2.90	3.40	2.70	5.50	3.00
Overcrowding	6.10	2.30	6.20	2.40	7.30	2.20
Dwelling	9.70	2.20	9.50	2.60	10.90	1.70
UBN (% of the sample)	55.40%		64.00%		61.40%	

Note
UBN: Unsatisfied Basic Needs. A home was considered to have UBN if at least one of the following indicators was identified: inappropriate dwelling; overcrowding (more than 3 people per room); head of household with incomplete secondary schooling with more than four dependents; presence of school-aged children not attending any educational system; absence of waste discharge system in household and absence of water supply pipes inside the house or land
Parental education. 0: no education; 1: incomplete primary school; 3: primary school degree; 6: incomplete high school; 9: high school degree or incomplete technical studies; 10: complete technical degree or incomplete college studies; 12: college degree or higher
Parental occupation. 0: unoccupied; 1: unstable worker; 2: unskilled laborer; 4: skilled laborer; 6: small autonomous producer; 7: administrative employee; 8: technical professional; 10: small business owner; 11: professional; 12: company manager
Overcrowding. People per room. 0: 6 or more; 3: 4–6; 6: 2–4; 9: 1–2
Dwelling. Values between 3 and 12. The score was calculated considering the type of dwelling, type of floor, water access, type of bathroom, type of roof, external walls and dwelling property

different cities of Argentina (i.e., Salta City, province of Buenos Aires (PBA), and Autonomous City of Buenos Aires (CABA)) (Table 1).

In order to identify different cognitive tasks' resolution profiles and their modulation by individual and contextual factors, we constructed a variable that considered cognitive performance trajectories in each task and program. This variable was called *H index* and was aimed to analyze the trial-by-trial performance trajectories of children. For this, we calculated the performance mean according to the age group of belonging for each trial into each task, within the context of each program. On the one hand, this allowed the different programs to be analyzed separately and to involve different populations in terms of their sociodemographic characterization. On the other hand, the literature indicated that, as development progresses, performance in cognitive tasks related to spatial working memory, flexibility, and planning tends to be more efficient (Best & Miller, 2010; Boudreau et al., 2018; Garon et al., 2008; Grammer et al., 2014; Zelazo et al., 2004). In addition, previous studies from our group considering the same population samples as those considered in the present study found a similar trend showing increasing performance with age (Fracchia et al., 2020; Hermida, 2012; Segretin et al., 2014). Once we estimated the mean performance in each trial, we compared it with the raw score of each participant. Finally, a cumulative sum method (i.e., CUSUM) was applied to characterize the trajectories of children's performance in each task.

The mentioned method consisted in a sequential data analysis, from which the *H index* indicator was selected (Grunkemeier et al., 2009; Siddiqui & Izawa, 2015). This index results from the accumulated sum of correct answers and contemplates the mean of the population performance, applying the formula [H = Precedent + (Present -X population)], where *H* is equal to the sum of the performance in the previous trial (i.e., precedent) with the difference between the performance level in the present trial (i.e., present), minus the population mean for that same trial (i.e., X population). Given the *H index* values obtained in all trials, individual cognitive performance curves were generated for each task. Based on this variable, the individual trajectories were classified into different resolution profiles (Fig. 1). Specifically, the trial-by-trial H index of each participant was compared with the sample median for each trial. Finally, a classification of resolution profiles was generated according to the highest percentage of trials in which the H index was below, equal to, or above the sample median. Once we generated the two variables of interest, and with the aim of analyzing the level of prediction of individual (e.g., sex) and contextual factors (e.g., home conditions, economic resource) on the classification of the resolution profiles of each task, we implemented an analysis of the associations between the variables of interest (i.e., ordinal logistic regression).

Results and Conclusions

Program number 1 was carried out in Salta City and involved 382 children from 3 to 5 years of age. The assessment procedure included (1) the Corsi Blocks task (Corsi) (Pickering, 2001), which was used to tap spatial working memory processes; (2) the Tower of London Task (TOL), which was designed to assess planning processes (Kaller et al., 2012; Shallice, 1982); and (3) the FIST task, which assessed cognitive flexibility (Jacques & Zelazo, 2001) (Fig. 2).

Analyses showed that *sex, household conditions, family composition, positive events,* and, marginally, the *economic situation of the household* predicted the classification of children in working memory resolution profiles. On the other hand, the cognitive flexibility task's classification was predicted by the *sex* variable. In relation to the planning task, no significant associations were found.

Program number 2 was carried out in PBA and included 288 children between 3 and 5 years old. These participants were administered Corsi, TOL, and FIST. The results showed a significant effect of the *economic resources* factor on the working memory resolution profiles. However, for the cognitive flexibility and planning tasks, no significant effects were found.

Program number 3 was carried out in the CABA and included 296 boys and girls aged 4 and 5 years old, who were administered Corsi and TOL. In relation to the working memory task, results of the prediction analysis indicated significant effects for *household conditions, family composition,* and *household needs*. Regarding the planning task, no significant associations were found.

Finally, program number 4 was implemented in CABA and included 46 5-year-olds children who were administered Corsi. The results showed a significant association between the working memory resolution profiles and *overcrowding* and *years of preschool attendance*.

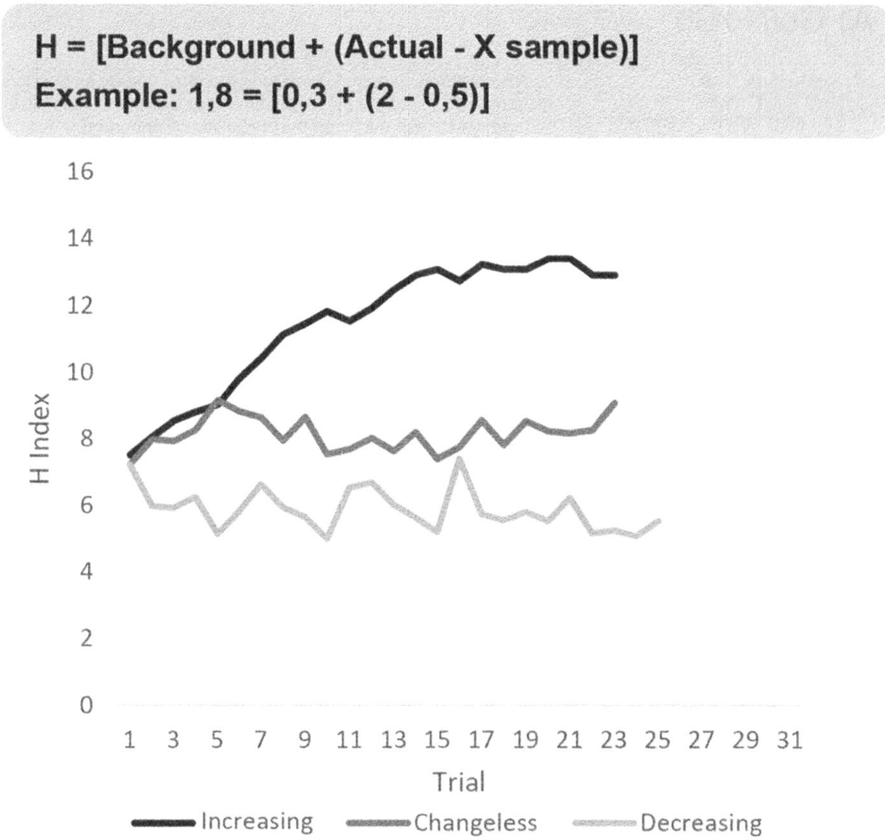

Fig. 1 Examples of the three different performance profiles in the TOL task

Working Memory

First, *girls* showed better performance profiles in the task. It should be mentioned that although there are a series of studies that show better performance of female participants, most evidence claims that male participants achieved better performances (Hyde, 2014; Voyer et al., 2017).

Second, *household conditions* also predicted the profile classification. In program 1, and contrary to the literature, children who achieved a better profile classification lived in more stressful contexts associated with worst housing conditions, building characteristics, and higher levels of overcrowding. In contrast, in program 3, the opposite association was verified, that is, more efficient profiles were associated with better housing conditions, lower levels of overcrowding, higher levels of access to parents' education and occupation, and lower scores in maternal anxiety and depression. In the literature, structural housing problems have been associated with the negative effects on children's cognitive development, as well as with the

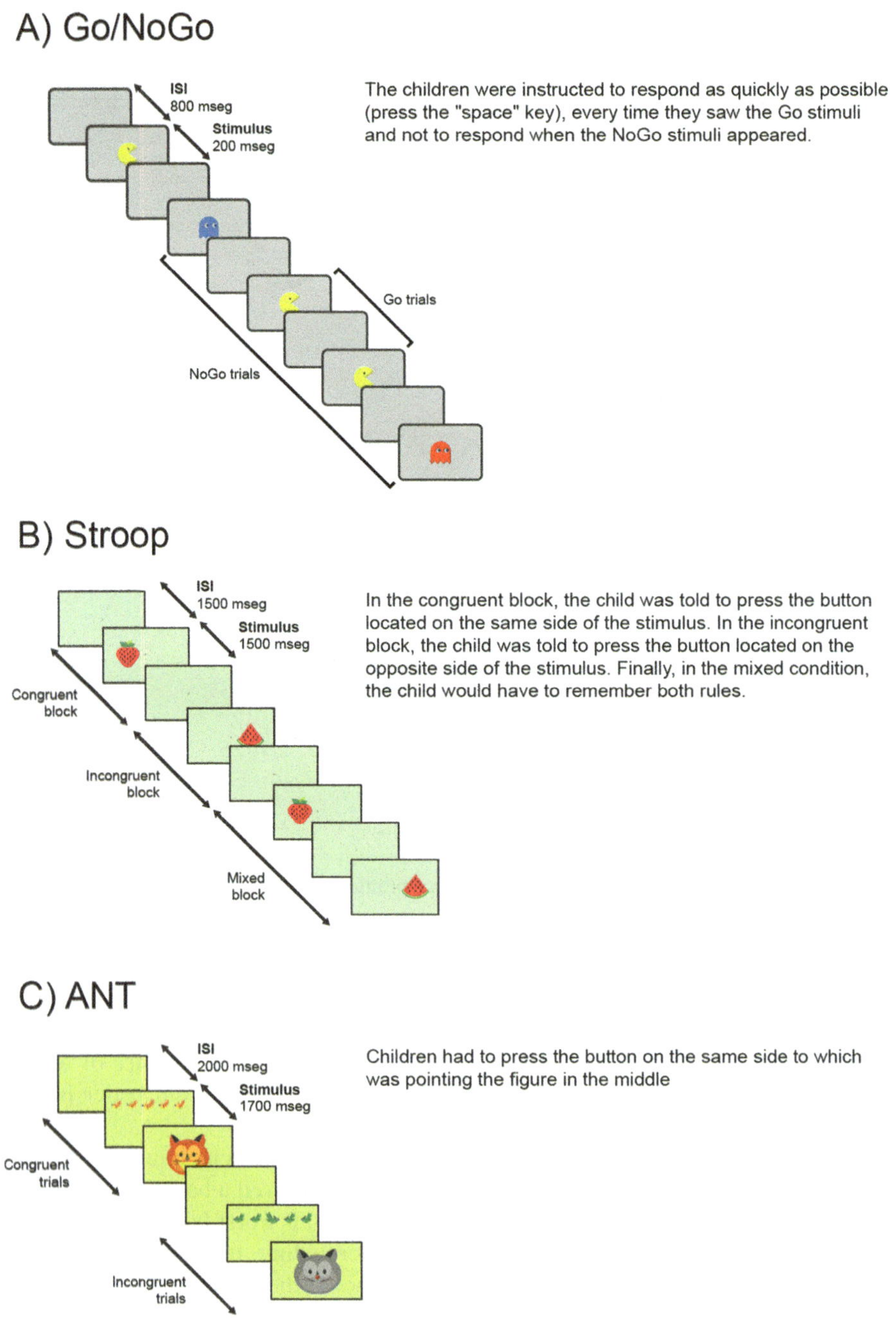

Fig. 2 Cognitive tasks involved in the intervention programs

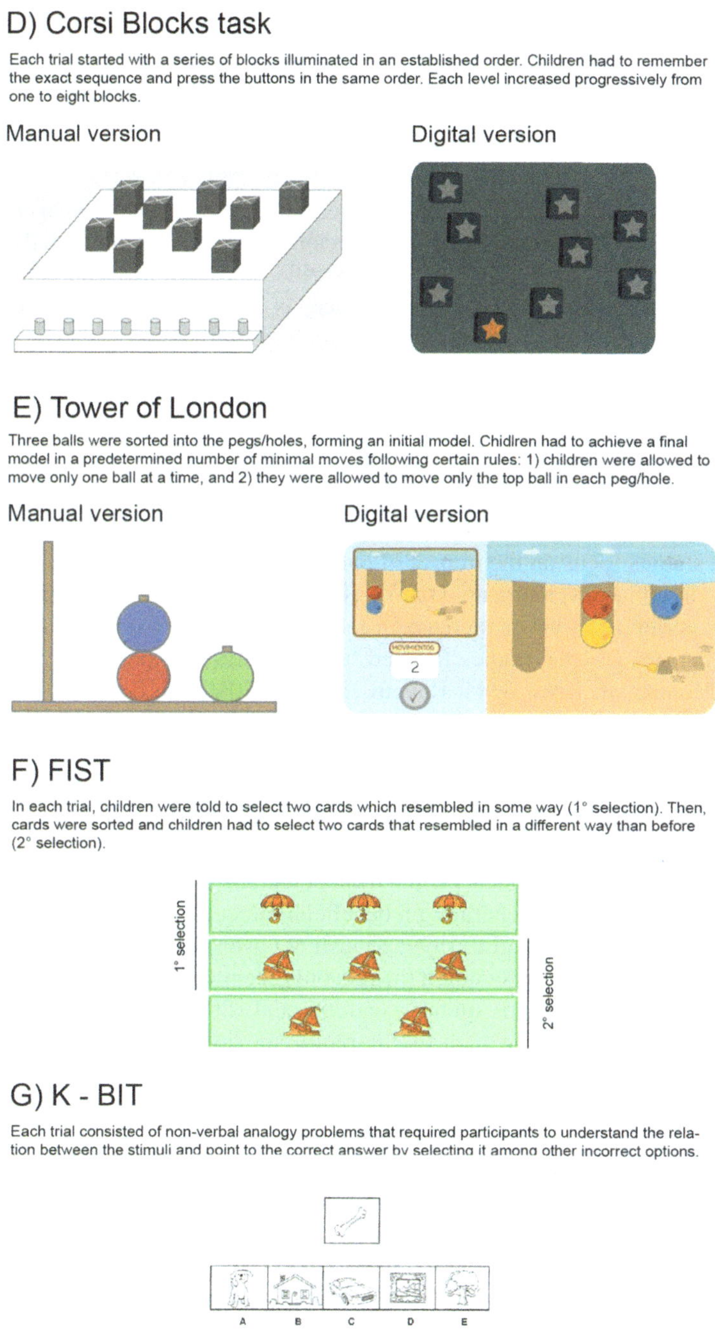

Fig. 2 (continued)

social and intellectual well-being of children (Bradley & Corwyn, 2002; Brooks-Gunn & Duncan, 1997; Flouri et al., 2014; Guo & Harris, 2000; Huston & Bentley, 2010; Lipina et al., 2013). It should be considered that (a) in program 1 the housing characteristics scores were high and optimal for child development and (b) that previous studies with the same population did not find such contradictory results (Segretin et al., 2014).

Third, the presence of *both parents at home* and better *parental occupation levels* were associated with more efficient profiles. In a previous study carried out with the same population sample, the same association was verified for flexibility and planning processes (Segretin et al., 2014). Although there is a controversy about the associations between family composition and children's cognitive performance, a possible explanation could advocate for an indirect association mediated by other relevant factors (e.g., belonging to a household single parent could be associated with lower income as a consequence of a single productive adult) (Santelices Álvarez et al., 2015; Hobcraft & Kiernan, 2010; Rhoades et al., 2011; Sarsour et al., 2011; Shonkoff & Philips, 2000).

In relation to *parental occupation,* literature shows two central aspects. On the one hand, certain studies focus on the impact of parent's schedules and the amount of working hours in child development. That is, children of mothers who worked full-time hours experienced a lower increase in learning (Han & Fox, 2011; Xie, 2016). On the other hand, other studies focused on the analysis of parental work types (Parcel & Menaghan, 1990, 1994; Repetti & Wang, 2010). In addition, less favorable work environments, as well as the presence of more work stressors, were associated with lower levels of parental commitment and sensitivity (Benjamin Goodman et al., 2008).

Fourth, and in contrast with the literature, results suggest an association between the perception of fewer *positive life events* and better resolution profiles. Several studies show that social support protects people from potentially negative influences of stressful events, in addition to generating a beneficial effect on health and well-being (Martín & Dávila, 2008; Mikulic, 1999; Klausli & Owen, 2009). However, it is important to consider that this association (a) was only identified within program 1 and (b) was not verified in previous studies (Segretin et al., 2014).

In the fifth place, and in relation with the literature, an association was found between greater *household economic resources* and more efficient profiles. Various studies support an association between children's cognitive performance and family economic resources (Johnson et al., 2016; Lawson, Hook & Farah, 2018; Lipina et al., 2013).

In the sixth place, a lower *presence of subsidies* was associated with more efficient profiles. Regarding the link between the perception of social and economic support and cognitive development, studies have shown an association between increased social support and better cognitive performance, which could be explained by a reduction in stress related to financial problems, as well as greater availability of resources and access to services (Morris et al., 2017; Offer et al., 2010; Park et al., 2010). It should be considered that most families in the sample received at least one subsidy, reducing its variability.

Finally, and contrary to the literature, it was found that a lower *number of years of preschool attendance* was associated with better profiles. In particular, early schooling in contexts of poverty has been associated with better academic achievement and positive results in other domains of life (Blair & Raver, 2014; Padilla & Ryan, 2018). In relation to this finding, it is important to consider that, on the one hand, most of the participants in the sample had between 2 and 3 years of previous school attendance, which implies a high attendance for preschoolers. On the other hand, it could be hypothesized that children living in more vulnerable households, and having poorer social support networks, were those in need for attend earlier to school.

Cognitive Flexibility

Girls showed better performance profiles. However, the literature supports an absence of sex differences in cognitive flexibility performances (Jacques & Zelazo, 2001; Lewis-Morrarty et al., 2012).

Summary of the Evidence

These results contribute to increasing knowledge about the relationships between individual performance trajectories in tasks with executive demands and individual and contextual factors. Consequently, they represent a contribution to the development of learning and psychosocial processes throughout the life course. In particular, the present section represents an advance in terms of the implementation of a new analysis method aimed at identifying individual differences modulating the trajectories of cognitive performance in comparison with a reference population (*H index*) (Jaeggi et al., 2011). This methodology differs from traditional ones, which usually include raw scores for the characterization of cognitive performance. Profiles could be considered an indicator of resolution strategies built by children from an active position in the tasks performance.

On the other hand, we identified that the association between sociodemographic variables and the resolution profiles varied between processes and children. In particular, there was a strong pattern of association between these variables in the working memory task, which could not be verified within flexibility and planning performances. Such associations could be interpreted from the conception of working memory as a more basic process (Garon et al., 2008). Likewise, another interesting contribution consisted in approaching poverty from a multidimensional and complex perspective. The previous contributes to deepening the knowledge about the complex and mediated relationships between poverty and child development. Therefore, it is necessary to continue exploring these issues, avoiding population-based generalizations.

Individualized and Adaptive Cognitive Intervention

Between the years 2017 and 2018, we conducted a cognitive training intervention in Argentina with 5-year-old children from poor homes (see Table 2 for details). Children were distributed into intervention (INT) and control (CON) groups and classified as high- and low-performing groups (HPG and LPG, respectively) based on the children's baseline performance in three tasks considered to be related to inhibitory control (Stroop-like), working memory (Corsi), and planning (TOL). Differential intervention menus were designed for each group. The intervention consisted of a total of 12 sessions in which children performed three different activities (four sessions for each activity). Then, the same cognitive tasks used in the pretest assessment were administered to each group in the posttest evaluation. The final sample was composed of 82 children (38 female; mean age, 5.32; SD, 0.34). For more information on attrition rate and reasons refer to Pietto and colleagues (2021). All the implemented procedures complied with national and international research norms and were reviewed and approved by the IRB (CEMIC, Protocols N°682, 961).

Assessment Procedures

All activities were administered in three 30-minute sessions in an ad hoc testing room from the kindergarten for the exclusive use of the team for research purposes and were performed on a tablet or a computer placed at a distance of about 30 cm from the child.

Table 2 Sociodemographic characteristics of the sample

	M	SD
SES variables		
Parental education	7.01	2.54
Parental occupation	3.91	1.74
Overcrowding	7.61	1.94
Dwelling	10.45	1.66
UBN (% of the sample)	95.46%	

Note

UBN: Unsatisfied Basic Needs. A home was considered to have UBN if at least one of the following indicators was identified: inappropriate dwelling; overcrowding (more than 3 people per room); head of household with incomplete secondary schooling with more than four dependents; presence of school-aged children not attending any educational system; absence of waste discharge system in household; absence of water supply pipes inside the house or land, and living unfavored neighborhoods

In each session, the following tasks were administered: Session 1, Go/NoGo task (Pietto et al., 2018); Session 2, Child-ANT (Rueda et al., 2004) and Corsi (Pickering, 2001); Session 3, Stroop-like (Hearts and Flowers, Davidson et al., 2006) and TOL (Kaller et al., 2012; Shallice, 1982); and Session 4, K-BIT (Kaufman, 1983) (Fig. 1). While all tasks were used to assess behavioral performance, the Go/NoGo task was also used to evaluate neural activity using a portable EEG (www.emotiv.com).

Intervention Procedures

After the pretest assessment, we implemented the cognitive intervention activities with two different intervention menus for each activity (one for each performance group, see below). Each intervention menu consisted of 12 weekly training sessions of 15 minutes each.

Intervention Group Activities

Participants in the training group performed three activities, each of them was administered in two consecutive sessions in the following order: *inhibitory control, working memory,* and *planning activities.* Then, this cycle was run one more time.

The activities followed two main principles: (1) the inclusion of new challenges with increasing difficulty and (2) repeated practice (Diamond, 2012). Each activity started at a particular difficulty level for each intervention menu and increased in difficulty with different criteria for each intervention menu. If children encountered difficulties in accomplishing the task, research assistants gave clues and encouraged them to continue practicing. Besides, if the task was misunderstood, the trainers repeated the instructions as many times as necessary.

The inhibitory control activity consisted of a Stroop-like task in which difficulty level differed between performance groups. In the first place, congruent and incongruent conditions for the LPG included a larger number of trials compared to the HPG. Further, in the congruent condition, the stimulus appeared more frequently on the opposite side of the button that the participant had to press. This was done to extend the time during which the LPG was exposed to inhibitory control trials before facing the mixed condition, giving a smoother difficulty slope to the LPG than to the HPG. Finally, the LPG had longer response times than the HPG (e.g., while the former had 9000 ms available to respond in the first level, the latter had only 4000 ms).

The working memory activity (Goldin et al., 2013) was designed to measure working memory for visual patterns and was based on the Self Ordered Pointing Task (SOPT) (Cragg & Nation, 2007; Luciana & Nelson, 1998). The difficulty increased as the children won more trials and differed between performance groups. The LPG was exposed to simpler items whereas the HPG was exposed to more complex ones. In addition, both groups were taught to use a mnemonic strategy at the highest levels. However, the intervention menu for the LPG was designed to give children more opportunities to practice this strategy than the HPG menu.

The planning activity was adapted from the Dog-Cat-Mouse task (Klahr, 1985). The difficulty of the task was given by the length of the required plan as well as the complexity of such a plan. For the HPG, difficulty levels included only the number of movements required for plan completion. For the LPG, difficulty parameters also included other parameters that smoothed the passage between trials.

Control Group Activities

Like the INT group, participants assigned to the CON group performed a series of activities throughout 12 sessions of 15 minutes. Unlike the INT group, participants in the CON group played three different games available for free download from Google Play Store (*Bubble shooter, Painting* game, *Dots)*. We did not expect these activities to generate training effects in participants.

Results and Conclusions

Inhibitory Control

In the case of the tasks with inhibitory control demands, (i.e., Stroop, ANT, and Go/NoGo) results showed intervention-related changes in performance and neural activity. In the behavioral level, only ANT showed changes in performance. In particular, within LPG, only LPG-INT showed an increase in performance, whereas within HPG both groups showed changes, with HPG-INT showing greater effect sizes. With regard to time variables, LPG groups showed increases in the *executive network*. Nonetheless, LPG-INT showed greater changes.

Attention was addressed as a measure of cognitive intervention impact in several studies with children coming from poor (Blair & Raver, 2014; Diamond et al., 2007; Goldin et al., 2014; Segretin et al., 2016b) and non-poor homes (Pozuelos et al., 2019; Rueda et al., 2012). With regard to the individual differences' analyses, results for ANT could be in line with studies including children from non-poor homes in tasks with executive demands (Guye et al., 2017; Karbach et al., 2017).

At the neural level, the N2, ERN, and event-related theta oscillations showed larger amplitudes in trials involving conflict (i.e., with NoGo stimuli and error response). This is consistent with prior studies using conflict-inducing tasks in preschool children (Abdul Rahman et al., 2017; Abundis-Gutiérrez et al., 2014; Canen & Brooker, 2017; Checa et al., 2014; Grammer et al., 2014; Lahat et al., 2010; Torpey et al., 2012). Training-related effects were also observed on both stimulus-locked and response-locked activity. Particularly, increased mid-frontal ΔN2 was observed in the INT-LPG and INT-HPG following training. Additionally, it was observed an increased mid-frontal ΔERN in the INT-HPG, which was accompanied by an effect in the power of the theta band (i.e., larger differentiation in power between correct and incorrect trials following training). The training effects observed on the stimulus-locked activity (i.e., ΔN2) are similar to those reported in previous intervention studies with preschool children (Liu et al., 2015; Pozuelos et al., 2019; Rueda et al., 2005; Rueda et al., 2012), whereas the pattern of brain activation observed in ΔERN resembles a more adult-like activation (Hoyniak, 2017; Lo, 2018).

Training-related effects in stimulus-locked (i.e., N2) and response-locked activity (i.e., ERN and theta oscillations) could reflect changes on conflict monitoring and/or control signal specification processes, at pre-response and post-response stages, respectively (Cavanagh & Frank, 2014; Downes et al., 2017; Iannaccone et al., 2015; Nigbur et al., 2011; Lo, 2018; Van Noordt et al., 2016; Yamanaka & Yamamoto, 2010; Yeung & Cohen, 2006).

Working Memory

HPG CON showed a decrease in the *total time* in the Corsi. Conversely, none of the groups showed changes in *performance*. It is relevant to mention that both *performance* and *total times* were not homogeneous between 2017 and 2018 samples.

Unlike recent literature on the impact of cognitive interventions in working memory demanding tasks with children from poor (Foy & Mann, 2014; Segretin et al., 2016a, and non-poor homes (Diamond & Ling, 2019), our results did not show changes in the Corsi. The lack of changes between phases could be related to the instruments chosen for assessment and training. It could be possible that it was not appropriate (1) for the participant's age or (2) as a measure of working memory transfer.

Planning

Both LPG showed increases in *performance*; conversely, within HPG, only HPG INT did. These results are similar to those reported in the literature, where performance in planning-demanding tasks showed intervention-related changes in children from poor (Korzeniowski et al., 2017; Segretin et al., 2014) and non-poor homes (Bergman Nutley et al., 2011; Goldin et al., 2013). Similar to our results, Korzeniowski and colleagues (2017) showed that children with higher initial performance values showed higher intervention-related changes in a planning task.

Fluid Reasoning

Trained (i.e., INT-HPG, INT-LPG) and untrained (i.e., CON-LPG) children showed an increase in the raw score of matrices subscale; however, children in the INT-HPG also showed an increase in the proportion of correct responses. The proportion of correct responses involved both raw score and errors, indicating that children in the INT-HPG not only achieved a greater number of correct trials but also committed fewer errors in the post-intervention stage. This result replicates previous evidence indicating that fluid reasoning abilities may be improved with training in basic (e.g., inhibitory control, working memory) and more complex cognitive skills (i.e., planning and problem-solving), which are suggested as significant contributors to fluid intelligence (Pozuelos et al., 2019; Rueda et al., 2005).

Apart from the apparent overlap between fluid and executive processing both at the construct and neural levels (Liu et al., 2015; Rueda, 2018), the pattern of results could be explained by the characteristics of the intervention. Specifically, evidence suggests that it could be easier for participants with higher cognitive performance levels to acquire and implement the strategies proposed by the trainers (Karbach et al., 2017).

Summary of the Evidence

Results showed that the effects of the intervention followed different patterns for each experimental and performance group depending on the cognitive task considered and the intervention menus implemented. Specifically, whereas children with lower initial performance showed higher changes in ANT, children with higher initial performance levels showed bigger changes in TOL and K-BIT.

As mentioned previously, only few intervention studies considered the effect of individual differences in initial performances. Some of them reported a *compensatory account* by which children with lower initial performances showed larger intervention-related changes (Karbach et al., 2017; Könen & Karbach, 2015). Others reported a *magnification account* by which children with higher initial performance values were the ones to show larger intervention-related changes (Karbach et al., 2017; Könen & Karbach, 2015; Lövdén et al., 2012). Authors refer that such effects could be attributed to different processes and intervention approaches. Our results show different effects depending on the processes considered. Analyses showed that children with lower levels of initial *performance* showed more changes in ANT but children with higher initial levels of *performance* showed greater changes in TOL and K-BIT. This pattern of results could be related to the distinction between simple and integrated processes. In particular, literature referred to the development of self-regulatory and EFs propose that those involved in attention, inhibitory control, and cognitive flexibility usually show an earlier development whereas complex processes such as planning and reasoning usually develop later on the basis of simpler processes (Garon et al. 2008; Nigg, 2017; Bailey & Jones, 2019).

In this way, intervention-related changes found in our analyses could be associated not only with the type of activities offered but also with the type of processes involved. Changes of *performance* in simpler processes such as the attentional were associated with the LPG while intervention-related changes in complex processes such as planning and reasoning were associated with the HPG. Future studies could follow this hypothesis in its theoretical, developmental, and empirical implications.

Our results in inhibitory control tasks show different effects depending on the level of analysis considered. Children in the INT-HPG showed changes only at the neural level, whereas children in the INT-LPG also showed changes at the behavioral level. The fact that for the INT-HPG, inhibitory control changes were verified only at the neural level could be due to the fact that children' functioning was already high, making it difficult to reach higher performance levels (Karbach et al., 2017). Contrary, given that children in the INT-LPG were performing at lower levels, they might have had higher possibility for improvements at two levels of analysis. Another hypothesis could relate the differential pattern of results with the specific training menus designed for the inhibitory control activity of each group. In this sense, children with better performance in inhibitory control may have needed more practice at the hardest levels of the inhibitory control activity. Despite the fact that complexity of the training activity designed for the high-performing group was higher than the one designed for the low-performing group, it could not have been complex enough to promote performance flexibility and consequently enable eventual training effects on behavior.

Finally, even though our results could point to the presence of both types of intervention-related effects for different processes, in the present study all children in the INT took part in activities involving all three processes. Moreover, all activities involved some level of scaffolding and/or strategy instruction in addition to the systematic training of the processes. In this way, the design of the study does not allow to tell conclusively if the differential effects of the intervention are related to the intervention groups, the intervention menus, or the targeted processes.

Discussion

Early development implies a group of basic cognitive processes such as executive functions, which make up a central aspect of child development. They are involved in the regulation of thoughts and behaviors engaged in daily activities. Research in the area also indicates that these cognitive processes encompass different qualitative and quantitative mechanisms that vary significantly between individuals (Espinet et al., 2012; Fracchia et al., 2016; Scruggs & Mastropieri, 2013). This suggests that boys and girls may not be equally susceptible to events in their developmental contexts, which also influences the diversity of strategies that they implement during the performance of cognitive tasks (Dunlosky et al., 2013). Furthermore, certain aspects of task performance and cognitive processing could be modulated by different individual (e.g., age, sex) and environmental (e.g., poverty during childhood) factors (Boone et al., 2018; Kaller et al., 2012; Lipina et al., 2013; Lipina & Segretin, 2015; Otto et al., 2013; Wolbers & Hegarty, 2010). That is, the protective and risk factors present in childhood could modulate the emergence and evolution of various aspects of these associated strategies and processes (D'angiulli et al., 2012; Evans & Kim, 2013; Fracchia et al., 2016; Sharkins et al., 2016).

In this context, it is important to note that the source of individual differences may vary due to various factors such as age, sociodemographic context, personality, and the processes addressed (Katz et al., 2016; Segretin et al., 2014). These results support the need for greater efforts to study how the individual characteristics of the participants interact with the intervention proposals. Potentially, this could involve expanding the universe of possible effects of the interventions beyond the effects of compensation and magnification. In other words, for different samples and intervention proposals, the changes generated by the interventions could show trajectories that do not coincide with these effects. In this sense, it is important to consider child development as a complex and multidimensional process (Spencer, Perone, & Buss, 2011). This implies considering, among other things, the modulation of the sociodemographic context in each aspect of child development. Therefore, Ecological Psychology supports a conception of development, considering it situated in a sociocultural context, which in turn modulates individual development (Barker, 1968; Bronfenbrenner, 1979). From this perspective, each person is significantly influenced by the interaction of a series of interdependent systems, throughout his life cycle. The environment is understood in both a dynamic and changing manner whose interrelations are necessary in order to understand development. Some of these levels influence the person more directly or closely, while others do it more indirectly or remotely (Bronfenbrenner & Ceci, 1994; Bronfenbrenner & Evans, 2000; Lerner, 2018). Considering the aforementioned, we can hypothesize that the different performance patterns found could be linked to the influence exerted from the diverse sociodemographic contexts in which we worked. Finally, in line with relational meta-theoretical approaches would imply thinking of performance trajectories as the result of dynamic and bidirectional interactions between individuals and various contexts (McClelland et al., 2015).

References

Abdul Rahman, A., Carroll, D. J., Espy, K. A., & Wiebe, S. A. (2017). Neural correlates of response inhibition in early childhood: Evidence from a Go/No-Go task. *Developmental Neuropsychology, 42*(5), 336–350.

Abundis-Gutiérrez, A., Checa, P., Castellanos, C., & Rueda, M. R. (2014). Electrophysiological correlates of attention networks in childhood and early adulthood. *Neuropsychologia, 57*, 78–92.

Baddeley, A. D., & Hitch, G. J. (1994). Developments in the concept of working memory. *Neuropsychology, 8*(4), 485.

Bailey, R., & Jones, S. M. (2019). An integrated model of regulation for applied settings. *Clinical Child and Family Psychology Review, 22*(1), 2–23. https://doi.org/10.1007/s10567-019-00288-y
Banich, M. T. (2009). Executive function: The search for an integrated account. *Current Directions in Psychological Science, 18*(2), 89–94.
Barker, R. G. (1968). *Ecological psychology: Concepts and methods for studying the environment of human behavior*. Stanford, CA: Stanford University Press.
Benjamin Goodman, W., Crouter, A. C., Lanza, S. T., & Cox, M. J. (2008). Paternal work characteristics and father-infant interactions in low-income, rural families. *Journal of Marriage and Family, 70*(3), 640–653.
Bergman Nutley, S., Söderqvist, S., Bryde, S., Thorell, L. B., Humphreys, K., & Klingberg, T. (2011). Gains in fluid intelligence after training non-verbal reasoning in 4-year-old children: A controlled, randomized study. *Developmental Science, 14*(3), 591–601.
Best, J. R., & Miller, P. H. (2010). A developmental perspective on executive function. *Child Development, 81*(6), 1641–1660.
Blair, C., & Raver, C. C. (2014). Closing the achievement gap through modification of neurocognitive and neuroendocrine function: Results from a cluster randomized controlled trial of an innovative approach to the education of children in kindergarten. *PLoS One, 9*(11), e112393.
Boone, A. P., Gong, X., & Hegarty, M. (2018). Sex differences in navigation strategy and efficiency. *Memory & Cognition, 46*(6), 909–922.
Boudreau, A. M., Dempsey, E. E., Smith, I. M., & Garon, N. (2018). A novel working memory task for preschoolers: Sensitivity to age differences from 3-5 years. *Child Neuropsychology, 24*(6), 799–822.
Bradley, R. H., & Corwyn, R. F. (2002). Socioeconomic status and child development. *Annual Review of Psychology, 53*(1), 371–399.
Bronfenbrenner, U. (1979). The ecology of human development. Harvard University Press.
Bronfenbrenner, U., & Ceci, S. J. (1994). Nature-nuture reconceptualized in developmental perspective: A bioecological model. *Psychological Review, 101*(4), 568.
Bronfenbrenner, U., & Evans, G. W. (2000). Developmental science in the 21st century: Emerging questions, theoretical models, research designs and empirical findings. *Social Development, 9*(1), 115–125.
Brooks-Gunn, J., & Duncan, G. J. (1997). The effects of poverty on children. *The Future of Children, 7*, 55–71.
Canen, M. J., & Brooker, R. J. (2017). ERN, theta power, and risk for anxiety problems in preschoolers. *Biological Psychology, 123*, 103–110.
Cavanagh, J. F., & Frank, M. J. (2014). Frontal theta as a mechanism for cognitive control. *Trends in Cognitive Sciences, 18*(8), 414–421.
Checa, P., Castellanos, M. C., Abundis-Gutiérrez, A., & Rosario Rueda, M. (2014). Development of neural mechanisms of conflict and error processing during childhood: Implications for self-regulation. *Frontiers in Psychology, 5*, 326.
Collins, A., & Koechlin, E. (2012). Reasoning, learning, and creativity: Frontal lobe function and human decision-making. *PLoS Biology, 10*(3), e1001293.
Cragg, L., & Nation, K. (2007). Self-ordered pointing as a test of working memory in typically developing children. *Memory, 15*(5), 526–535.
D'angiulli, A., Lipina, S. J., & Olesinska, A. (2012). Explicit and implicit issues in the developmental cognitive neuroscience of social inequality. *Frontiers in Human Neuroscience, 6*, 254.
D'Esposito, M., & Postle, B. R. (2015). The cognitive neuroscience of working memory. *Annual Review of Psychology, 66*, 115–142.
Dajani, D. R., & Uddin, L. Q. (2015). Demystifying cognitive flexibility: Implications for clinical and developmental neuroscience. *Trends in Neurosciences, 38*(9), 571–578.
Davidson, M. C., Amso, D., Anderson, L. C., & Diamond, A. (2006). Development of cognitive control and executive functions from 4 to 13 years: Evidence from manipulations of memory, inhibition, and task switching. *Neuropsychologia, 44*(11), 2037–2078.

Diamond, A., & Ling, D. S. (2016). Conclusions about interventions, programs, and approaches for improving executive functions that appear justified and those that, despite much hype, do not. *Developmental Cognitive Neuroscience, 18*, 34–48.

Diamond, A., & Ling, D. S. (2019). *Review of the evidence on, and fundamental questions about, efforts to improve executive functions, including working memory*. En Cognitive and Working Memory Training (Número January). https://doi.org/10.1093/oso/9780199974467.003.0008.

Diamond, A., Barnett, W. S., Thomas, J., & Munro, S. (2007). Preschool program improves cognitive control. *Science (New York, N.Y.), 318*(5855), 1387.

Diamond, A. (2012). Activities and programs that improve children's executive functions. *Current Directions in Psychological Science, 21*(5), 335–341.

Diamond, A. (2013). Executive functions. *Annual Review of Psychology, 64*, 135–168.

Doebel, S. (2020). Rethinking executive function and its development. *Perspectives on Psychological Science, 15*(4), 942–956.

Downes, M., Bathelt, J. O. E., & De Haan, M. (2017). Event-related potential measures of executive functioning from preschool to adolescence. *Developmental Medicine & Child Neurology, 59*(6), 581–590.

Dunlosky, J., Rawson, K. A., Marsh, E. J., Nathan, M. J., & Willingham, D. T. (2013). Improving students' learning with effective learning techniques: Promising directions from cognitive and educational psychology. *Psychological Science in the Public Interest, 14*(1), 4–58.

Espinet, S. D., Anderson, J. E., & Zelazo, P. D. (2012). Reflection training improves executive function in preschool-age children: Behavioral and neural effects. *Developmental Cognitive Neuroscience, 4*, 3–15. https://doi.org/10.1016/j.dcn.2012.11.009

Evans, G. W., & Kim, P. (2013). Childhood poverty, chronic stress, self-regulation, and coping. *Child Development Perspectives, 7*(1), 43–48.

Farah, M. J. (2018). Socioeconomic status and the brain: Prospects for neuroscience-informed policy. *Nature Reviews Neuroscience, 19*, 428.

Flouri, E., Midouhas, E., & Joshi, H. (2014). Family poverty and trajectories of children's emotional and behavioural problems: The moderating roles of self-regulation and verbal cognitive ability. *Journal of Abnormal Child Psychology, 42*(6), 1043–1056.

Foy, J. G., & Mann, V. A. (2014). Adaptive cognitive training enhances executive control and visuospatial and verbal working memory in beginning readers. *International Education Research, 2*(2), 19–43.

Fracchia, C. S., Giovannetti, F., Gili, J., Lopez-Rosenfeld, M., Hermida, M. J., Prats, L. M., … Lipina, S. J. (2016). Individuality and self-regulation in preschoolers. *Diskurs Kindheits-und Jugendforschung, 11*(4), 462–468.

Fracchia, C. S., Segretin, M. S., Hermida, M. J., Prats, L. M., & Lipina, S. J. (2020). Mediating role of poverty in the association between environmental factors and cognitive performance in preschoolers. *Revista Argentina de Ciencias del Comportamiento (RACC), 12*(2), 24–38.

Friedman, N. P., & Miyake, A. (2017). Unity and diversity of executive functions: Individual differences as a window on cognitive structure. *Cortex, 86*, 186–204.

Garon, N., Bryson, S. E., & Smith, I. M. (2008). Executive function in preschoolers: A review using an integrative framework. *Psychological Bulletin, 134*(1), 31.

Gassman-Pines, A., Gibson-Davis, C. M., & Ananat, E. O. (2015). How economic downturns affect children's development: An interdisciplinary perspective on pathways of influence. *Child Development Perspectives, 9*(4), 233–238.

Goldin, A. P., Hermida, M. J., Shalom, D. E., Costa, M. E., Lopez-Rosenfeld, M., Segretin, M. S., … Sigman, M. (2014). Far transfer to language and math of a short software-based gaming intervention. *Proceedings of the National Academy of Sciences, 111*(17), 6443–6448.

Goldin, A. P., Segretin, M. S., Hermida, M. J., Paz, L., Lipina, S. J., & Sigman, M. (2013). Training planning and working memory in third graders. *Mind, Brain, and Education, 7*(2), 136–146.

Grammer, J. K., Carrasco, M., Gehring, W. J., & Morrison, F. J. (2014). Age-related changes in error processing in young children: A school-based investigation. *Developmental Cognitive Neuroscience, 9*, 93–105.

Grunkemeier, G. L., Jin, R., & Wu, Y. (2009). Cumulative sum curves and their prediction limits. *The Annals of Thoracic Surgery, 87*(2), 361–364.
Guo, G., & Harris, K. M. (2000). The mechanisms mediating the effects of poverty on children's intellectual development. *Demography, 37*(4), 431–447.
Guye, S., De Simoni, C., & von Bastian, C. C. (2017). Do individual differences predict change in cognitive training performance? A Latent Growth Curve Modeling approach. *Journal of Cognitive Enhancement, 1*(4), 374–393. https://doi.org/10.1007/s41465-017-0049-9
Han, W. J., & Fox, L. E. (2011). Parental work schedules and children's cognitive trajectories. *Journal of Marriage and Family, 73*(5), 962–980.
Hermida, M. J. (2012). *Neurociencia cognitiva y educación: Dispositivos de enseñanza para niños y niñas en riesgo social por pobreza*. Tesis Doctoral (Universidad Nacional de San Luis).
Hobcraft, J. N., & Kiernan, K. E. (2010). *Predictive factors from age 3 and infancy for poor child outcomes at age 5 relating to children's development, behaviour and health: Evidence from the Millennium Cohort Study*. University of York.
Hoyniak, C. (2017). Changes in the NoGo N2 event-related potential component across childhood: A systematic review and meta-analysis. *Developmental Neuropsychology, 42*(1), 1–24.
Huston, A. C., & Bentley, A. C. (2010). Human development in societal context. *Annual Review of Psychology, 61*, 411–437.
Hyde, J. S. (2014). Gender similarities and differences. *Annual Review of Psychology, 65*, 373–398.
Iannaccone, R., Hauser, T. U., Staempfli, P., Walitza, S., Brandeis, D., & Brem, S. (2015). Conflict monitoring and error processing: New insights from simultaneous EEG–fMRI. *NeuroImage, 105*, 395–407.
Jacques, S., & Zelazo, P. D. (2001). The Flexible Item Selection Task (FIST): A measure of executive function in preschoolers. *Developmental Neuropsychology, 20*(3), 573–591.
Jaeggi, S. M., Buschkuehl, M., Jonides, J., & Shah, P. (2011). Short-and long-term benefits of cognitive training. *Proceedings of the National Academy of Sciences, 108*(25), 10081–10086.
Johnson, S. B., Riis, J. L., & Noble, K. G. (2016). State of the art review: Poverty and the developing brain. *Pediatrics, 137*(4), e20153075.
Jurado, M. B., & Rosselli, M. (2007). The elusive nature of executive functions: A review of our current understanding. *Neuropsychology Review, 17*(3), 213–233.
Kaller, C. P., Rahm, B., Spreer, J., Weiller, C., & Unterrainer, J. M. (2011). Dissociable contributions of left and right dorsolateral prefrontal cortex in planning. *Cerebral Cortex, 21*(2), 307–317.
Kaller, C. P., Unterrainer, J. M., & Stahl, C. (2012). Assessing planning ability with the Tower of London task: Psychometric properties of a structurally balanced problem set. *Psychological Assessment, 24*(1), 46.
Karbach, J., Könen, T., & Spengler, M. (2017). Who benefits the most? Individual differences in the transfer of executive control training across the lifespan. *Journal of Cognitive Enhancement, 1*(4), 394–405.
Katz, B., Jones, M. R., Shah, P., Buschkuehl, M., & Jaeggi, S. M. (2016). Individual differences and motivational effects. In *Cognitive training* (pp. 157–166). Springer, Cham.
Katz, B., Jones, M. R., Shah, P., Buschkuehl, M., & Jaeggi, S. M. (2021). Individual differences in cognitive training research. *Cognitive Training*, 107–123. https://doi.org/10.1007/978-3-030-39292-5_8
Kaufman, A. S. (1983). *Kaufman assessment battery for children*. Wiley.
Klahr, D. (1985). Solving problems with ambiguous subgoal ordering: Preschoolers' performance. *Child Development, 56*, 940–952.
Könen, T., & Karbach, J. (2015). The benefits of looking at intraindividual dynamics in cognitive training data. *Frontiers in Psychology, 6*, 615.
Korzeniowski, C., Ison, M. S., & Difabio, H. (2017). Group cognitive intervention targeted to the strengthening of executive functions in children at social risk. *International Journal of Psychological Research, 10*(2), 34–45.

Lahat, A., Todd, R., Mahy, C. E., Lau, K., & Zelazo, P. D. (2010). Neurophysiological correlates of executive function: A comparison of European-Canadian and Chinese-Canadian 5-year-olds. *Frontiers in Human Neuroscience, 3*, 72.

Lawson, G. M., Hook, C. J., & Farah, M. J. (2018). A meta-analysis of the relationship between socioeconomic status and executive function performance among children. *Developmental Science, 21*(2), e12529.

Lerner, R. M. (2018). *Concepts and theories of human development*. Routledge.

Lewis-Morrarty, E., Dozier, M., Bernard, K., Terracciano, S. M., & Moore, S. V. (2012). Cognitive flexibility and theory of mind outcomes among foster children: Preschool follow-up results of a randomized clinical trial. *Journal of Adolescent Health, 51*(2), S17–S22.

Li, M., Riis, J. L., Ghazarian, S. R., & Johnson, S. B. (2017). Income, family context and self-regulation in 5-year-old children. *Journal of Developmental and Behavioral Pediatrics: JDBP, 38*(2), 99.

Lipina, S. J., & Segretin, M. S. (2015). 6000 días más: evidencia neurocientífica acerca del impacto de la pobreza infantil. *Psicología Educativa, 21*(2), 107–116.

Lipina, S. J., Segretin, M. S., Hermida, M. J., Prats, L., Fracchia, C., López-Camelo, J., & Colombo, J. A. (2013). Linking childhood poverty and cognition: Individual and environmental predictors of non-verbal executive control in an Argentine sample. *Developmental Science, 16*, 697–707.

Lipina, S. J., Sigman, M., & Slezak, D. F. (2017). *Aportes de la ciencia cognitiva y la neurociencia al desarrollo de las TIC. In Pensar las TIC desde la ciencia cognitiva y la neurociencia* (pp. 15–32). Gedisa.

Lipina, S. J. (2016). *Pobre cerebro: los efectos de la pobreza sobre el desarrollo cognitivo y emocional, y lo que la neurociencia puede hacer para prevenirlos*. Siglo Veintiuno Editores.

Liu, Q., Zhu, X., Ziegler, A., & Shi, J. (2015). The effects of inhibitory control training for preschoolers on reasoning ability and neural activity. *Scientific Reports, 5*(1), 1–11.

Lo, S. L. (2018). A meta-analytic review of the event-related potentials (ERN and N2) in childhood and adolescence: Providing a developmental perspective on the conflict monitoring theory. *Developmental Review, 48*, 82–112.

Lövdén, M., Brehmer, Y., Li, S. C., & Lindenberger, U. (2012). Training-induced compensation versus magnification of individual differences in memory performance. *Frontiers in Human Neuroscience, 6*, 141.

Luciana, M., & Nelson, C. A. (1998). The functional emergence of prefrontally-guided working memory systems in four-to eight-year-old children. *Neuropsychologia, 36*(3), 273–293.

Martín, E., & Dávila, L. M. (2008). Redes de apoyo social y adaptación de los menores en acogimiento residencial. *Psicothema*, 229–235.

McClelland, M. M., John Geldhof, G., Cameron, C. E., & Wanless, S. B. (2015). Development and self-regulation. In *Handbook of child psychology and developmental science* (pp. 1–43).

McCormack, T., & Atance, C. M. (2011). Planning in young children: A review and synthesis. *Developmental Review, 31*(1), 1–31.

Mikulic, I. M. (1999). La evaluación Psicológica de los Recursos Sociales y los Estresores de vida. Aportes del Inventario LISRES. Psychological Assessment of Social Resources and Life Stressors. Contributions from the LISRES Inventory.

Miyake, A., Friedman, N. P., Emerson, M. J., Witzki, A. H., Howerter, A., & Wager, T. D. (2000). The unity and diversity of executive functions and their contributions to complex "frontal lobe" tasks: A latent variable analysis. *Cognitive Psychology, 41*(1), 49–100.

Morris, A. S., Robinson, L. R., Hays-Grudo, J., Claussen, A. H., Hartwig, S. A., & Treat, A. E. (2017). Targeting parenting in early childhood: A public health approach to improve outcomes for children living in poverty. *Child Development, 88*(2), 388–397.

Müller, U., & Kerns, K. (2015). The development of executive function. In L. S. Liben, U. Müller, & R. M. Lerner (Eds.), Handbook of child psychology and developmental science: Cognitive processes (pp. 571–623). John Wiley & Sons, Inc.. https://doi.org/10.1002/9781118963418.childpsy214

Munakata, Y., Snyder, H. R., & Chatham, C. H. (2012). Developing cognitive control: Three key transitions. *Current Directions in Psychological Science, 21*(2), 71–77.

Nigbur, R., Ivanova, G., & Stürmer, B. (2011). Theta power as a marker for cognitive interference. *Clinical Neurophysiology, 122*(11), 2185–2194.

Nigg, J. T. (2017). Annual Research Review: On the relations among self-regulation, self-control, executive functioning, effortful control, cognitive control, impulsivity, risk-taking, and inhibition for developmental psychopathology. *Journal of Child Psychology and Psychiatry, 58*(4), 361–383.

Noble, K. G., & Giebler, M. A. (2020). The neuroscience of socioeconomic inequality. *Current Opinion in Behavioral Sciences, 36*, 23–28.

Offer, S., Sambol, S., & Benjamin, O. (2010). Learning to negotiate network relations: Social support among working mothers living in poverty. *Community, Work & Family, 13*(4), 467–482.

Otto, A. R., Raio, C. M., Chiang, A., Phelps, E. A., & Daw, N. D. (2013). Working-memory capacity protects model-based learning from stress. *Proceedings of the National Academy of Sciences, 110*(52), 20941–20946.

Klausli, J. F., & Owen, M. T. (2009). Stable maternal cohabitation, couple relationship quality, and characteristics of the home environment in the child's first two years. *Journal of Family Psychology, 23*(1), 103.

Padilla, C. M., & Ryan, R. M. (2018). School readiness among children of Hispanic immigrants and their peers: The role of parental cognitive stimulation and early care and education. *Early Childhood Research Quarterly, 52*, 154.

Parcel, T. L., & Menaghan, E. G. (1990). Maternal working conditions and children's verbal facility: Studying the intergenerational transmission of inequality from mothers to young children. *Social Psychology Quarterly, 53*, 132–147.

Parcel, T. L., & Menaghan, E. G. (1994). Early parental work, family social capital, and early childhood outcomes. *American Journal of Sociology, 99*(4), 972–1009.

Park, S. Y., Glidden, L. M., & Shin, J. Y. (2010). Structural and functional aspects of social support for mothers of children with and without cognitive delays in Vietnam. *Journal of Applied Research in Intellectual Disabilities, 23*(1), 38–51.

Pickering, S. J. (2001). The development of visuo-spatial working memory. *Memory, 9*(4–6), 423–432.

Pietto, M. L., Giovannetti, F., Segretin, M. S., Belloli, L. M. L., Lopez-Rosenfeld, M., Goldin, A. P., … Lipina, S. J. (2018). Enhancement of inhibitory control in a sample of preschoolers from poor homes after cognitive training in a kindergarten setting: Cognitive and ERP evidence. *Trends in Neuroscience and Education, 13*, 34–42.

Posner, M. I., & Rothbart, M. K. (2007). Research on attention networks as a model for the integration of psychological science. *Annual Review of Psychology, 58*, 1–23.

Posner, M. I., Rothbart, M. K., & Rueda, M. R. (2013). Developing attention and selfregulation in infancy and childhood. In *Neural circuit development and function in the healthy and diseased brain* (pp. 395–411).

Pozuelos, J. P., Combita, L. M., Abundis, A., Paz-Alonso, P. M., Conejero, Á., Guerra, S., & Rueda, M. R. (2019). Metacognitive scaffolding boosts cognitive and neural benefits following executive attention training in children. *Developmental Science, 22*(2), e12756.

Raver, C. C., Blair, C., & Willoughby, M. (2013). Poverty as a predictor of 4-year-olds' executive function: New perspectives on models of differential susceptibility. *Developmental Psychology, 49*(2), 292.

Repetti, R., & Wang, S.-w. (2010). Parent employment and chaos in the family. In G. W. Evans & T. D. Wachs (Eds.), Chaos and its influence on children's development: An ecological perspective (pp. 191–208). American Psychological Association. https://doi.org/10.1037/12057-012

Rhoades, B. L., Greenberg, M. T., Lanza, S. T., & Blair, C. (2011). Demographic and familial predictors of early executive function development: Contribution of a person-centered perspective. *Journal of Experimental Child Psychology, 108*(3), 638–662.

Roy, A. L., & Raver, C. C. (2014). Are all risks equal? Early experiences of poverty-related risk and children's functioning. *Journal of Family Psychology, 28*(3), 391.
Rueda, M. R. (2018). Attention in the heart of intelligence. *Trends in Neuroscience and Education, 13*, 26–33.
Rueda, M. R., Checa, P., & Cómbita, L. M. (2012). Enhanced efficiency of the executive attention network after training in preschool children: Immediate changes and effects after two months. *Developmental Cognitive Neuroscience, 2*, S192–S204.
Rueda, M. R., Fan, J., McCandliss, B. D., Halparin, J. D., Gruber, D. B., Lercari, L. P., & Posner, M. I. (2004). Development of attentional networks in childhood. *Neuropsychologia, 42*(8), 1029–1040.
Rueda, M. R., Rothbart, M. K., McCandliss, B. D., Saccomanno, L., & Posner, M. I. (2005). Training, maturation, and genetic influences on the development of executive attention. *Proceedings of the National Academy of Sciences, 102*(41), 14931–14936.
Santelices Álvarez, M. P., Besoain, C., & Escobar, M. J. (2015). Parenthood, maternal work and child development: A Chilean study in children attending daycares in poverty context. *Universitas Psychologica, 14*(2), 675–684.
Sarsour, K., Sheridan, M., Jutte, D., Nuru-Jeter, A., Hinshaw, S., & Boyce, W. T. (2011). Family socioeconomic status and child executive functions: The roles of language, home environment, and single parenthood. *Journal of the International Neuropsychological Society, 17*(1), 120–132.
Schelble, J. L., Therriault, D. J., & Miller, M. D. (2012). Classifying retrieval strategies as a function of working memory. *Memory & Cognition, 40*(2), 218–230.
Scionti, N., Cavallero, M., Zogmaister, C., & Marzocchi, G. M. (2020). Is cognitive training effective for improving executive functions in preschoolers? A systematic review and meta-analysis. *Frontiers in Psychology, 10*, 2812.
Scruggs, T. E., & Mastropieri, M. A. (2013). Individual differences and learning challenges. *Theory Into Practice, 52*(sup1), 63–72.
Segretin, M. S., Hermida, M. J., Prats, L. M., Fracchia, C. S., Colombo, J. A., & Lipina, S. J. (2016a). Estimulación de procesos cognitivos en niños de 4 años: comparaciones entre formatos individual y grupal de intervención. *Revista Argentina de Ciencias del Comportamiento (RACC), 8*(3), 48–60.
Segretin, M. S., Hermida, M. J., Prats, L., Fracchia, C. S., Ruetti, E., & Lipina, S. J. (2016b). Childhood poverty and cognitive development in Latin America in the 21st century. *New Directions for Child and Adolescent Development, 2016*(152), 9–29. https://doi.org/10.1002/cad.20162
Segretin, M. S., Lipina, S. J., Hermida, M. J., Sheffield, T. D., Nelson, J. M., Espy, K. A., & Colombo, J. A. (2014). Predictors of cognitive enhancement after training in preschoolers from diverse socioeconomic backgrounds. *Frontiers in Psychology, 5*, 205.
Shallice, T. (1982). Specific impairments of planning. *Philosophical Transactions of the Royal Society of London B, Biological Sciences, 298*(1089), 199–209.
Sharkins, K. A., Newton, A. B., Albaiz, N. E. A., & Ernest, J. M. (2016). Preschool children's exposure to media, technology, and screen time: Perspectives of caregivers from three early childcare settings. *Early Childhood Education Journal, 44*(5), 437–444.
Shonkoff, J. P., & Philips, D. A. (2000). *Development from neurons to neighborhoods: The science of early childhood development rethinking nature and nurture. Development from neurons to neighborhoods: The science of early childhood development* (pp. 39–56). National Academies Press.
Siddiqui, K. M., & Izawa, J. I. (2015). Systematic methods for measuring outcomes: How they may be used to improve outcomes after radical cystectomy. *Arab Journal of Urology, 13*(2), 122–127.
Smith, E. E., & Jonides, J. (1999). Storage and executive processes in the frontal lobes. *Science, 283*(5408), 1657–1661.

Spencer, M. B., & Swanson, D. P. (2013). Opportunities and challenges to the development of healthy children and youth living in diverse communities. *Development and Psychopathology, 25*(4pt2), 1551–1566.

Spencer, J. P., Perone, S., & Buss, A. T. (2011). Twenty years and going strong: A dynamic systems revolution in motor and cognitive development. *Child Development Perspectives, 5*(4), 260–266.

Torpey, D. C., Hajcak, G., Kim, J., Kujawa, A., & Klein, D. N. (2012). Electrocortical and behavioral measures of response monitoring in young children during a Go/No-Go task. *Developmental Psychobiology, 54*(2), 139–150.

Unterrainer, J. M., & Owen, A. M. (2006). Planning and problem solving: From neuropsychology to functional neuroimaging. *Journal of Physiology-Paris, 99*(4–6), 308–317.

Van Noordt, S. J., Campopiano, A., & Segalowitz, S. J. (2016). A functional classification of medial frontal negativity ERPs: Theta oscillations and single subject effects. *Psychophysiology, 53*(9), 1317–1334.

von Eye, A., Bergman, L. R., & Hsieh, C.-A. (2015). Person-oriented methodological approaches. In W. F. Overton, P. C. M. Molenaar, & R. M. Lerner (Eds.), Handbook of child psychology and developmental science: Theory and method (pp. 789–841). John Wiley & Sons, Inc.. https://doi.org/10.1002/9781118963418.childpsy121

Voyer, D., Voyer, S. D., & Saint-Aubin, J. (2017). Sex differences in visual-spatial working memory: A meta-analysis. *Psychonomic Bulletin & Review, 24*(2), 307–334.

Walker, S. P., Wachs, T. D., Grantham-McGregor, S., Black, M. M., Nelson, C. A., Huffman, S. L., … Gardner, J. M. M. (2011). Inequality in early childhood: Risk and protective factors for early child development. *The Lancet, 378*(9799), 1325–1338.

Wolbers, T., & Hegarty, M. (2010). What determines our navigational abilities? *Trends in Cognitive Sciences, 14*(3), 138–146.

Xie, S. (2016). Dual-earning parents' work-family balance and time with children: the moderating effects of gender and age (Master's thesis).

Yamanaka, K., & Yamamoto, Y. (2010). Single-trial EEG power and phase dynamics associated with voluntary response inhibition. *Journal of Cognitive Neuroscience, 22*(4), 714–727.

Yeung, N., & Cohen, J. D. (2006). The impact of cognitive deficits on conflict monitoring: Predictable dissociations between the error-related negativity and N2. *Psychological Science, 17*(2), 164–171.

Yoshikawa, H., Aber, J. L., & Beardslee, W. R. (2012). The effects of poverty on the mental, emotional, and behavioral health of children and youth: Implications for prevention. *American Psychologist, 67*(4), 272.

Zelazo, P. D., Crack, F. I. M., & Booth, L. (2004). Executive function across the life span. *Acta Psychologica, 115*, 167–183.

Technology on Our Side: Using Technology for Transferring Cognitive Science to Education

Camila Zugarramurdi, Maria Soledad Assis, and Juan Valle-Lisboa

Introduction

The advent of digital technologies has transformed human life, and it has promised to transform education. Nevertheless, the contributions to education made by many relevant and widespread technologies, like radio, television, and lately PCs, tablets, and smartphones, are strongly dependent on the existence of a carefully designed curriculum respecting what is known about human learning (Hirsh-Pasek et al., 2015). There are many successful applications of different digital technologies to education, both to the academic and the socio-emotional realms (Borzekowski et al., 2019; Goldin et al., 2014; Judd & Klingberg, 2021; Mares & Pan, 2013; Nwaerondu & Thompson, 1987; Watkins & Dehaene, 2021; Watson & McIntyre, 2020; Wilson et al., 2006). As with the development of many other applications, a user-centered approach to Educational technologies is needed. In the educational milieu, the users are students, teachers, parents, and policymakers, and the technological applications need to be developed taking into account how students learn and what can be done

C. Zugarramurdi (✉) · M. S. Assis
Centro de Investigación Básica en Psicología, Facultad de Psicología, Universidad de la República, Montevideo, Uruguay

Centro Interdisciplinario de Cognición para la Enseñanza y el Aprendizaje, Universidad de la República, Montevideo, Uruguay
e-mail: camilazuga@gmail.com

J. Valle-Lisboa
Centro de Investigación Básica en Psicología, Facultad de Psicología, Universidad de la República, Montevideo, Uruguay

Centro Interdisciplinario de Cognición para la Enseñanza y el Aprendizaje, Universidad de la República, Montevideo, Uruguay

Sección Biofísica y Biología de Sistemas, Facultad de Ciencias, Universidad de la República, Montevideo, Uruguay

M. V. Alves et al. (eds.), *Cognitive Sciences and Education in Non-WEIRD Populations*, https://doi.org/10.1007/978-3-031-06908-6_17

about it, something that puts the topics within the realm of educational and cognitive psychology or, more broadly speaking, in the field of applied cognitive sciences known as "the learning sciences." There are many possible uses of technology in education according to which users the technologies are tailored to. For instance, the Plan Ceibal in Uruguay uses the adaptive math platform (PAM in Spanish) to help teachers create math-related activities for children and teenagers that can be used within the classrooms or at home, thanks to the distribution of laptops and tablets to all children attending public schools. These tools have been shown to be effective for part of the population (Perera & Aboal, 2018). Some other digital tools are designed to test and follow each child's progress and find the stumbling block of their learning process and can be used for both high stakes or low stakes testing. There are also the so-called educational applications that are mainly targeted to children and supposedly can help them develop some skills and competences.

Given that teacher training is expensive and takes time, the use of technology as a means to give access to education has been specially sought in third-world countries, leading sometimes to naive proposals (Arora, 2010; Mitra, 2003). This has heralded the need to have widely applicable theories of learning and cognition that can point to evidence-based educational interventions (Dehaene, 2020; Golinkoff & Hirsh-Pasek, 2016; Meltzoff et al., 2009). These theories have wide principles and specific details. The wide principles are sometimes presented as pillars (Dehaene, 2020) or easy-to-remember lists (Golinkoff & Hirsh-Pasek, 2016) but for some authors they revolve around attention, active learning, significant feedback, and consolidation, whereas others would also include social learning, meaningfulness, and joyful play. The relevance of these principles cannot be overstated, and we believe that these aspects are quite generally applicable, despite the fact that most of them have been studied in the WEIRD (Western, educated, industrialized, rich, and democratic) population (Henrich et al., 2010).

Nevertheless, there is another, more specific knowledge that particularly refers to learning specific topics like language, mathematics, science, history, or literature. These theories are in a sense more vulnerable to the WEIRDness objection. In particular, as it is discussed elsewhere in this volume, most of cognitive science and neuroscience theories have been developed based on studies of a biased sample of humanity, the so-called WEIRD population. Both in terms of ethnic and social diversity, Latin America, Africa, Asia, and Oceania have diverse populations, with different cultures and socioeconomic situations. All these dimensions affect the way information is processed and acquired, something that needs to be taken into account when designing educational interventions, especially those based on digital technologies.

In this chapter, we present three studies in order to illustrate the need to consider the specific cultural underpinnings of learning when devising and using cognitive-based technological applications. The ANS-*Puma* study presents a tablet-based intervention on mathematical skills, the *Lexiland* one presents a digital assessment of prereading skills, and the *Kalulu* study presents a tablet-based intervention on literacy acquisition (Fig. 1 and Table 1). In each study, we review how and why the particular aspects of diverse populations should be taken into account when designing the studies and interpreting the results. We believe that by considering these theoretical bases, the promise of technology for education can be fulfilled.

Fig. 1 Screenshots from the digital Apps presented in the studies. Left column: PUMA (Prueba Uruguaya de Matemáticas) presentation screen assessment (top) and Approximate Number System (ANS) assessment (bottom). Middle column: Lexiland presentation screen (top) and phonological awareness game within Lexiland (bottom). Right column: Kalulu presentation screen (top) and grapheme to phoneme mapping game within Kalulu (bottom)

Table 1 Reading, writing, and letter knowledge measured before (Pre) and after the intervention (Post)

Task	Group	Pre		N	Post		N
		Mean	SD		Mean	SD	
Writing	C	0.292	0.239	53	0.352	0.252	53
	T	0.278	0.218	43	0.347	0.231	44
Reading	C	0.455	0.371	54	0.551	0.387	52
	T	0.444	0.338	41	0.532	0.367	45
Letter knowledge	C	0.659	0.194	54	0.722	0.195	54
	T	0.659	0.205	43	0.702	0.223	43

Reading: read a list of 20 words, and we measure the proportion of correct words read. Writing: proportion of words read to the children that are correctly written. Letter knowledge is the proportion of letters from the alphabet that children can pronounce

The ANS-Puma Study: Interventions in Early Mathematics Across SES Levels

According to one current theory of mathematical cognition, there are several systems that embody mathematical knowledge and that scaffold the learning of formal mathematics (Mazzocco et al., 2011). In the field of arithmetics and early numerical competence, these so-called *core systems of knowledge* (Spelke & Kinzler, 2007) imply, at a minimum, the capacity to estimate in parallel small and even large numerosities, termed the approximate number system (ANS), and the capacity to follow a limited number of objects, the object tracking system (Le Corre & Carey, 2007; vanMarle et al., 2018). Substantial evidence suggests that both of these systems and maybe others as well are involved in the adult numerical and arithmetic capacity, although there is some controversy concerning the role these systems have in establishing the early numerical knowledge (Carey & Barner, 2019; Libertus et al., 2011; Szkudlarek & Brannon, 2017) and whether they can be stimulated to obtain better learning outcomes (Merkley et al., 2017; Park & Brannon, 2013; Valle-Lisboa et al., 2017; Wang et al., 2016).

How much of this theoretical framework applies to diverse populations, including those from lower SES countries? As it happens with many other school topics, children from lower SES social environments tend to perform poorer than their high or medium SES level peers, especially in symbolic mathematics (Sirin, 2005). If the theoretical framework applies to all populations, one would predict that those populations at greater risk of falling behind would benefit the most from interventions geared at enhancing any of the underlying systems and their connections, as enhancing the workings of these systems would increase the likelihood of developing a functioning symbolic system. With this idea in mind, we designed an intervention program to stimulate the precision of the ANS and evaluate its impact on early numerical and arithmetic competence, in a population of low and high-SES children.

The aim of the project was to determine whether a game-based, 2-week intervention on ANS could impact the learning of mathematics in children coming from low and high-SES schools. To this end we collected data from seven schools, three characterized as low-SES and four characterized as high-SES, based on a composite measure of unsatisfied basic needs of the households including maternal education, overcrowding, precarious housing, school attendance, and maternal education. A total of 454 first graders, with ages ranging from 6.42 to 8.76, took part in this study.

Both training and assessment were conducted within the classroom with the full class playing in parallel. Teachers were present but did not take part in the intervention. For the intervention program, children completed 6-minute ANS training sessions twice a week for 2 weeks for a total of four sessions. The ANS training task presented different numbers of dots on the two sides of the screen, and children had to decide which side of the screen presented a higher number of dots. Training was delivered through tablets, and data was recorded online.

The tasks we used were included in the tablet-based *Prueba Uruguaya de Matemática (PUMA),* which evaluates different mathematical skills related to number symbol knowledge. PUMA has shown a great internal validity and a strong correlation with TEMA-3 (Ginsburg & Baroody, 2003) assessments (López, De Leon, Maiche, this volume). All the tasks are presented as games. Besides formal mathematical tasks, we created games to evaluate *time discrimination,* i.e., the capacity to compare the duration of two time intervals and *digit span*, the amount of digits a child can retain in a short period of time. We also assessed *ANS acuity* by means of the standardized Panamath task (Halberda et al., 2008). The *ANS training task* consisted of asking children to decide which side of the screen presented a higher number of dots. Children completed four ANS training sessions with feedback, with a duration of 6 minutes each.

Unfortunately, data from the control group could not be completed, and thus the effects of the intervention program could not be effectively assessed. However, the collected data on the training across SES groups shed light on some of the inequities brought by SES levels, their underpinnings, and how training can affect them differentially.

Results showed that PUMA scores measuring formal math abilities differed significantly between the two SES levels, as did the ANS and time discrimination measured (for full statistical details see Valle-Lisboa et al., 2017). While, regrettably, this is not an unexpected result, we went further in trying to understand how the SES effect is mediated. In a previous work (Odic et al., 2016), we had found correlations between formal math performance and cognitive variables such as ANS and time discrimination that were partially independent of each other. This opens the question of whether ANS and/or time discrimination mediate the effect of SES on PUMA scores. Thus, we conducted a regression analysis to understand what variables explain the differences in PUMA scores. Crucially, when ANS acuity, digit span, and pre-/postassessment time were regressed together with time discrimination and SES, none time discrimination nor SES levels were significant predictors of PUMA scores. Thus, these cognitive factors somewhat mediate the effect of SES, and they change during the intervention.

The lack of a control group impedes the establishment of causal relationships. Nevertheless, by analyzing the results in the PUMA test with respect to the number of instances a child played the training games during the intervention, we evaluated whether there were any dose-related gains, as there was natural variation for external reasons (malfunctioning tablets, child absence due to sickness, etc.). The main result of this analysis showed that, besides pretest scores, doing more trials of the intervention games leads to a marginally significant increase of the PUMA score in the posttest ($p < 0.06$). Critically, the interaction of SES and number of training trials was significant, in particular with a stronger effect at the low-SES level.

Overall, the results suggest that the effects of SES can be mediated by basic cognitive abilities and that these abilities can be the target of interventions. Of course, the methodological limitations preclude us from making strong claims about the effectiveness of the interventions. In recent years there has been a set of studies questioning the effectiveness of ANS training, both empirical ones (Merkley et al., 2017) and theoretical ones (Carey & Barner, 2019). One response to these studies is to refine the theory (Spelke, 2017). In that new version of the theory, what is required to learn mathematics is not to strengthen or enhance the workings of the basic cognitive systems, but to link them to linguistic or other abilities to develop the full-fledged symbolic but grounded system of representations. In that sense, merely increasing the acuity of the ANS can only have a limited impact on symbolic mathematics. Consistent with this picture, in a recent study (Dillon et al., 2017) what worked best for enhancing mathematical knowledge was the intervention based on combining approximate and symbolic numerical representations. Likewise, several results show that the development of other notions linked to numerical knowledge, like the order representations, might have a greater impact furthering elementary mathematics learning (Lyons & Ansari, 2015). Likewise, we have recently shown that spatiotemporal conceptual knowledge predicts early mathematics above and beyond age, general vocabulary, and intelligence (Fitipalde et al., submitted). The differences in formal mathematics we found between children from low and high-SES schools can thus be traced to other more general abilities. There is a complementary interpretation of these results, though. Given the screening of the effect of SES on mathematics by ANS and digit span we reported, one might wonder whether the lack of effect some studies have found is a reflection of several of these studies being conducted in affluent countries. Then, the other response to these studies is that maybe in those societies the basic cognitive abilities are close to being saturated, but that there are other non-WEIRD countries where there might still be value in trying to promote the development of basic core cognitive abilities. This should be further studied in a preregistered manner in different populations.

The *Lexiland* Study: Learning to Read in a Transparent Orthography

Reading is a paramount skill for personal and professional development (Arnold et al., 2005). Reading, unlike language, is an acquired skill, and it requires teaching and practice for it to develop. Typically, this happens in formal school settings. The worldwide efforts to increase the number of children completing primary schooling have yielded good results, but this has not guaranteed that all of these children leave primary school reaching the expected minimum levels of proficiency in math and reading (Roser & Ortiz-Ospina, 2013). According to World Bank and UNESCO's reports, 53% of children in low- and middle-income countries fall within the category of what they call learning poverty (World Bank, 2019). Predictably, the COVID crisis has made the situation even worse, with estimates of the number of children in learning poverty escalating up to 70% (World Bank, 2021). A promising way to face this crisis is to use technology, for instance, by screening for reading difficulties early on, even before formal reading instruction begins, at times where interventions are most effective (Fletcher & Vaughn, 2009). This is possible since, for decades now, we have known that some of the prerequisites for reading acquisition start developing way before the beginning of formal reading instruction.

According to substantial evidence, one of the most relevant preliteracy skills needed for successful reading acquisition is phonological awareness, our ability to isolate and manipulate speech units (Melby-Lervåg et al., 2012). It is measured through tasks assessing the capacity to segment words into their constituent syllables or phonemes or those requiring blending phonemes to form a word, among others. In addition, children need to be familiar with the alphabetic principle, the concept that letters represent the sounds of oral language (Rayner et al., 2001). Next, children need to master letter knowledge (Foulin, 2005; Byrne & Fielding-Barnsley, 1989), that is, knowing, for each letter, which is its corresponding name or sound.

These three factors (phonological awareness, the alphabetic principle, and letter knowledge) underlie the first stage of reading acquisition, known as decoding (Hulme & Snowling, 2013). Decoding is the process by which children take each of the letters in a written word, convert it into its corresponding sound, and blend them together into a spoken word. Once they achieve this feat, they can access the word's meaning. The decoding process in its beginning is slow and effortful, thus children need to maintain each converted sound in their phonological short-term memory in order to be able to blend them into a word by the time they reach the last letter.

Two stages follow decoding in the reading acquisition process: fluency and comprehension (Nation, 2019). Fluency entails automatizing the decoding process, in order for it to be fast, precise, and effortless. It involves additional skills, mainly related to lexical access, pattern recognition, and visuo-attentional skills relevant for text processing (Kuhn et al., 2010). These factors are frequently measured through an experimental task known as rapid automatized naming (RAN). In this task, children are presented with a grid of objects, numbers, letters, or colors and are asked to name them as accurately and rapidly as possible. Taken together, phonological awareness, the alphabetic principle, letter knowledge, phonological short-term memory, and the RAN task—known as *preliteracy skills*—have been shown to be strong predictors of future reading outcomes. Since none of them requires actual reading, they can be assessed in the kindergarten years, before formal reading instruction begins.

These preliteracy skills can be used to predict and intervene promptly on those cases where it is needed (Fletcher & Vaughn, 2009). A limitation of most predictive studies is that they have been carried out either in the laboratory or by individual assessments in the school context, an approach which is not scalable, especially in under-resourced economies. The use of technology can aid in overcoming these limitations, by potentially allowing automatic autonomous assessment, carried out in groups in the school context.

With this scenario as a backdrop, in the *Lexiland* project, we aimed at developing a digital universal screener of reading difficulties targeted at children attending kindergarten. We developed an app—*Lexiland*—to assess preliteracy skills and validated its potential to predict future reading outcomes through a longitudinal follow-up study from kindergarten to second grade of 617 children from public schools in Montevideo, Uruguay. The sample was characterized as middle income (quintiles 3 and 4 of the income distribution). Children were assessed in groups of four to five at schools during school time. Since most schools did not have any extra room for assessment to take place, the assessment was done mainly in hallways or playgrounds. The research team provided the tablets with the loaded app, and two research assistants monitored data collection. All tasks' instructions and feedback (for practice trials) were prerecorded and delivered through the videogame through headphones, and research assistants were available for clarification upon demand. The initial sample of children was followed for 2 additional years, assessed at the end of first grade and the end of second grade, to evaluate their reading skills, including decoding, fluency, and comprehension.

The study resulted in both expected and unexpected findings. Among the expected, but yet novel, findings, we showed that by a brief—20 to 30 minute—assessment of only three preliteracy skills at the end of kindergarten (letter knowledge, phonological awareness, and short-term memory), we could predict future reading outcomes with high accuracy. Our model, tested through cross-validation, offered 90% sensitivity and almost 80% specificity (Zugarramurdi et al., 2022b). This high accuracy classification levels are equivalent to those reported in previous lab or individual assessment studies (Ozernov-Palchik & Gaab, 2016; Thompson et al., 2015) but remarkably in our case were produced by a group-based, in-school, autonomous assessment.

Among the unexpected findings, our study showed that the role that phonological awareness plays during reading acquisition is not as universal as previously thought. While we did find phonological awareness to predict future reading acquisition, we also found that, when considering it together with letter knowledge and verbal short-term memory, phonological awareness did not contribute any additional unique variance (Zugarramurdi et al., 2022a). These results contradict those reported for English-speaking children and add to the scarce evidence coming from non-English-based studies.

We believe that the differential contribution of phonological awareness is due to two factors. First, we found that phonological awareness at the phoneme level does not develop before reading acquisition in most children. It is possible that this is a result of the cultural environment, where the contribution of oral language skills, in general, and phonological awareness, in particular, to reading acquisition is not generally acknowledged. A large proportion of the evidence on reading acquisition comes from studies carried out in the UK, a country with a long established policy for teaching reading centered around phonics methods that assign great importance to the development of phonological awareness and letter knowledge (Machin et al., 2018; National Reading Panel, 2000) and therefore it is not surprising that phonological awareness is a strong unique predictor of future reading outcomes. On the contrary, in other countries, such as the USA or Uruguay, home environment and teaching practices are much more variable, and although the capacity to develop phonological awareness might be present, it is not consistently stimulated in schools or at home. Second, our results can be interpreted in terms of the orthographic specificities of the Spanish language. Orthographies can be characterized in terms of the consistency of the mapping between graphemes and phonemes, in a continuum from transparent to opaque. English, the most studied language in the reading acquisition literature (Share, 2008), locates in the opaque end of the continuum, where mappings are very irregular. That is, the same grapheme can have many very different associated phonemes, as exemplified by the different sounds of the letter /a/ in the words /cat/ and /table/. On the contrary, in Spanish (and similarly in Finnish or Italian), graphemes almost always map to the same sound, irrespective of the context in which it is embedded. In such transparent orthographies, learning to decode requires learning a few associations and mastering the ability to blend those sounds. Given that phonological awareness is not explicitly taught, when children are taught to read, they develop phonological awareness at the same time, which explains why it does not add unique variance to prediction. While some studies before ours have made similar points (see, e.g., Landerl et al., 2019), the bulk of the evidence has come from children learning to read in English, an opaque and atypical orthography, obscuring the diversity inherent to the reading acquisition process.

Thus, the Lexiland study highlights the great benefits of incorporating technology into education, while also providing novel infrequent evidence on the reading acquisition process from a less studied population and language.

The *Kalulu* Study: Lessons Learned From a Tablet-Based Intervention on Reading Instruction

The third study we review concerns the use of tablets in early reading instruction. The use of digital platforms is especially challenging during the end of kindergarten and beginning of first grade, where direct social interactions play a primary role in learning. In particular, learning to read depends on multiple factors, both individual and institutional, that need to be taken care of for an effective learning process. While educational institutions take all the factors into account, the traditional teaching/learning dynamic was not designed to rely upon technological tools (von Brevern, 2004).

The contributions from cognitive sciences can help in bridging this gap. There is substantial evidence about the initial steps of learning to read (see before and Cuetos, 2008; Defior et al., 2015; National Reading Panel, 2000; Perfetti & Bolger, 2004; Bradley & Bryant, 1983). In addition, several recent studies have specifically evaluated the effectiveness of incorporating technological tools for educational purposes in the domain of reading and literacy (Gaudreau et al., 2020; Kirkorian, 2018; Ojanen et al., 2015; Potier Watkins et al., 2020). These studies demonstrate that technology can help to optimize time management and organization of school work, as well as to design out effective interventions. Here, we describe the adaptation of a game—*Kalulu*—designed to enhance reading acquisition. *Kalulu*, originally developed in France, is based on accumulated evidence stemming from neuroscience research and its transfer to the field of education (Dehaene, 2011, 2014, 2015). It was designed to complement classroom teaching by using it either within the classroom or at home, and its effectiveness has recently been tested in a first experimental research study in France (Potier Watkins et al., 2020).

The game targets reading acquisition through the explicit and systematic teaching of letters and letter-sound correspondences. It presents a series of training screens, termed *lessons,* where each student receives a short explanation of the sounds and use of letters in different contexts. After this, several interactive games are presented that aim to strengthen the knowledge presented in each lesson. In this way, players learn the correspondence between graphemes and phonemes in their different forms (uppercase, lowercase, and italics) and then deepen this knowledge through the proposals of each interactive module.

Crucially, *Kalulu* presents letters in a systematic manner, from the simplest to the most complex, according to an automatic procedure that weights the consistency and frequency of different grapheme-phoneme mappings (see below). In addition to a carefully designed progression of stimuli, the game has the benefit of gradually adapting to the individual difficulties of each player, being slower when students need more time and increasing the pace for students who can handle more difficult games. Therefore, *Kalulu* offers a regulated, systematized teaching in ascending order of complexity to facilitate the decoding of words and stimulate the learning of reading in a simple and attractive way for the youngest children.

Since the progression of letters is such a critical factor and the complexity of grapheme-phoneme mappings is orthography-specific, a first step in the adaptation involved adapting the progression of stimuli to the characteristics of the Spanish orthography. For this purpose, an algorithm including a seq2seq transformer was used, which was specifically designed to map phonetic coding of words to their graphemic representation. The input to the transformer consisted of a list of 13,184 frequent Spanish words in children's books (Corral et al., 2009), with the syllabic structure and psycholinguistic properties obtained from ESPAL (Duchon et al., 2013). As a result, an ordered sequence of letters was obtained for Spanish, based on the consistency of the grapheme to phoneme mappings, from simpler to more complex ones (Potier Watkins et al., 2019), coupled with words and phrases that can be used for practice. The method produces a reasonable didactic progression. For instance, the vowels are taught in the first lessons due to their very consistent grapheme to phoneme mapping and their high frequency. On the contrary, the letter C is not taught in the first lessons because it is associated with two sounds in Spanish (/k/ and /s/); the letters X, K, and W are presented in the end due to the low frequency of use. Grapheme-phoneme mappings are presented in isolation and also in the context of frequent words.

In order to test the effectiveness of Kalulu, a randomized control study was carried out, with first grade children from five public low-SES schools in Montevideo. We choose to apply Kalulu in vulnerable contexts due to the fact that there is scientific evidence that indicates that the context can act as an influencing factor in children's reading performance (Diuk & Ferroni, 2012; Fish & Pinkerman, 2003). The study included 145 children (80 boys and 65 girls) ranging 6 to 7 years old, attending the last trimester of the school year. Children played Kalulu in tablets, individually, in the school context, in groups of five children playing in parallel, outside the classroom. Each child played twice a week, and completed at least ten 20-minute lessons. The training group played the Kalulu lessons described above, while the control group also played Kalulu but on a different path focused on mathematical skills (not described). Before and after the intervention, children were evaluated on preliteracy skills (letter knowledge and phonological awareness), verbal short-term memory, IQ, and word reading and writing, through *Lexiland* (see above and Zugarramurdi et al., 2022b). Groups were matched on phonological awareness and IQ, and no significant differences were found on letter knowledge, reading, or writing before study onset.

Results showed that all children improved in reading, writing, and letter knowledge between pre- and post-measures. However, no significant differences were found between treatment and control groups.

Since the initial lessons of Kalulu are highly focused on grapheme to phoneme mappings, and in trying to better understand the equivalent growth of literacy skills between groups, we further analyzed children's letter knowledge before the intervention began. The data showed that most children knew more than half of the letters from the alphabet. In particular, when looking at letters taught during the ten *Kalulu* lessons that children played, we found that, on average, children knew 75%, showing a ceiling effect. We believe these unexpected results stem from a discrepancy between curricula and in-classroom teaching practices. Explicitly teaching grapheme to phoneme mappings is not encouraged in the national curricula, and thus, before the study, we did not expect children to master *Kalulu's* lessons contents at the onset. However, as informally reported by the teachers taking part in the study, all participants were getting daily explicit training on grapheme-phoneme correspondences. These teachers had enough experience to know that such knowledge was essential for children to develop basic reading skills, irrespective of what the curricula suggest (or ignores). According to our previous experience with middle- and high-SES schools (Zugarramurdi et al., 2022a), it is mostly teachers from low-SES schools that insist on teaching letter knowledge and word synthesis, effectively implementing a form of synthetic phonics program. Moreover, since the intervention took place in the last trimester of the school year, children had had enough time to work on the lessons presented by the game in person with their teachers.

Surprisingly, even though children showed strong knowledge of letters and their sounds, performance in post-intervention word writing assessments showed results below 40% accuracy. Although teachers were able to work successfully on the teaching of letters, they reported that the times explicitly dedicated to the teaching of writing did not exceed 5 hours per week. In this sense, we believe that using tools such as *Kalulu* at the beginning of literacy acquisition would optimize the time devoted to the decoding process, allowing for an increase in the hours dedicated to teaching writing. This is due to the fact that *Kalulu* presents the information in an explicit and systematic way through the use of different interactive games that proved to be attractive for children, which further motivates their learning.

General Discussion and Conclusions

What the three studies presented here have in common is that they were based on the use of cognitive science to devise testing and intervention strategies using digital media. Both the mathematics intervention and the reading assessment were aimed at easing testing in school settings, one by deploying PUMA (see also López et al., this volume) and the other by creating Lexiland. The *Lexiland* study was indeed the development and validation of an assessment tool. The *Kalulu* study was intervention in the literacy acquisition process by the use of a principled way of presenting letters during the early steps of learning to read. Most of the testing tasks in this study were possible thanks to the presence of *Lexiland*, developed for the second study. *Lexiland* was also important because it was adapted to the characteristics of the local population. As explained before, one important aspect of early reading in Spanish, evidenced by the Lexiland study, which is not usually considered when dealing with more opaque writing systems, is that in a transparent orthography the most important early predictors of future reading are letter knowledge, memory, and rapid automatized naming, whereas phonological awareness, although it can be used to predict future learning to read, seems not to be a prerequisite. We believe that, in Spanish and other transparent orthographies, once letter knowledge and enough working memory are in place, learning to read is underway. To put it in another way, if a child knows the sounds of each letter and can hold them in memory, when they learn how to blend the different sounds, which is a phonological awareness task, they know how to decode. This contrasts with other orthographies, where letter knowledge, memory, and phonological awareness are not enough to be able to read most words, especially because what is measured as letter knowledge is not enough to define grapheme-phoneme mappings. Phonological awareness is thus a prerequisite in those orthographies. Although there are both transparent and opaque writing systems in WEIRD countries, most of the results in the mainstream literature come from English, whose writing is quite opaque. Thus, when we deployed our testing strategy, we remained open to the possibility that some predictors are more important than others. Notice that this does not question the main framework of the science of reading but parametrizes its application to the specifics of the culture, a conclusion that becomes especially evident when working outside WEIRD populations.

A similar parametrization applies to the PUMA-ANS study results. In effect, although the lack of a proper control group limits the interpretability of the results, the fact that there is a dose-response between training intensity and PUMA scores, and that most of the SES effects are screened by basic cognitive skills, points to an important explanatory value of these cognitive variables. What is noteworthy about the role of these variables is that most researchers in the field nowadays maintain that enhancing ANS is not enough (Dillon et al., 2017; Hyde et al., 2021) or even necessary (Carey & Barner, 2019; Lyons et al., 2014) to master elementary mathematical knowledge. This is interpreted as a result of limited association between symbolic and nonsymbolic representations of magnitudes. How can our results be interpreted? One possibility is that the range of situations that are explored in WEIRD countries is limited and so the development of ANS follows more or less the same pathway in those countries, where differences in ANS accuracy have little or no importance. In contrast, in non-WEIRD countries where poverty is more extreme, there might be cases of children with little opportunity to develop the approximate number system, making stimulation much more important. This is consistent with our results showing that children from low-SES schools gain more in math when trained with ANS stimulation. The result contrasts somewhat with those of a large study conducted in India (Dillon et al., 2017) where the authors showed no effect of non-symbolic math training in school scores. Notice that our intervention was short and that these authors use school mathematics as their dependent variable, which is a more distal measure compared to our PUMA scores. Also, the characteristics of the India population need not be equal to our Latin American schools; not all non-WEIRD countries are created equal. An attempt to partially replicate some positive findings of short interventions stimulating ANS on math in the same setting have failed to find significant results (Díaz-Simón, 2021) but these studies used extremely short interventions. The theoretical interpretations of the different sets of results remains open. Clearly, more work needs to be done.

Lastly, when we think about interventions in low-SES settings, it is important to consider which other social agents are at play in the field. Although in our country the reading curriculum is organized around an eclectic proposal mixing several aspects of a whole language approach with some phonological awareness ideas, and the inspector system strongly suggests that teachers follow these guidelines, several teachers and principals of the Kalulu study said they were following more closely a phonics-based curriculum, because, if not, children would not learn. In essence, our experience from higher SES schools was not applicable to this low-SES setting. In this sense, we only managed to repeat what they had already learned a few months before. Our next step is clear, we should start earlier and make the game more adaptive to what children already know, and not to what we expect them to know, not because they have some particular gift, but because we are not the only agents in the field.

To conclude, our three studies evidence three types of peculiarities that need to be taken into account when trying to apply cognitive science to non-WEIRD populations. In the first place, as in Lexiland, there are linguistic and more general cultural differences between different countries that need to be taken into account. Second, there might be situations where interventions in some countries use variables (like ANS accuracy) that are not elastic enough because in that particular place they are already saturated, but that in other places they might need to be enhanced; this should be analyzed in each particular case. Lastly, when applying interventions, especially in settings that receive attention from many different agents, what these other agents are doing needs to be put into the picture and dealt with appropriately. If these provisions are met, the possibility of using cognitive science to base the design of digitally based interventions is a principled way to use technology in early education settings and a particularly valuable assets in low-income countries where human and financial resources are scarce.

References

Arnold, E. M., Goldston, D. B., Walsh, A. K., Reboussin, B. A., Daniel, S. S., Hickman, E., & Wood, F. B. (2005). Severity of emotional and behavioral problems among poor and typical readers. *Journal of Abnormal Child Psychology, 33*(2), 205–217. https://doi.org/10.1007/s10802-005-1828-9

Arora, P. (2010). Hope-in-the-Wall? A digital promise for free learning. *British Journal of Educational Technology, 41*(5), 689–702. https://doi.org/10.1111/j.1467-8535.2010.01078.x

Borzekowski, D. L. G., Singpurwalla, D., Mehrotra, D., & Howard, D. (2019). The impact of Galli Galli Sim Sim on Indian preschoolers. *Journal of Applied Developmental Psychology, 64*, 101054. https://doi.org/10.1016/j.appdev.2019.101054

Bradley, L., & Bryant, P. E. (1983). Categorizing sounds and learning to read: A causal connection. *Nature, 301*(5899), 419–421.

Byrne, B., & Fielding-Barnsley, R. (1989). Phonemic awareness and letter knowledge in the child's acquisition of the alphabetic principle. *Journal of Educational Psychology, 81*(3), 313–321. https://doi.org/10.1037/0022-0663.81.3.313

Carey, S., & Barner, D. (2019). Ontogenetic origins of human integer representations. *Trends in Cognitive Sciences, 23*(10), 823–835. https://doi.org/10.1016/j.tics.2019.07.004

Corral, S., Ferrero, M., & Goikoetxea, E. (2009). LEXIN: a lexical database from Spanish kindergarten and first-grade readers. *Behavior Research Methods, 41*, 1009–1017. https://doi.org/10.3758/BRM.41.4.1009

Cuetos Vega, F. (2008). *Psicología de La Lectura*. Wolters Kluwer.

Defior, S., Jiménez-Fernández, G., Calet, N., & Serrano, F. (2015). Aprendiendo a Leer y Escribir En Español: Además de La Fonología, ¿qué Otros Procesos? *Estudios de Psicología, 36*(3), 571–591.

Dehaene, S. (2020). *How we learn: Why brains learn better than any machine . . . for now*.

Dehaene, S. (2011). The massive impact of literacy on the brain and its consequences for education. *Human Neuroplasticity and Education*, (October 2010), 19–32. http://www.pas.va/content/dam/accademia/pdf/sv117.pdf#page=17

Dehaene, S. (2014). Siglo XX *El Cerebro Lector*.

Dehaene, S. (2015). *Aprender a Leer : De Las Ciencias Cognitivas Al Aula*. http://www.sigloxxieditores.com.ar/fichaLibro.php?libro=978-987-629-505-5.

Díaz-Simón, N. (2021). Histéresis en tareas de entrenamiento del Sistema Numérico Aproximado: relación con el aprendizaje de la matemática simbólica. *Tesis de Maestría en Ciencias Cognitivas*.

Dillon, M. R., Kannan, H., Dean, J. T., Spelke, E. S., & Duflo, E. (2017). Cognitive science in the field: A preschool intervention durably enhances intuitive but not formal mathematics. *Science, 357*(6346), 47–55. https://doi.org/10.1126/science.aal4724

Diuk, B., & Ferroni, M. (2012). Dificultades de lectura en contextos de pobreza: ¿un caso de Efecto Mateo? *Psicologia Escolar e Educacional, 16*(2), 209–217. https://doi.org/10.1590/S1413-85572012000200003

Duchon, A., Perea, M., Sebastián-Gallés, N., Martí, A., & Carreiras, M. (2013). EsPal: One-stop shopping for Spanish word properties. *Behavior Research Methods, 45*(4), 1246–1258.

Fish, M., & Pinkerman, B. (2003). Language skills in low-SES rural Appalachian children: Normative development and individual differences, infancy to preschool. *Journal of Applied Developmental Psychology, 23*(5), 539–565.

Fitipalde, D., Maiche, A., & Valle-Lisboa, J. (submitted). Evidencia de la asociación entre conceptos témporo-espaciales y habilidades matemáticas en niños preescolares. *Infancia y Aprendizaje*.

Fletcher, J. M., & Vaughn, S. (2009). Response to intervention: Preventing and remediating academic difficulties. *Child Development Perspectives, 3*(1), 30–37. https://doi.org/10.1111/j.1750-8606.2008.00072.x

Foulin, J. N. (2005). Why is letter-name knowledge such a good predictor of learning to read? *Reading and Writing, 18*(2), 129–155. https://doi.org/10.1007/s11145-004-5892-2

Gaudreau, C., et al. (2020). Preschoolers benefit equally from video chat, pseudo-contingent video, and live book reading: Implications for storytime during the Coronavirus pandemic and beyond. *Frontiers in Psychology, 11*(September), 2158.

Ginsburg, H. P., & Baroody, A. J. (2003). *Test of early mathematics ability* (3rd ed.). PRO-ED.

Goldin, A. P., Hermida, M. J., Shalom, D. E., Costa, M. E., Lopez-Rosenfeld, M., Segretin, M. S., Fernández-Slezak, D., Lipina, S. J., & Sigman, M. (2014). Far transfer to language and math of a short software-based gaming intervention. *Proceedings of the National Academy of Sciences*, 201320217.

Golinkoff, R. M., & Hirsh-Pasek, K. (2016). *Becoming brilliant: What science tells us about raising successful children*. Washington DC: American Psychological Association.

Halberda, J., Mazzocco, M. M. M., & Feigenson, L. (2008). Individual differences in non-verbal number acuity correlate with maths achievement. *Nature, 455*(7213), 665–668. https://doi.org/10.1038/nature07246

Henrich, J., Heine, S. J., & Norenzayan, A. (2010). The weirdest people in the world? *Behavioral and Brain Sciences, 33*(2-3), 61–83.

Hirsh-Pasek, K., Zosh, J. M., Golinkoff, R. M., Gray, J. H., Robb, M. B., & Kaufman, J. (2015). Putting education in "Educational" apps: Lessons from the science of learning. *Psychological Science in the Public Interest, 16*(1), 3–34. https://doi.org/10.1177/1529100615569721

Hulme, C., & Snowling, M. J. (2013). Learning to read: What we know and what we need to understand better. *Child Development Perspectives, 7*(1), 1–5. https://doi.org/10.1111/cdep.12005

Hyde, D. C., Mou, Y., Berteletti, I., Spelke, E. S., Dehaene, S., & Piazza, M. (2021). Testing the role of symbols in preschool numeracy: An experimental computer-based intervention study. *PLoS One, 16*(11), e0259775. https://doi.org/10.1371/journal.pone.0259775

Judd, N., & Klingberg, T. (2021). Training spatial cognition enhances mathematical learning in a randomized study of 17,000 children. *Nature Human Behaviour*, 1–7. https://doi.org/10.1038/s41562-021-01118-4

Kirkorian, H. L. (2018). When and how do interactive digital media help children connect what they see on and off the screen? *Child Development Perspectives, 12*(3), 210–214.

Kuhn, M. R., Schwanenflugel, P. J., Meisinger, E. B., Levy, B. A., & Rasinski, T. V. (2010). Aligning theory and assessment of reading fluency: Automaticity, prosody, and definitions of fluency. *Reading Research Quarterly, 45*(2), 230–251. https://doi.org/10.1598/rrq.45.2.4

Landerl, K., Freudenthaler, H. H., Heene, M., De Jong, P. F., Desrochers, A., Manolitsis, G., Parrila, R., & Georgiou, G. K. (2019). Phonological awareness and rapid automatized naming as longitudinal predictors of reading in five alphabetic orthographies with varying degrees of consistency. *Scientific Studies of Reading, 23*(3), 220–234. https://doi.org/10.1080/10888438.2018.1510936

Le Corre, M., & Carey, S. (2007). One, two, three, four, nothing more: An investigation of the conceptual sources of the verbal counting principles. *Cognition, 105*(2), 395–438. https://doi.org/10.1016/j.cognition.2006.10.005

Libertus, M. E., Feigenson, L., & Halberda, J. (2011). Preschool acuity of the approximate number system correlates with school math ability. *Developmental Science, 14*(6), 1292–1300. https://doi.org/10.1111/j.1467-7687.2011.01080.x

Lyons, I. M., & Ansari, D. (2015). Numerical order processing in children: From reversing the distance-effect to predicting arithmetic. *Mind, Brain, and Education, 9*(4), 207–221.

Lyons, I. M., Price, G. R., Vaessen, A., Blomert, L., & Ansari, D. (2014). Numerical predictors of arithmetic success in grades 1–6. *Developmental Science, 17*(5), 714–726.

Machin, S., McNally, S., & Viarengo, M. (2018). Changing how literacy is taught: Evidence on synthetic phonics. *American Economic Journal: Economic Policy, 10*(2), 217–241. https://doi.org/10.1257/pol.20160514

Mares, M.-L., & Pan, Z. (2013). Effects of Sesame Street: A meta-analysis of children's learning in 15 countries. *Journal of Applied Developmental Psychology, 34*(3), 140–151. https://doi.org/10.1016/j.appdev.2013.01.001

Mazzocco, M. M. M., Feigenson, L., & Halberda, J. (2011). Preschoolers' precision of the approximate number system predicts later school mathematics performance. *PLoS One, 6*(9), e23749. https://doi.org/10.1371/journal.pone.0023749

Melby-Lervåg, M., Lyster, S.-A., & Hulme, C. (2012). Phonological skills and their role in learning to read: A meta-analytic review. *Psychological Bulletin, 138*(2), 322–352. https://doi.org/10.1037/a0026744

Meltzoff, A. N., Kuhl, P. K., Movellan, J., & Sejnowski, T. J. (2009). Foundations for a new science of learning. *Science (New York, N.Y.), 325*(5938), 284–288. https://doi.org/10.1126/science.1175626

Merkley, R., Matejko, A. A., & Ansari, D. (2017). Strong causal claims require strong evidence: A commentary on Wang and colleagues. *Journal of Experimental Child Psychology, 153*, 163–167. https://doi.org/10.1016/j.jecp.2016.07.008

Mitra, S. (2003). Minimally invasive education: A progress report on the "hole-in-the-wall" experiments: Colloquium. *British Journal of Educational Technology, 34*(3), 367–371. https://doi.org/10.1111/1467-8535.00333

Nation, K. (2019). Children's reading difficulties, language, and reflections on the simple view of reading. *Australian Journal of Learning Difficulties, 24*(1), 47–73. https://doi.org/10.1080/19404158.2019.1609272

National Reading Panel. (2000). Teaching children to read: An evidence-based assessment of the scientific research literature on reading and its implications for reading instruction. In *NIH Publication No. 00-4769* (Vol. 7). https://doi.org/10.1002/ppul.1950070418

Nwaerondu, N. G., & Thompson, G. (1987). The use of educational radio in developing countries: Lessons from the past. *International Journal of E-Learning & Distance Education / Revue Internationale Du e-Learning et La Formation à Distance, 2*(2), 43–54.

Odic, D., Lisboa, J. V., Eisinger, R., Olivera, M. G., Maiche, A., & Halberda, J. (2016). Approximate number and approximate time discrimination each correlate with school math abilities in young children. *Acta Psychologica, 163*, 17–26. https://doi.org/10.1016/j.actpsy.2015.10.010

Ojanen, E., Ronimus, M., Ahonen, T., Chansa-Kabali, T., February, P., Jere-Folotiya, J., Kauppinen, K.-P., Ketonen, R., Ngorosho, D., Pitkänen, M., Puhakka, S., Sampa, F., Walubita, G., Yalukanda, C., Pugh, K., Richardson, U., Serpell, R., & Lyytinen, H. (2015). GraphoGame – a catalyst for multi-level promotion of literacy in diverse contexts. *Frontiers in Psychology, 6*. https://doi.org/10.3389/fpsyg.2015.00671

Ozernov-Palchik, O., & Gaab, N. (2016). Tackling the "dyslexia paradox": Reading brain and behavior for early markers of developmental dyslexia. *Wiley Interdisciplinary Reviews: Cognitive Science, 7*(2), 156–176. https://doi.org/10.1002/wcs.1383

Park, J., & Brannon, E. M. (2013). Training the approximate number system improves math proficiency. *Psychological Science, 24*(10), 2013–2019. https://doi.org/10.1177/0956797613482944

Perera, M., & Aboal, D. (2018). The impact of a mathematics computer-assisted learning platform on students' mathematics test scores. https://digital.fundacionceibal.edu.uy/jspui/handle/123456789/225

Perfetti, C., & Bolger, D. (2004). The brain might read that way. In *Scientific studies of reading* (pp. 293–304).

Potier Watkins, C., Dehaene, O., & Dehaene, S. (2019). Automatic construction of a phonics curriculum for reading education using the Transformer Neural Network. In S. Isotani, E. Millán, A. Ogan, P. Hastings, B. McLaren, & R. Luckin (Eds.), *Artificial intelligence in education* (pp. 226–231). Springer International Publishing.

Potier Watkins, C., Caporal, J., Merville, C., Kouider, S., & Dehaene, S. (2020). Accelerating reading acquisition and boosting comprehension with a cognitive science-based tablet training. *Journal of Computers in Education, 7*, 183–212. https://doi.org/10.1007/s40692-019-00152-6

Potier Watkins, C., & Dehaene, S. (2021). Can a game application that boosts phonics knowledge in kindergarten advance 1st grade reading? *PsyArXiv*. https://doi.org/10.31234/osf.io/pwumg

Rayner, K., Foorman, B. R., Perfetti, C., Pesetsky, D., & Seidenberg, M. S. (2001). How psychological science informs the teaching of reading. *Psychological Science, 2*(2 Suppl), 31–74. http://www.ncbi.nlm.nih.gov/pubmed/11878018

Roser, M., & Ortiz-Ospina, E. (2013). Primary and secondary education. *Published online at OurWorldInData.org*. Retrieved from: https://ourworldindata.org/primary-and-secondary-education, November 23, 2021.

Share, D. L. (2008). On the anglocentricities of current reading research and practice: The Perils of Overreliance on an " Outlier ". *Orthography., 134*(4), 584–615. https://doi.org/10.1037/0033-2909.134.4.584

Sirin, S. R. (2005). Socioeconomic status and academic achievement: A meta-analytic review of research. *Review of Educational Research, 75*(3), 417–453. https://doi.org/10.3102/00346543075003417

Spelke, E. S. (2017). Core knowledge, language, and number. *Language Learning and Development, 13*(2), 147–170. https://doi.org/10.1080/15475441.2016.1263572

Spelke, E. S., & Kinzler, K. D. (2007). Core knowledge. *Developmental Science, 1*, 89–96. https://doi.org/10.1111/j.1467-7687.2007.00569.x

Szkudlarek, E., & Brannon, E. M. (2017). Does the approximate number system serve as a foundation for symbolic mathematics? *Language Learning and Development, 13*(2), 171–190.

Thompson, P. A., Hulme, C., Nash, H. M., Gooch, D., Hayiou-Thomas, E., & Snowling, M. J. (2015). Developmental dyslexia: Predicting individual risk. *Journal of Child Psychology and Psychiatry, 56*(9), 976–987. https://doi.org/10.1111/jcpp.12412

Valle-Lisboa, J., Cabana, Á., Eisinger, R., Mailhos, Á., Luzardo, M., Halberda, J., & Maiche, A. (2017). Cognitive abilities that mediate SES's effect on elementary mathematics learning: The Uruguayan tablet-based intervention. *Prospects, 46*, 301–316. https://doi.org/10.1007/s11125-017-9392-y

vanMarle, K., Chu, F. W., Mou, Y., Seok, J. H., Rouder, J., & Geary, D. C. (2018). Attaching meaning to the number words: Contributions of the object tracking and approximate number systems. *Developmental Science, 21*(1), e12495.

von Brevern, H. (2004). Cognitive and logical rationales for e-learning objects. *Journal of Educational Technology & Society, 7*(4), 2–25. http://www.jstor.org/stable/jeductechsoci.7.4.2

Wang, J. (Jenny), Odic, D., Halberda, J., & Feigenson, L. (2016). Changing the precision of preschoolers' approximate number system representations changes their symbolic math performance. *Journal of Experimental Child Psychology, 147*, 82–99. https://doi.org/10.1016/j.jecp.2016.03.002

Watson, J., & McIntyre, N. (2020). Educational television: A rapid evidence review. *Zenodo*. https://doi.org/10.5281/ZENODO.4556935

Wilson, A. J., Revkin, S. K., Cohen, D., Cohen, L., & Dehaene, S. (2006). An open trial assessment of «The Number Race», an adaptive computer game for remediation of dyscalculia. *Behavioral and Brain Functions: BBF, 2*, 20. https://doi.org/10.1186/1744-9081-2-20

World Bank. (2019, Oct). *Learning poverty*. https://www.worldbank.org/en/topic/education/brief/learning-poverty.

World Bank. (2021, Oct). *Pandemic threatens to drive unprecedented number of children into learning poverty*. https://www.worldbank.org/en/news/press-release/2021/10/29/world-bank-pandemic-threatens-to-drive-unprecedented-number-of-children-into-learning-poverty

Zugarramurdi, C., Fernández, L., Lallier, M., Valle-Lisboa, J., & Carreiras, M. (2022a). Mind the orthography: revisiting the contribution of pre-reading phonological awareness to reading acquisition. *Developmental Psychology, 58*(6), 1003–1016. https://doi.org/10.1037/DEV0001341

Zugarramurdi, C., Fernández, L., Lallier, M., Carreiras, M., & Valle-Lisboa, J. (2022b). A tablet-based universal screener for in-classroom reading assessment. *Journal of Educational Computing Research*. https://doi.org/10.1177/07356331221074300

Early Language Intervention in School Settings: What Works for Whom?

Juliana Perina Gândara, Heloisa Gonçalves da Silveira, Helena Sousa, Mirela Cunha Cardoso Ramacciotti, and Marina Leite Puglisi

Early intervention has been widely proven to be socially and economically effective, with an estimated rate of return to investment in human capital of about 3$ to 17$ for each dollar invested (Heckman, 2006; Knudsen et al., 2006). There is less evidence, however, on which intervention programs work for whom. Addressing this question represents a crucial step in order to bridge scientific knowledge into health and education public policies.

This chapter considers language interventions that have been implemented in school settings in WEIRD (Western, educated, industrialized, rich, democratic) and non-WEIRD countries and explores if their success is associated with environmental factors, child characteristics, school features, or the nature of the intervention itself. Possible explanations and implications will be discussed.

Language Development and Life Skills

It is well established in the scientific literature that oral language skills are associated with literacy development and academic achievement (Hoff, 2013; Lonigan & Shanahan, 2010; Schoon et al., 2010; Snow, 2016; Snowling & Hulme, 2021), having long-lasting consequences on social, emotional, and mental health (Hoff, 2006; Justice et al., 2018; Yew & O'Kearney, 2013). Based on these findings, researchers have argued that language skills constitute one of the solid foundation blocks for children's development, especially in the early years (Law et al., 2017; Snowling & Hulme, 2011, 2021).

J. P. Gândara
Centro Baobá, São Paulo, São Paulo, Brazil

H. G. da Silveira · H. Sousa · M. C. C. Ramacciotti · M. L. Puglisi (✉)
Departamento de Fonoaudiologia – Universidade Federal de São Paulo, São Paulo, SP, Brazil
e-mail: puglisi.marina@gmail.com

M. V. Alves et al. (eds.), *Cognitive Sciences and Education in Non-WEIRD Populations*, https://doi.org/10.1007/978-3-031-06908-6_18

Language skills are not only important for learning concepts and building strong social relations but are also considered a cornerstone for reading development. According to the Simple View of Reading,[1] the phonological component of language – especially phoneme awareness – is a strong predictor of learning to decode words, while vocabulary, grammar, and broader language abilities are the precursors of reading comprehension (Catts, 2018; Fricke et al., 2013). Each strand develops relatively independently (Lonigan, 2006), but both are required for proficient reading and should be fostered way before children enter school (Lonigan & Shanahan, 2010).

Unfortunately, although oral language and preliteracy skills are usually targeted in the school curricula since the early childhood education and care (ECEC), teachers are often not properly trained to deliver language activities using the most effective techniques (Lonigan, 2006). This is worrying given that a considerable number of children rely on highly structured and rich linguistic school environments to optimally develop their language skills. These are usually children living in socioeconomically and culturally disadvantaged conditions and/or children with learning language difficulties or language disorders.

Prevalence studies have consistently shown that around 7% of school-aged children have significant language disorders of unknown origin (i.e., developmental language disorder – DLD) (Norbury et al., 2016; Tomblin, Records, et al., 1997). This is a high rate even compared to other well-known neurodevelopmental disorders (McGregor et al., 2020). A smaller but still high prevalence has been found for dyslexia (around 5%), a neurodevelopmental condition that affects the development of decoding skills and can co-occur with DLD, among others (Scerri & Schulte-Körne, 2010).

A longitudinal study by Snowling et al. (2016) showed that children who presented poor oral language skills at age 5 (when entering compulsory education) had literacy and academic difficulties at age 8, showing that early oral language development influences later academic outcomes. Thus, to become a skilled reader, a number of developmental precursors are necessary, including language abilities that children may struggle with (Lonigan, 2006; Lundberg, 2009) for several reasons (from socioeconomic variables to language proficiency and emotional disposition).

We will next explore the wide variety of factors that influence oral language development, imposing different challenges and opportunities for children's language, literacy, and academic achievements. These factors may have a biological or environmental nature; some constitute risk factors for language and reading disorders, while others may influence the course of language development without necessarily imposing risks (Bishop et al., 2017; Choudhury & Benasich, 2003; Conti-Ramsden & Durkin, 2016; Eghbalzad et al., 2021; Lundberg, 2009; Rudolph, 2017).

[1] The Simple View of Reading is a theoretical framework developed by Gough and Tunmer in 1986. It states that proficient reading depends on two basic components: decoding and language comprehension skills

Biological Factors That Influence Language and Literacy Development

Biological factors have been extensively studied and more is known today about what puts children at disadvantage for both oral language and literacy development (Rudolph, 2017). These factors include family history of neurodevelopmental disorders, sex differences, and pre-, peri-, or neonatal problems (Bishop et al., 2017; Conti-Ramsden & Durkin, 2016; Weindrich et al., 1998). There is strong evidence that genetics play an important role in language development as research shows that language and literacy disorders run in families (Bishop, 2002, 2006; Choudhury & Benasich, 2003; Tomblin, Smith, & Zhang, 1997). That means most children with language and literacy impairment are likely to have first-degree relatives affected as well.

The reputed prevalence of language impairment among boys, on the other hand, is controversial. Whereas some evidence shows that more boys are referred to specialized services (Conti-Ramsden & Botting, 1999), epidemiological studies found a more balanced proportion of language deficits for boys and girls (Tomblin, Smith, & Zhang, 1997). Such discrepancy may be rooted on behavioral problems being more common among boys, which usually puts them on the spotlight earlier (Yew & O'Kearney, 2013). Case in point, a meta-analysis showed that biological sex met the criteria for clinical significance as a risk factor for language impairment (Rudolph, 2017). The same study also showed the relevance of pre- and perinatal indicators (Apgar score and birth order) for language development. Other studies have also underscored premature birth (Brósch-Fohraheim et al., 2019) and neonatal complications (Whitehouse et al., 2014) as risk factors.

Environmental Factors That Influence Language Development

There are a great number of environmental factors relevant to language development and literacy. The most consistently investigated are parental educational level, socioeconomic status (SES), and multilingual contexts (Bishop et al., 2016; Conti-Ramsden & Durkin, 2016; Dicataldo et al., 2020; Eghbalzad et al., 2021). Recently, the COVID-19 pandemic has evidenced the effects of such environmental factors, with implications for early language as well as academic development (Charney et al., 2021).

Environmental variables may tip the scales for language and academic achievements. Parental educational level and socioeconomic status (SES) are some of the factors that bear on the quality of the linguistic stimuli children are being provided with (Eghbalzad et al., 2021; Justice et al., 2020; Kucirkova et al., 2016; Pace et al., 2017; Puglisi et al., 2017). As reported, less educated parents and impoverished families would usually provide less quantity as well as lower quality of linguistic input, thus impacting children's language development with rippling effects on literacy and academic performances (Dollaghan et al., 1999; Hirsh-Pasek et al., 2015; Hoff & Tian, 2005; Law et al., 2011; Spencer et al., 2017).

Research on multilingual environments shows they may pose an initial disadvantage for children's language and literacy development when compared to monolingual children, especially when the language spoken at home is different from the language used in the school setting (Hoff, 2013, 2021; Hoff & Elledge, 2005). Although multilingualism per se is not a risk factor for language impairment, studies show it may influence the quality of the linguistic input children receive in the language spoken at school (for a review, see Hoff, 2021). Hence, multilingualism does provide a different context for language learning, posing important variations in opportunities and experiences, amount and proportion of exposure, and environmental support. Such variations may affect the course of language development and academic achievement (Chan & Sylva, 2015; Hoff, 2013, 2021; Sorenson Duncan & Paradis, 2018).

Protective Factors for Language and Literacy Development: Early Language Intervention

Inasmuch as roadblocks to language development may vary, experimental and clinical evidence converge on how protective factors may upend children's developmental course improving their language outcomes (Catts, 2017; Conti-Ramsden & Durkin, 2016). Prosociality and sociability[2] have been recently studied as potential protective factors (Conti-Ramsden & Durkin, 2016; Toseeb et al., 2017). The benefits and effects of early interventions have also been extensively researched and recognized as a way to fill in the gaps (Catts, 2017; Collisson et al., 2016; Dobinson & Dockrell, 2021; Fricke et al., 2013; Greenwood et al., 2020). In this matter, early high-quality education bears strong evidence for positive impact on the development of all children but especially those from disadvantaged backgrounds (for a review, see Sylva, 2014). Hence, over the past decade, emphasis on the use of evidence-based programs to promote children's development and prevent disorders is of note.

Whereas research on the effectiveness of some interventions for oral language and literacy skills improvement is mounting (Dobinson & Dockrell, 2021; Greenwood et al., 2020; Snowling & Hulme, 2011), wide is the range of targets and implementation variables that may account for differences in their specific outcomes (for a review, see Williams & Beidas, 2019). Some programs may focus on foundation skills, such as phonological awareness, vocabulary, and narrative, while others emphasize preliteracy and reading skills. Some interventions are conducted individually or in small groups, and others promote whole-class activities; some are

[2] Prosociality refers to behaviors that are intended to benefit others, caring for their needs and welfare. Sociability is defined as the quality of being sociable, i.e., being friendly

conducted by teachers and/or assistants or speech and language pathologists in the school setting, while others are delivered by parents or caregivers at home. The method and amount of training for program delivery also vary, as well as the amount and intensity of the intervention.

Overall, early intervention programs are recognized and validated as a way to foster children's language and literacy skills, but scientific evidence on what works for whom is still lacking, especially in school settings and, particularly, in low- and middle-income countries.

Therefore, here we scrutinize the available scientific evidence on the success of early language interventions in school settings. Specifically, we explore the variables and characteristics of programs implemented in various countries (both WEIRD and non-WEIRD) to identify contextual differences accounting for effective interventions, i.e., those that foster children's language development and literacy readiness, especially in disadvantaged environments.

Review Methods

This chapter aims to provide a comprehensive review of contextual factors that may be related to the success of language interventions in school settings. We used PICOS search strategy to retrieve studies of interest based on the following question: how and to what extent is the success of educational language interventions associated to contextual factors?

Data Sources

We conducted a formal literature search using PICOS criteria, as presented in Table 1. The terms used in each search were differently combined, so that strategies were appropriate for each database.

The search was performed at the following databases: Cochrane, JSTOR, PUBMED, SciELO, ERIC, SpringerLink, and SAGE Journals. Because we were especially interested in obtaining studies conducted with non-WEIRD populations, we also checked the reference lists of the studies retrieved from the databases that could meet our criteria.

Study Selection and Inclusion and Exclusion Criteria

We started by analyzing the titles and abstracts of all articles identified by the search to remove clearly irrelevant papers to the purpose of this review. To be included, papers should have tested the effects of language interventions delivered in school

Table 1 Terms used for PICOS strategy search

P	Population	(infant OR child OR child* OR "child, preschool" OR students OR school OR "early childhood education" OR "early childhood education center" OR "elementary school" OR "early education" OR "elementary education" OR preschool) NOT ("language disord*" OR autism OR "hearing loss" OR adolescent OR math* OR mathematics OR arithmetic OR bullying OR "sexual abuse" OR social OR emotion* OR attention OR physical OR anxiety OR perception)
I	Intervention	"Early intervention, educational" OR "school intervention" OR "school program" OR "educational program" OR "language intervention" OR "language program" OR "preliteracy skills" OR program OR preliteracy OR language OR intervention
C	Comparison	Not specified
O	Outcome	language OR "semantics" OR "language tests" OR "phonological awareness" OR phonology OR "oral language" OR reading OR preliteracy OR "early literacy skills" OR test
S	Study type	"Clinical trial" OR "randomized controlled trial" OR "controlled clinical trial" OR "meta-analysis" OR systematic review

settings. Exclusion criteria involved interventions in different settings, with a different scope, and with insufficient information on methods or results. After the initial selection of papers based on title and abstract, duplicate papers were removed, leaving the remaining articles to be retrieved for full-text review. The selected papers were then rated using the GRADE (Grading of Recommendations Assessment, Development, and Evaluation; Guyatt et al., 2008) approach, which is a methodology that assesses the certainty of evidence. Only studies rated with high or moderate grades were included in our analyses.

The flowchart (Fig. 1) shows the number of studies retrieved and excluded at each point, as well as the final number of studies included in this review.

Our search showed a disproportionate overall number of intervention studies from WEIRD and non-WEIRD countries, as shown in Fig. 2. There were less studies that included non-WEIRD populations ($N = 12$, as opposed to $N = 34$ from WEIRD countries), with a lower proportion of them being graded with high and moderate quality (50%, as opposed to 59%), suggesting a lack of quality evidence in these contexts. In the next section, we present the findings from all papers that met inclusion criteria, from WEIRD and non-WEIRD countries.

Findings: What Works for Whom and in Which Context?

Results showed that most intervention programs had overall positive effects. The interaction between specific variables and the extent of the effects, however, varied. In order to present and discuss these findings in light of the existing literature, we divided this section in topics to tap into each contextual factor of interest: (1) environmental, (2) child, (3) school, and (4) intervention. For each analysis, we

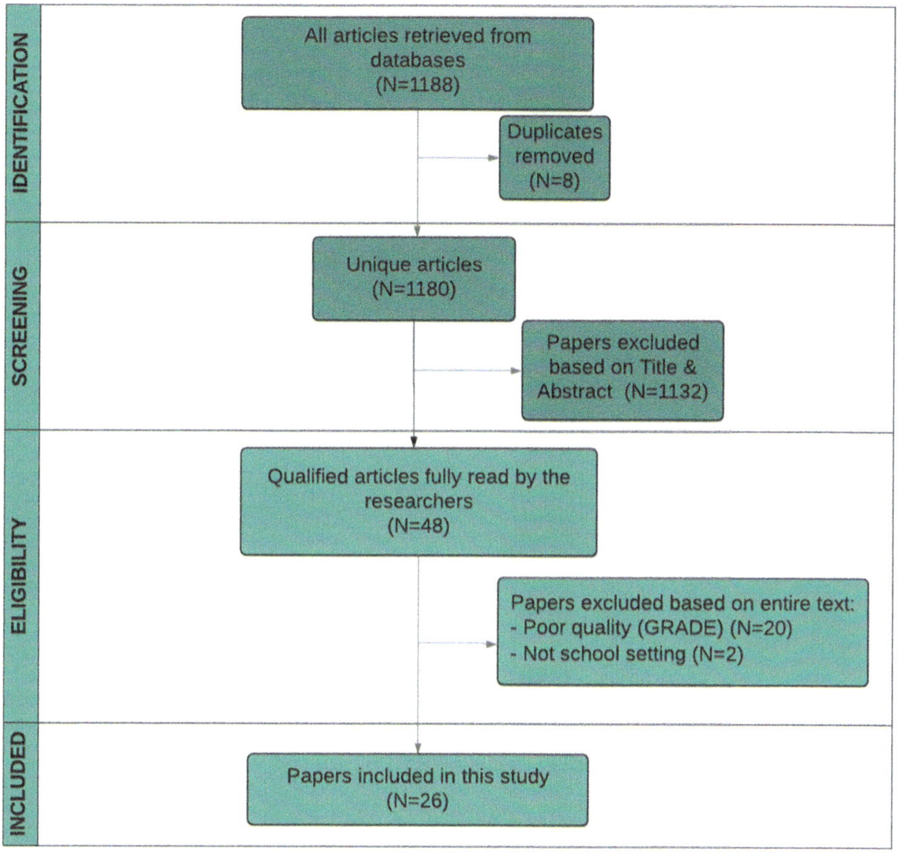

Fig. 1 Flowchart showing the number of studies included and excluded at ach point

identified the effects of the intervention on phonological awareness (PA), oral language (OL), and reading skills (read). Figure 3 summarizes the proportion of studies with positive effects for each variable of the contextual factors (1), (2), and (3), which are discussed below.

Environmental (Social and Linguistic) Factors

We identified interventions that were delivered in different SES (low- and mixed-SES) and linguistic (mono- and multilingual) environments. Figure 3 shows the proportion of studies that found significant positive effects for three different outcomes: phonological awareness, oral language, and reading.

We found more studies analyzing intervention effects in low-SES samples (N = 15, 58%) and only three studies gathering subjects from mixed-SES (two

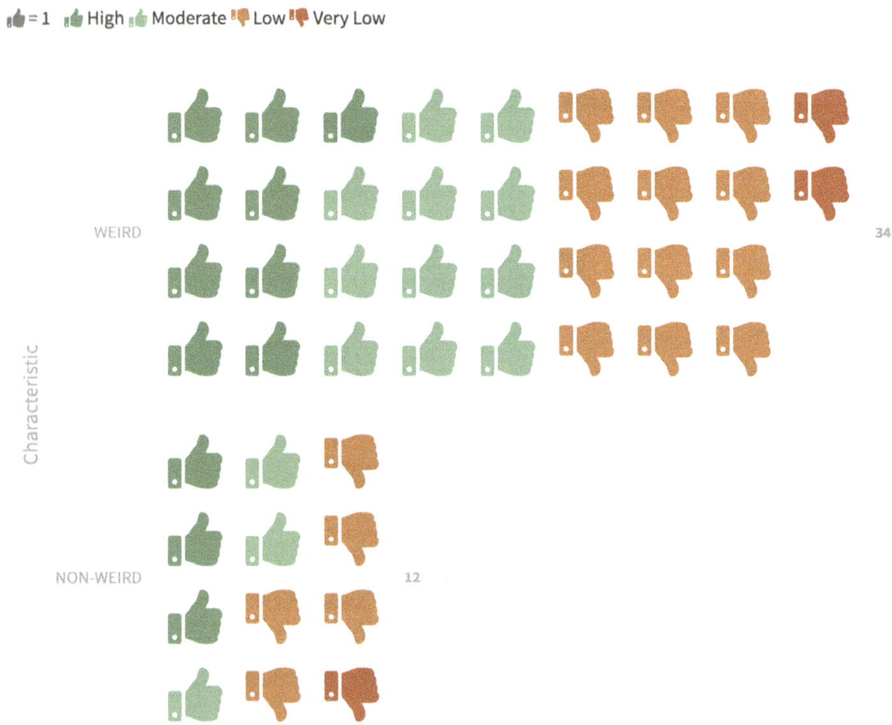

Fig. 2 Quality of evidence of studies including WEIRD versus non-WEIRD populations

low- and middle-SES; one low- and high-SES). A considerable number of studies, however, did not specify participants' socioeconomic characteristics ($N = 8$, 31%; not included in Fig. 3), posing a bias to this analysis. We found that in both low- and mixed-SES conditions, positive outcomes were observed in the majority (65–100%) of the studies. At first glance, it could seem that the proportion of positive findings was higher in the mixed-SES samples, but the limited number of studies (small dot sizes) requires caution in comparing findings.

There is a body of studies showing that children from low-SES usually perform poorly on language tasks (Dicataldo et al., 2020; Hirsh-Pasek et al., 2015; Hoff, 2006, 2013; Hoff & Tian, 2005; Law et al., 2011; Pace et al., 2017; Sirin, 2005). This is usually associated with the quality of the stimuli they are receiving, both at home and at school. A lower-than-ideal stimuli quality is, in turn, a product of a series of conditions related to cultural and social aspects, from parental level of education to the family's cultural habits and logistical possibilities in daily routines, as well as various dimensions of the classroom language environment, including characteristics of the school and the teacher's communication features (Justice et al., 2018;Sylva, 2014 ; Sylva et al., 2013). Nevertheless, studies have shown that poorer language and cognitive performances by low-SES children could be counterbalanced by high-quality education at an early stage (Sylva, 2014;

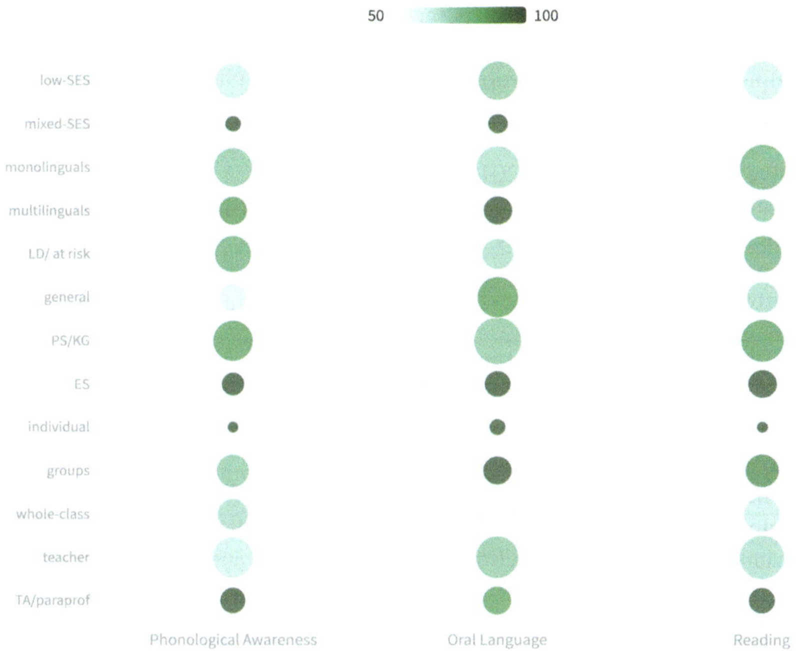

Fig. 3 Percentage of studies that found positive effects in phonological awareness, oral language, and reading skills, for environmental, child, and school factors

Sylva et al., 2013). This effect may be even stronger for children with better implicit learning skills, as suggested by Eghbalzad et al. (2021). In this study, we did not find that the intervention effects were higher for low-SES samples, but instead that the intervention was beneficial for children in all studies, regardless of their SES. However, we would need more studies analyzing the intervention effects in mid- and high-SES populations to properly address this question.

Regarding language diversity, there were only eight studies analyzing the effect of language interventions in multilingual environments, which showed convergent findings: all of them yielded positive effects in at least one outcome (oral language was the most benefited, with significant effects in all studies in which these skills were measured). However, it is of note that most of these studies were carried out in samples of bilingual immigrant children living in monolingual countries, which is a specific subgroup that is more susceptible to social vulnerability and is not fully representative of multilingual samples.

Many comparison studies have shown that multilingual children may be at disadvantage regarding language development when they enter school (Hoff, 2013, 2021; Hoff & Elledge, 2005). This characteristic, however, is related to the quality and the amount of input these children receive in each language, which is, in turn, related to other environmental factors, such as the effects of SES and other conditions of social vulnerability (e.g., immigrants and refugees) (Hoff, 2013; Hoff et al., 2012; Thordardottir, 2011). For instance, findings from multiple studies conducted in the United States suggest a "school readiness gap" between low-income bilingual children and monolingual middle-class children (Castro et al., 2011). It is a fact that language minority homes offer a variability of conditions (some only hear their heritage language at home, and others are bilinguals from the start, although the balance between languages may vary), and all these differences may pose both risk and protective factors (Hoff, 2006, 2013).

Regardless of the differences observed in oral language abilities, multilingualism does not make children slow learners (Hoff, 2021; Marchman et al., 2010). In fact, there is strong evidence that multilinguals may show advancements in other areas (e.g., executive functions) once they experienced a rich linguistic environment in the target language (Hoff, 2013, 2021). As with SES and other environmental factors, children's individual learning abilities may also constitute a protective factor in a context of linguistic diversity (Hoff, 2021), and high-quality education have a fundamental role in bridging the gap for their language and academic outcomes (Hoff, 2013).

In sum, most studies included in our sample showed positive results for the outcomes of interest (PA, OL, read) in all environmental conditions. This means that most intervention programs were effective both for disadvantaged children (i.e., children from low-SES and those from multilingual environments) and children with more favorable environmental backgrounds. The implications of the improvements, however, may be interpreted differently. For disadvantaged students, the positive effects suggest that most interventions have the potential to close the gaps imposed by the unfavorable conditions, providing them with more opportunities to succeed academically. For students with unidentified environmental disadvantages, the interventions may have boosted the development of linguistic abilities that are prerequisites to literacy and formal academic instruction.

As argued by Hoff (2013), the initial differences caused by poorer linguistic environments should be recognized as deficits rather than differences that should be embraced. As such, the implementation of effective interventions may leverage the ground for all children to achieve their maximum potential. Although there is scientific evidence that creates expectation for positive effects for language interventions in low-SES and multilingual environments, systematic reviews are needed to address this question more robustly.

Child Factors

Children's characteristics also varied between interventions. We identified interventions that targeted different populations based on their language status (language disorder/at risk for language disorder – LD/at risk and "general," which means children were not selected for their language status) and educational level (preschool/kindergarten and elementary school[3]). Figure 3 shows the proportion of studies that found significant positive effects for three different outcomes: phonological awareness, oral language, and reading.

We found that intervention effects did not vary based on the target population. The intervention effects for all outcomes (phonological awareness, oral language, and reading) were positive for all groups, regardless of children's language status (60–83%).

The group of children with or at risk for language disorders included both individuals with confirmed diagnosis and those who had never seen a specialist but performed poorly on language tests. As discussed previously, children may struggle with language and literacy development for several reasons (Lonigan, 2006; Lundberg, 2009), from individual predisposition to poor stimulating environments. Our clinical sample might have reflected this heterogeneity, including both children with persistent language learning difficulties (language disorders) and those with poor language development (language delay). Different responses to intervention, however, are expected for each of these subgroups.

Children with language disorders usually require intensive and ostensive training delivered with multisensory learning cues to compensate their long-lasting language difficulties (Ebbels, 2014; Ebbels, McCartney, et al., 2019, Ebbels, Wright, et al., 2017). Children with language delay, on the other hand, need an enriched environment to overcome their initial difficulties, since their language learning capacity is usually preserved. These are the ones who probably benefit the most from good quality interventions (Dodge, 2020). From the 12 papers that studied clinical samples included in this review, 11 recruited at-risk children and only one study involved children with diagnosed language disorder (over half of these included children from low-SES). We can hypothesize, thus, that most children in our clinical groups meet criteria for language delay instead of disorder, which helps explain their good response to intervention.

Regarding children's educational level, all interventions delivered in elementary school showed positive effects, while those targeting preschool and kindergarten years were still effective but to a lesser extent (a little over 80% for each outcome). It is important to emphasize, however, that there were fewer studies focusing on elementary school children, as shown in Fig. 3.

Even so, at this point, the importance of early language intervention for literacy readiness is beyond question (Fricke et al., 2013; Snowling & Hulme, 2021; Terrell & Watson, 2018). There is enough evidence showing that individual differences in oral language skills for school entry-level children can predict literacy and later reading performance (Snowling, Bishop, et al., 2000, Snowling, Nash, et al., 2019). Indeed, our results showed that most interventions studies actually produced positive effects for all relevant outcomes, regardless of the educational level targeted.

---

[3] Preschool, also known as early childhood education and care, refers to noncompulsory education for under school-aged children. Kindergarten, also known as pre-elementary or primary school entry level, refers to the compulsory, first stage of formal education. Elementary school, also known as primary or grade school, refers to compulsory first level education for ages 5–10

Nevertheless, the finding that for children in preschool and kindergarten not all interventions were effective suggests that the implementation of structured language programs may be more difficult in the early years, especially for complex programs, which pose additional difficulties in maintaining quality and consistency over time (Snow & Matthews, 2016).

School Factors

We found interventions that have been delivered in different school settings (whole classroom x small groups x individual) and by professionals with different backgrounds (regular teachers, teacher assistants, and paraprofessionals, which were students, tutors, or professionals with nontechnical backgrounds). Figure 3 shows the proportion of studies that found significant positive effects for three different outcomes: phonological awareness, oral language, and reading.

Regarding school settings, only two studies presented programs delivered in individual sessions; 11 reported interventions in small groups, and for other 11 interventions were delivered for the whole class (other two studies did not mention this information). Positive effects on phonological awareness were frequent for the single intervention delivered in individual sessions, as well as for interventions adopting group (75%) and whole-classroom arrangements (71%). When the outcome was oral language, positive effects were robustly found for individual and group interventions (100%), but only 50% of the studies with whole-classroom settings were effective. A similar trend was observed for outcomes in reading skills: the only study with individual and most of the studies with group (87%) interventions had positive effects, while whole-classroom arrangements also had positive findings but to a lesser extent (67%).

The fact that individual sessions were the best setting to improve children's phonological awareness, reading and language skills is not surprising, given that, in that context, the professional delivering the intervention can precisely recognize and address children's needs. However, this conclusion must be taken cautiously because of the very few quality studies with this school setting. Likewise, evidence showing small group settings are effective to promote language and literacy skills for either targeted (Snowling & Hulme, 2011) and universal interventions (Cohen-Mimran et al., 2014) come as no surprise. The fact that classroom-based interventions did not improve children's oral language and reading outcomes as much as the other settings may also be related to the amount of support and participation needed in activities that target – directly or indirectly – these abilities. Within the classroom context, it is considerably more difficult for the person delivering the program to involve all students and recognize their individual needs (ref).

For professional background, we also found different patterns according to the outcome of interest. Curiously, interventions delivered by teacher assistants and paraprofessionals had a slightly higher impact on all skills (100% for phonological awareness, 83% for oral language, and 100% for reading) than interventions delivered by regular teachers (67%, 77%, and 71% respectively). It is important to note, however, that there was a much higher number of studies involving teachers (18 studies) than teacher assistants and paraprofessionals (8 studies), which may have posed some bias in our results.

The literature shows that language interventions can be delivered successfully by trained professionals, being them teaching assistants (Bowyer-Crane et al., 2008) or specialist teachers (Snowling & Hulme, 2011). However, Lonigan (2006) found that programs evaluated as teacher-implemented rather than researcher-implemented yielded significantly smaller gains on children's skills. He argues that many early childhood educators would consider some of the phonological awareness activities outside the range of developmentally appropriate practice for preschool children. If these beliefs are in fact true for experienced teachers, we could speculate whether paraprofessionals and teacher assistants are naturally more dependent on norm-following and "sticking to the rules" of intervention guidelines due to their novel status as applicants. As far as speculations go, of note here is research lacking clear parameters for such granular analysis.

Intervention Factors

Interventions differed in scope (which abilities were at focus) and intensity (length of the session, frequency of sessions per week, and overall duration in weeks). Figure 4 shows the proportion of studies that found significant effects in each variable related to these factors, for three different outcomes: phonological awareness, oral language, and reading.

We identified interventions that targeted different language skills: phonological awareness, oral language, and/or reading. Intervention scope was divided into three categories: (1) "outcome as the single target" refers to interventions in which the outcome (either PA, OL, or read) was the only skill trained directly, (2) "outcome as one of the targets" refers to interventions in which the outcome was one of the skills trained directly, and (3) "outcome not a target" refers to interventions in which the outcome has not been directly trained but could have been influenced indirectly.

Phonological awareness seemed to be more effectively improved with interventions targeting them directly in combination with other skills (89%) or indirectly (75%). Interestingly, fewer studies showed improvements in these skills when targeted in isolation (only 50%). Oral language, on the other hand, had a more even improvement, regardless of how it was targeted: the greater proportion of positive effects were observed when it was one of the targets (89%), but significant improvements were also observed when it was directly targeted in isolation (75%) and when it was not directly targeted (75%). Showing a different pattern, reading skills improved the most as single targets (100%) but also ameliorated as one of several targets (78%) and, to a lesser extent, as an outcome not directly targeted (60%).

In general, interventions were effective to improve trained skills – in isolation or combined with other targets. The exception was the effectiveness of programs

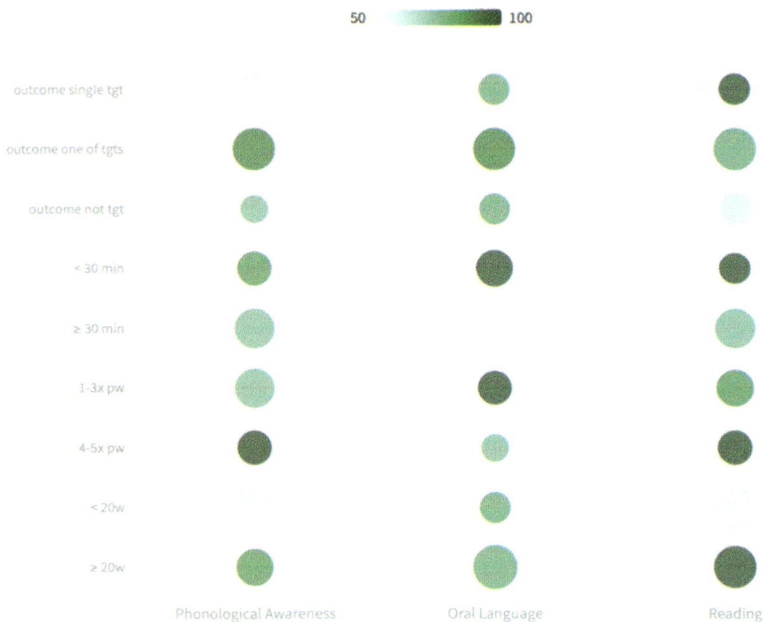

Fig. 4 Percentage of studies that found positive effects in phonological awareness, oral language and reading skills, for intervention factors

focusing on phonological awareness skills as single targets. A more refined analysis showed that, in our sample, phonological awareness was better improved when trained together with reading ($N = 6$, 100%) than with oral language skills only ($N = 2$, 0%). We speculate, therefore, that combining phonological awareness training with reading may more beneficial to children than targeting them alone.

Regarding transfer effects to nontargeted areas, we surprisingly found significant findings, especially for oral language and phonological awareness. There is a consistent body of evidence showing indeed that intervention studies rarely find what is called "transfer of knowledge." For example, in a large-scale longitudinal study training dialogic reading in 4-year-olds, positive effects on vocabulary did not generalize to reading scores at the end of second grade (Lonigan, 2006). The author justifies that promoting oral language does not impact phonological awareness or print knowledge. Similarly, Bowyer-Crane et al. (2008) found that training phonological awareness and reading fostered decoding skills whereas the oral language program improved vocabulary and grammatical skills. The same group of researchers concluded in a further study that neither program produced statistically significant improvements in secondary outcomes (Fricke et al., 2017).

Given the evidence that emergent literacy skills are relatively modular and that interventions usually improve more consistently the trained skills, Lonigan (2006) advocates that it may be advantageous to incorporate in preschool curricula all language and emergent literacy skills that are crucial for learning to read and write. He even demonstrates, for example, that phonological awareness interventions that have included letter knowledge training produced larger gains than phonological awareness training alone (Lonigan, 2006). Our data is in line with these findings, since interventions that simultaneously targeted multiple skills were the ones that yielded more positive effects for all outcomes, with a more consistent body of evidence.

With regard to intervention intensity, we identified programs with different lengths for sessions (less or more than 30 minutes), for weekly frequency (1–2; 3–4, or 5 times per week) and overall weekly duration (less than 10 weeks, between 10 and 20 weeks, and more than 20 weeks). Figure 4 shows the proportion of studies that found significant effects for phonological awareness, oral language, and reading.

Phonological awareness improved both through interventions that had shorter (less than 30 minutes) and longer (more than 30 minutes) sessions (83% and 75%, respectively), lower (1–3 times per week) and higher (4–5 times per week) frequency (75% and 100%, respectively) but overall longer (≥ 20 weeks), rather than shorter (≤ 20 weeks) duration (86% and 57%, respectively).

Reading skills had a similar pattern, benefiting both from interventions with shorter (100%) and, to a lesser extent, longer (75%) sessions. The frequency of the sessions also did not differentiate the proportion of positive findings, as both higher and lower frequencies showed consistent positive results (100% and 86%, respectively). Reading outcomes, however, consistently required longer (100%) rather than shorter (57%) intervention program (20 or more weeks were mostly effective).

Differently, oral language skills improved more consistently in studies with interventions that had shorter (100%) rather than longer (50%) sessions. They also required a much less intense training than phonological awareness, with 1–3 sessions per week (100%) yielding more consistent positive effects, although daily sessions were also effective for most studies (75%). The duration of the program did not matter as much for these skills, as studies with both shorter and longer durations had mostly positive effects (80%).

For intervention intensity, at first glance, the more seems to be the merrier. However, it is important to scrutinize the variables used to measure the intervention intensity and the effects on each type of outcome. Our analyses showed that most intervention regimens produced positive effects on all measured outcomes. The duration of sessions (less than versus equal or more than 30 minutes) was not a variable that played a significant part in producing more improvements for phonological awareness or reading, but shorter sessions were better for oral language outcomes. Frequency of the sessions were also did not differentiate the effectiveness of the programs for any measured outcomes. The duration of the intervention, on the other hand, yielded different effects for phonological awareness and reading outcomes, in favor of longer interventions (equal to or more than 20 weeks). Thus, it is reasonable to suggest that shorter sessions (low dose) distributed with higher or lower frequency for longer periods of time produce better outcomes for phonological awareness, oral language, and reading skills.

Regarding variables used to measure the intervention intensity, Schmitt et al. (2017) were the first and only ones to find that more intensive treatments measured as time were not significantly associated with improved outcomes. They demonstrated that best combinations of features to improve the language abilities of clinical groups were treatments delivered in high frequency and low dose or low frequency and high dose. These two conditions promoted better outcomes than the ones involving high frequency, high dose or low frequency, low dose. Similarly, Smith-Lock et al. (2013) found that, under a constant overall number of sessions, children receiving more distributed treatment over time (weekly) had more gains in grammar than massed treatment sessions (daily). It is important therefore to analyze intensity parameters separately rather than cumulatively.

Regarding the overall duration of language interventions, there is evidence showing that two versions of the same language program (one lasting for 20 weeks and the other lasting for 30 weeks) improved children's oral language skills, but the shorter version showed small effect size ($d = 0.21$) while the longer version had moderate effect size ($d = 0.30$) (Fricke et al., 2017).

Summary

In summary, this chapter explored for whom and in which contexts language interventions seem to be more effective. Although our intention was to analyze the effectiveness of language interventions delivered in school settings comparing WEIRD and non-WEIRD populations, we could only find a few studies from non-WEIRD countries, most of which rated with poor quality of evidence. Therefore, the results of all good-quality studies (both with WEIRD and non-WEIRD populations) were taken together in our analyses. We emphasize the need for more research with good methodological quality in non-WEIRD countries for further conclusions on possible effectiveness differences as compared to WEIRD populations.

We found that, in general, factors related to the quality of the provided interventions (i.e., methodological variables, such as who delivered the intervention, which abilities were stimulated, and intensity of the intervention) produced more consistent effects than factors related to the individuals (i.e., which are risk factors, such as background SES, multilingualism, and previously detected language disorders).

It is relevant to point out, though, the limitations of these analyses. This chapter provided a scope rather than a systematic review. Although our search retrieved 46 studies that initially met the criteria for our analyses, almost half of them were excluded from our sample for poor quality ratings (20 studies). Furthermore, not all of the contextual factors were explored in all studies included, which restricts the robustness of our findings. We are aware of these limitations and highlight that the intention of this chapter is to raise the contextual factors that seem to play an important role on the effect of language interventions. Future studies are needed to approach the topic in a more systematic way.

References

Bishop, D. (2002). The role of genes in the etiology of specific language impairment. *Journal of Communication Disorders, 35*(4), 311–328. https://doi.org/10.1016/s0021-9924(02)00087-4

Bishop, D. V. (2006). Developmental cognitive genetics: How psychology can inform genetics and vice versa. *Quarterly Journal of Experimental Psychology, 59*(7), 1153–1168. https://doi.org/10.1080/17470210500489372

Bishop, D. V., Snowling, M. J., Thompson, P. A., & Greenhalgh, T. (2016). CATALISE: A multinational and multidisciplinary Delphi consensus study. Identifying language impairments in children. *PLoS One, 11*(7), e0158753. https://doi.org/10.1371/journal.pone.0158753

Bishop, D. V., Snowling, M. J., Thompson, P. A., & Greenhalgh, T. (2017). Phase 2 of CATALISE: A multinational and multidisciplinary Delphi consensus study of problems with language development: Terminology. *Journal of Child Psychology and Psychiatry, 58*(10), 1068–1080. https://doi.org/10.1111/jcpp.12721

Bowyer-Crane, C., Snowling, M. J., Duff, F. J., Fieldsend, E., Carroll, J. M., Miles, J., Götz, K., & Hulme, C. (2008). Improving early language and literacy skills: Differential effects of an oral language versus a phonology with reading intervention. *Journal of Child Psychology and Psychiatry, 49*(4), 422–432. https://doi.org/10.1111/j.1469-7610.2007.01849.x

Brósch-Fohraheim, N., Fuiko, R., Marschik, P. B., & Resch, B. (2019). The influence of preterm birth on expressive vocabulary at the age of 36 to 41 months. *Medicine, 98*(6), e14404. https://doi.org/10.1097/md.0000000000014404

Castro, D. C., Páez, M. M., Dickinson, D. K., & Frede, E. (2011). Promoting language and literacy in young dual language learners: Research, practice, and policy. *Child Development Perspectives, 5*(1), 15–21. https://doi.org/10.1111/j.1750-8606.2010.00142.x

Catts, H. W. (2017). Early identification of reading disabilities. In I. K. Cain, D. L. Compton, & R. Parrila (Eds.), *Theories of Reading development* (pp. 311–332). John Benjamins Publishing.

Catts, H. W. (2018). The simple view of reading: Advancements and false impressions. *Remedial and Special Education, 39*(5), 317–323. https://doi.org/10.1177/0741932518767563

Chan, L. L., & Sylva, K. (2015). Exploring emergent literacy development in a second language: A selective literature review and conceptual framework for research. *Journal of Early Childhood Literacy, 15*(1), 3–36. https://doi.org/10.1177/1468798414522824

Charney, S. A., Camarata, S. M., & Chern, A. (2021). Potential impact of the COVID-19 pandemic on communication and language skills in children. *Otolaryngology–Head and Neck Surgery, 165*(1), 1–2. https://doi.org/10.1177/0194599820978247

Choudhury, N., & Benasich, A. A. (2003). A family aggregation study: The influence of family history and other risk factors on language development. *Journal of Speech, Language, and Hearing Research, 46*(2), 261–272. https://doi.org/10.1044/1092-4388(2003/021)

Cohen-Mimran, R., Reznik-Nevet, L., & Korona-Gaon, S. (2014). An activity-based language intervention program for kindergarten children: A retrospective evaluation. *Early Childhood Education Journal, 44*(1), 69–78. https://doi.org/10.1007/s10643-014-0676-z

Collisson, B. A., Graham, S. A., Preston, J. L., Rose, M. S., McDonald, S., & Tough, S. (2016). Risk and protective factors for late talking: An epidemiologic investigation. *The Journal of Pediatrics, 172*, 168–174.e1. https://doi.org/10.1016/j.jpeds.2016.02.020

Conti-Ramsden, G., & Botting, G. C. (1999). Characteristics of children attending language units in England: A national study of 7-year-olds. *International Journal of Language & Communication Disorders, 34*(4), 359–366. https://doi.org/10.1080/136828299247333

Conti-Ramsden, G., & Durkin, K. (2016). What factors influence language impairment? Considering resilience as well as risk. *Folia Phoniatrica et Logopaedica, 67*(6), 293–299. https://doi.org/10.1159/000444750

Dicataldo, R., Florit, E., & Roch, M. (2020). Fostering broad oral language skills in preschoolers from low SES background. *International Journal of Environmental Research and Public Health, 17*(12), 4495. https://doi.org/10.3390/ijerph17124495

Dobinson, K. L., & Dockrell, J. E. (2021). Universal strategies for the improvement of expressive language skills in the primary classroom: A systematic review. *First Language, 41*(5), 527–554. https://doi.org/10.1177/0142723721989471

Dodge, K. A. (2020). Annual research review: Universal and targeted strategies for assigning interventions to achieve population impact. *Journal of Child Psychology and Psychiatry, 61*(3), 255–267. https://doi.org/10.1111/jcpp.13141

Dollaghan, C. A., Campbell, T. F., Paradise, J. L., Feldman, H. M., Janosky, J. E., Pitcairn, D. N., & Kurs-Lasky, M. (1999). Maternal education and measures of early speech and language. *Journal of Speech, Language, and Hearing Research, 42*(6), 1432–1443. https://doi.org/10.1044/jslhr.4206.1432

Ebbels, S. (2014). Effectiveness of intervention for grammar in school-aged children with primary language impairments: A review of the evidence. *Child Language Teaching and Therapy, 30*(1), 7–40. https://doi.org/10.1177/0265659013512321

Ebbels, S. H., McCartney, E., Slonims, V., Dockrell, J. E., & Norbury, C. F. (2019). Evidence-based pathways to intervention for children with language disorders. *International Journal of Language & Communication Disorders, 54*(1), 3–19. https://doi.org/10.1111/1460-6984.12387

Ebbels, S. H., Wright, L., Brockbank, S., Godfrey, C., Harris, C., Leniston, H., Neary, K., Nicoll, H., Nicoll, L., Scott, J., & Marić, N. (2017). Effectiveness of 1: 1 speech and language therapy for older children with (developmental) language disorder. *International Journal of Language & Communication Disorders, 52*(4), 528–539. https://doi.org/10.1111/1460-6984.12297

Eghbalzad, L., Deocampo, J. A., & Conway, C. M. (2021). How statistical learning interacts with the socioeconomic environment to shape children's language development. *PLoS One, 16*(1), e0244954. https://doi.org/10.1371/journal.pone.0244954

Fricke, S., Bowyer-Crane, C., Haley, A. J., Hulme, C., & Snowling, M. J. (2013). Efficacy of language intervention in the early years. *The Journal of Child Psychology and Psychiatry, 54*(3), 280–290.

Fricke, S., Burgoyne, K., Bowyer-Crane, C., Kyriacou, M., Zosimidou, A., Maxwell, L., Lervåg, A., Snowling, M. J., & Hulme, C. (2017). The efficacy of early language intervention in mainstream school settings: A randomized controlled trial. *Journal of Child Psychology and Psychiatry and Allied Disciplines, 58*(10), 1141–1151. https://doi.org/10.1111/jcpp.12737

Greenwood, C. R., Schnitz, A. G., Carta, J. J., Wallisch, A., & Irvin, D. W. (2020). A systematic review of language intervention research with low-income families: A word gap prevention perspective. *Early Childhood Research Quarterly, 50*(1), 230–245. https://doi.org/10.1016/j.ecresq.2019.04.001

Guyatt, G. H., Oxman, A. D., Kunz, R., Vist, G. E., Falck-Ytter, Y., & Schünemann, H. J. (2008). What is "quality of evidence" and why is it important to clinicians? *BMJ, 336*(7651), 995–998. https://doi.org/10.1136/bmj.39490.551019.be

Heckman, J. J. (2006). Skill formation and the economics of investing in disadvantaged children. *Science, 312*(5782), 1900–1902. https://doi.org/10.1126/science.1128898

Hirsh-Pasek, K., Adamson, L. B., Bakeman, R., Owen, M. T., Golinkoff, R. M., Pace, A., Yust, P. K., & Suma, K. (2015). The contribution of early communication quality to low-income children's language success. *Psychological Science, 26*(7), 1071–1083. https://doi.org/10.1177/0956797615581493

Hoff, E. (2006). How social contexts support and shape language development. *Developmental Review, 26*(1), 55–88. https://doi.org/10.1016/j.dr.2005.11.002

Hoff, E. (2013). Interpreting the early language trajectories of children from low-SES and language minority homes: Implications for closing achievement gaps. *Developmental Psychology, 49*(1), 4–14. https://doi.org/10.1037/a0027238

Hoff E. (2021). Chapter four-why bilingual development is not easy. In J. J. Lockman (Ed.), *Advances in child development and behavior* (Vol. 61, pp. 129–167). https://doi.org/10.1016/bs.acdb.2021.03.002.

Hoff, E., Core, C., Place, S., Rumiche, R., Señor, M., & Parra, M. (2012). Dual language exposure and early bilingual development. *Journal of Child Language, 39*(1), 1–27. https://doi.org/10.1017/s0305000910000759

Hoff E, & Elledge C. (2005). Bilingualism as one of many environmental variables that affect language development. In *Proceedings of the 4th International Symposium on Bilingualism* (pp. 1034–1040).

Hoff, E., & Tian, C. (2005). Socioeconomic status and cultural influences on language. *Journal of Communication Disorders, 38*(4), 271–278. https://doi.org/10.1016/j.jcomdis.2005.02.003

Justice, L. M., Jiang, H., & Strasser, K. (2018). Linguistic environment of preschool classrooms: What dimensions support children's language growth? *Early Childhood Research Quarterly, 42*, 79–92. https://doi.org/10.1016/j.ecresq.2017.09.003

Justice, L. M., Jiang, H., Bates, R., & Koury, A. (2020). Language disparities related to maternal education emerge by two years in a low-income sample. *Maternal and Child Health Journal, 24*(11), 1419–1427. https://doi.org/10.1007/s10995-020-02973-9

Knudsen, E., Heckman, J., Cameron, J., & Shonkoff, J. (2006). Economic, neurobiological and behavioral perspectives on building America's future workforce. *PNAS, 103*(27), 10155–10162. https://doi.org/10.1073/pnas.0600888103

Kucirkova, N., Dale, P. S., & Sylva, K. (2016). Parents reading with their 10-month-old babies: Key predictors for high-quality reading styles. *Early Child Development and Care, 188*(2), 195–207. https://doi.org/10.1080/03004430.2016.1211117

Law, J., McBean, K., & Rush, R. (2011). Communication skills in a population of primary school-aged children raised in an area of pronounced social disadvantage. *International Journal of Language & Communication Disorders, 46*(6), 657–664. https://doi.org/10.1111/j.1460-6984.2011.00036.x

Law, J., Charlton, J., & Asmussen, K. (2017). *Language as a child wellbeing indicator [technical report]*. Early Intervention Foundation. https://www.eif.org.uk/report/language-as-a-child-wellbeing-indicator

Lonigan, C. J. (2006). Development, assessment, and promotion of preliteracy skills. *Early Education and Development, 17*(1), 91–114. https://doi.org/10.1207/s15566935eed1701_5

Lonigan, C. J., & Shanahan, T. (2010). Developing early literacy skills: Things we know we know and things we know we don't know. *Educational Researcher, 39*(4), 340–346. https://doi.org/10.3102/0013189x10369832

Lundberg, I. (2009). Early precursors and enabling skills of reading acquisition. *Scandinavian Journal of Psychology, 50*(6), 611–616. https://doi.org/10.1111/j.1467-9450.2009.00778.x

Marchman, V. A., Fernald, A., & Hurtado, N. (2010). How vocabulary size in two languages relates to efficiency in spoken word recognition by young Spanish–English bilinguals. *Journal of Child Language, 37*(4), 817–840. https://doi.org/10.1017/s0305000909990055

McGregor, K. K., Goffman, L., Van Horne, A. O., Hogan, T. P., & Finestack, L. H. (2020). Developmental language disorder: Applications for advocacy, research, and clinical service. *Perspectives of the ASHA Special Interest Groups, 5*(1), 38–46. https://doi.org/10.1044/2019_persp-19-00083

Norbury, C. F., Gooch, D., Wray, C., Baird, G., Charman, T., Simonoff, E., Vamvakas, G., & Pickles, A. (2016). The impact of nonverbal ability on prevalence and clinical presentation of language disorder: Evidence from a population study. *Journal of Child Psychology and Psychiatry, 57*(11), 1247–1257. https://doi.org/10.1111/jcpp.12573

Pace, A., Luo, R., Hirsh-Pasek, K., & Golinkoff, R. M. (2017). Identifying pathways between socioeconomic status and language development. *Annual Review of Linguistics, 3*(1), 285–308. https://doi.org/10.1146/annurev-linguistics-011516-034226

Puglisi, M. L., Hulme, C., Hamilton, L. G., & Snowling, M. J. (2017). The home literacy environment is a correlate, but perhaps not a cause, of variations in children's language and literacy development. *Scientific Studies of Reading, 21*(6), 498–514. https://doi.org/10.1080/10888438.2017.1346660

Rudolph, J. M. (2017). Case history risk factors for specific language impairment: A systematic review and meta-analysis. *American Journal of Speech-Language Pathology, 26*(3), 991–1010. https://doi.org/10.1044/2016_ajslp-15-0181

Scerri, T. S., & Schulte-Körne, G. (2010). Genetics of developmental dyslexia. *European Child & Adolescent Psychiatry, 19*(3), 179–197. https://doi.org/10.1007/s00787-009-0081-0

Schmitt, M. B., Justice, L. M., & Logan, J. A. (2017). Intensity of language treatment: Contribution to children's language outcomes. *International Journal of Language & Communication Disorders, 52*(2), 155–167. https://doi.org/10.1111/1460-6984.12254

Schoon, I., Parsons, S., Rush, R., & Law, J. (2010). Childhood language skills and adult literacy: A 29-year follow-up study. *Pediatrics, 125*(3), e459–e466. https://doi.org/10.1542/peds.2008-2111

Sirin, S. R. (2005). Socioeconomic status and academic achievement: A meta-analytic review of research. *Review of Educational Research, 75*(3), 417–453. https://doi.org/10.3102/00346543075003417

Smith-Lock, K., Leitão, S., Lambert, L., Prior, P., Dunn, A., Cronje, J., Newhouse, S., & Nickels, L. (2013). Daily or weekly? The role of treatment frequency in the effectiveness of grammar treatment for children with specific language impairment. *International Journal of Speech-Language Pathology, 15*(3), 255–267. https://doi.org/10.3109/17549507.2013.777851

Snow, C. E., & Matthews, T. J. (2016). Reading and language in the early grades. *The Future of Children, 26*(2), 57–74. https://doi.org/10.1353/foc.2016.0012

Snow, P. C. (2016). Elizabeth usher memorial lecture: Language is literacy is language - positioning speech-language pathology in education policy, practice, paradigms and polemics. *International Journal of Speech-Language Pathology, 18*(3), 216–228. https://doi.org/10.3109/17549507.2015.1112837

Snowling, M., Bishop, D. V., & Stothard, S. E. (2000). Is preschool language impairment a risk factor for dyslexia in adolescence? *Journal of Child Psychology and Psychiatry, 41*(5), 587–600. https://doi.org/10.1111/1469-7610.00651

Snowling, M. J., Duff, F. J., Nash, H. M., & Hulme, C. (2016). Language profiles and literacy outcomes of children with resolving, emerging, or persisting language impairments. *Journal of Child Psychology and Psychiatry, 57*(12), 1360–1369. https://doi.org/10.1111/jcpp.12497

Snowling, M. J., & Hulme, C. (2011). Evidence-based interventions for reading and language difficulties: Creating a virtuous circle. *British Journal of Educational Psychology, 81*(1), 1–23. https://doi.org/10.1111/j.2044-8279.2010.02014.x

Snowling, M. J., & Hulme, C. (2021). Annual research review: Reading disorders revisited – The critical importance of oral language. *Journal of Child Psychology and Psychiatry, 62*(5), 635–653. https://doi.org/10.1111/jcpp.13324

Snowling, M. J., Nash, H. M., Gooch, D. C., Hayiou-Thomas, M. E., & Hulme, C. (2019). Developmental outcomes for children at high risk of dyslexia and children with developmental language disorder. *Child Development, 90*(5), e548–e564. https://doi.org/10.1111/cdev.13216

Sorenson Duncan, T., & Paradis, J. (2018). How does maternal education influence the linguistic environment supporting bilingual language development in child second language learners of English? *International Journal of Bilingualism, 24*(1), 46–61. https://doi.org/10.1177/1367006918768366

Spencer, S., Clegg, J., Stackhouse, J., & Rush, R. (2017). Contribution of spoken language and socio-economic background to adolescents' educational achievement at age 16 years. *International Journal of Language & Communication Disorders, 52*(2), 184–196. https://doi.org/10.1111/1460-6984.12264

Sylva, K. (2014). The role of families and pre-school in educational disadvantage. *Oxford Review of Education, 40*(6), 680–695. https://doi.org/10.1080/03054985.2014.979581

Sylva, K., Sammons, P., Chan, L. L., Melhuish, E., Siraj-Blatchford, I., & Taggart, B. (2013). The effects of early experiences at home and pre-school on gains in English and mathematics in primary school: A multilevel study in England. *Zeitschrift für Erziehungswissenschaft, 16*(2), 277–301. https://doi.org/10.1007/s11618-013-0364-6

Terrell, P., & Watson, M. (2018). Laying a firm foundation: Embedding evidence-based emergent literacy practices into early intervention and preschool environments. *Language, Speech, and Hearing Services in Schools, 49*(2), 148–164. https://doi.org/10.1044/2017_lshss-17-0053

Thordardottir, E. (2011). The relationship between bilingual exposure and vocabulary development. *International Journal of Bilingualism, 15*(4), 426–445. https://doi.org/10.1177/1367006911403202

Tomblin, J. B., Records, N. L., Buckwalter, P., Zhang, X., Smith, E., & O'Brien, M. (1997). Prevalence of specific language impairment in kindergarten children. *Journal of Speech, Language, and Hearing Research, 40*(6), 1245–1260. https://doi.org/10.1044/jslhr.4006.1245

Tomblin, J., Smith, E., & Zhang, X. (1997). Epidemiology of specific language impairment: Prenatal and perinatal risk factors. *Journal of Communication Disorders, 30*(4), 325–344. https://doi.org/10.1016/s0021-9924(97)00015-4

Toseeb, U., Pickles, A., Durkin, K., Botting, N., & Conti-Ramsden, G. (2017). Prosociality from early adolescence to young adulthood: A longitudinal study of individuals with a history of language impairment. *Research in Developmental Disabilities, 62*, 148–159. https://doi.org/10.1016/j.ridd.2017.01.018

Weindrich, D., Jennen-Steinmetz, C., Laucht, M., Esser, G., & Schmidt, M. (1998). At risk for language disorders? Correlates and course of language disorders in preschool children born at risk. *Acta Paediatrica, 87*(12), 1288–1294. https://doi.org/10.1111/j.1651-2227.1998.tb00953.x

Whitehouse, A. J., Shelton, W. M., Ing, C., & Newnham, J. P. (2014). Prenatal, perinatal, and neonatal risk factors for specific language impairment: A prospective pregnancy cohort study. *Journal of Speech, Language, and Hearing Research, 57*(4), 1418–1427. https://doi.org/10.1044/2014_jslhr-l-13-0186

Williams, N. J., & Beidas, R. S. (2019). Annual research review: The state of implementation science in child psychology and psychiatry: A review and suggestions to advance the field. *Journal of Child Psychology and Psychiatry, 60*(4), 430–450. https://doi.org/10.1111/jcpp.12960

Yew, S. G., & O'Kearney, R. (2013). Emotional and behavioural outcomes later in childhood and adolescence for children with specific language impairments: Meta-analyses of controlled prospective studies. *Journal of Child Psychology and Psychiatry, 54*(5), 516–524. https://doi.org/10.1111/jcpp.12009

Developing Higher Order Thinking in Elementary School Science: A Narrative Around a Research Group Quest

Melina Furman, Mariana Luzuriaga, and Inés Taylor

The Question Where Our Quest Begins

Imagine the children of today in 10, 20, 50 years' time... What kind of adults do you hope they are?

This is one of the main questions that guide our work and commitment to improving education. At conferences, training sessions, and interviews, we repeatedly ask school principals, teachers, and parents to reflect on their answers to that question.

Frequent responses include "curious," "creative," "resilient," "kind," "thoughtful," "critical thinkers," and "problem-solvers." Thus, there is a consensus that we aspire for children to have the necessary knowledge and tools to function in the world, to develop a taste for lifelong learning, and to have an awareness of the "common good." All of these speak to the need for developing higher order thinking skills, which encompass a continuum of cognitive processes, such as understanding, applying, analyzing, evaluating, and creating, across factual, conceptual, procedural, and metacognitive knowledge dimensions (Anderson & Krathwohl, 2001; Bloom et al., 1956; Forehand, 2005), key in developing twenty-first-century citizenship (Scott, 2015).

The next questions we ask are "how do we achieve it?" and "what must we do now to prepare our students for the future we envision?."

One way of doing so is through science education, which has long been considered a field of opportunity to develop higher order thinking skills. In particular, an important goal of science education is the development of scientific thinking,

M. Furman (✉)
School of Education, Universidad de San Andres, Buenos Aires, Argentina

Consejo Nacional de Investigaciones Científicas y Técnicas, Buenos Aires, Argentina

M. Luzuriaga · I. Taylor
School of Education, Universidad de San Andres, Buenos Aires, Argentina

M. V. Alves et al. (eds.), *Cognitive Sciences and Education in Non-WEIRD Populations*, https://doi.org/10.1007/978-3-031-06908-6_19

defined as the skill set of "ways of knowing" involved in science, which include higher order thinking skills, such as inductive and deductive reasoning, interpreting and constructing data and models to explain phenomena, designing valid experiments, building evidence-based arguments, and finding and assessing ways to solve socioscientific problems (Klahr et al., 2011).

Although there is evidence that children have some rudiments of scientific thinking skills even at a very young age (Gopnik et al., 2001; Gweon et al., 2010; Gweon & Shultz, 2011), they do not fully develop spontaneously (Kuhn, 2010). Instead, they need to be intentionally taught and put into practice in a systematic and sustained way overtime (Duschl et al., 2007).

An effective science education for all is perhaps even more important within non-WEIRD (Western, Educated, Industrialized, Rich and Democratic) populations, both as a means to foster and increase citizens' informed participation in critical issues, such as health and environmental matters, and potentially promoting vocational orientations in the fields of science, technology, engineering, and mathematics (STEM), which are fundamental in knowledge-based economies. There is evidence that, in Argentina, the context of this study, as in many other countries, only few students pursue career choices related to STEM, especially among women and populations from socioeconomically vulnerable contexts (Albornoz et al., 2009; Szenkman & Lotitto, 2020).

Consequently, science teaching reform efforts were implemented in several countries over the last decades. As a recent study on the state of affairs regarding science education in primary schools in Latin America shows, starting in the second half of the 2000s, several countries, such as Chile, Colombia, Ecuador, México, Paraguay, and Uruguay, have renewed their science curriculum and implemented professional development programs, all aimed at enhancing science teaching with a particular focus on developing scientific thinking skills (Furman, 2020). This is also the case in Argentina (Tedesco, 2009).

However, low standardized test results and research studies consistently suggest that what happens within classroom contexts is still far from what is expected. Argentine and Latin American students, in general, achieve very low levels of performance in science and particularly struggle to solve questions relating to higher order thinking skills, such as interpreting data and drawing conclusions, analyzing and solving science-related problematic situations, and assessing or proposing experimental designs to answer inquiry questions (Meschengieser & Otero, 2016; UNESCO, 2015). Furthermore, there is great inequity in learning opportunities, with a consistent pattern showing students from wealthier families attaining higher levels of learning achievement than their less privileged peers in many countries (OECD, 2019; Vegas et al., 2014).

Faced with this scenario, we wonder: how can these societies facing structural challenges aspire to thrive when so few students seem to be equipped with the necessary skills? More specifically, the prevalence of such low learning outcomes opens up questions regarding current science teaching practices: are they encouraging children to become the thinkers and problem-solvers we aspire them to be?

Unfortunately, it appears that we aren't. Research on teaching practices, although limited, indicates that teachers across many Latin American countries show persistent difficulties in handling conceptual content and enacting reform-based teaching practices (Furman, 2020; Kisilevsky et al., 2019). In turn, this is consistent with these countries having poor teacher training systems in general (Bruns & Luque, 2015) and in science in particular (Furman & Luzuriaga, 2017; Maiztegui et al., 2000), which seems to negatively affect the quality of science instruction.

Thus, some of the main questions our research projects address are as follows: What do science teachers need to be able to improve their "business as usual" practices? How can we help them move toward enhancing science education and thus student learning of scientific thinking skills?

This is where the "CABA study," which will be described throughout this chapter, came in. Our quest to better understand the current state of scientific thinking skills development in relation to science teaching practices, and how to strengthen them, led us to develop a research project set in a representative sample of schools in the City of Buenos Aires (CABA), Argentina. Through this study we sought to characterize science teaching practices and student learning outcomes, both in general and as a result of teachers receiving specific professional development (the full design and results can be found in Albornoz et al., 2019).

In particular, we set out to rigorously evaluate if providing teachers with educative curriculum materials (ECM), specially designed to foster scientific thinking skills in students, is an effective way to improve science teaching and learning in the local context. ECM are detailed lesson plans, including reading materials, worksheets, and other resources for students, aimed at supporting teachers in the organization, content, and pedagogy of given topics (Ball & Cohen, 1996; Davis & Krajcik, 2005).

ECM became a widespread, regular source of consultation for teachers as well as a key component in curriculum reforms and teacher training policies in general as in science education, endorsed by research that brought attention to their potential, particularly in contexts with low-quality indicators in teacher education and student learning (Bruns & Luque, 2015; Mourshed et al., 2010).

However, while studies show mixed results on the effects of providing teachers with ECM on student learning and teaching practices (Davis et al., 2016), there is a dearth of evidence to this regard in the region. Considering that, within these contexts of resource scarcity, providing ECM represents a significant investment, we identified a pressing need to contribute with large-scale, high-quality, rigorous studies on the implementation and effects of science ECM in student learning of higher order scientific thinking skills as well as in teaching practices.

In this chapter we present a narrative of how this research project unfolded, from our initial questions to its design, the findings and new questions that emerged along the way, to the lessons learned at the end of the process. Although the overall findings for the wider randomized controlled trial were published (Albornoz et al., 2019), as are several more in-depth papers considering different aspects of the pedagogical intervention (Furman et al., 2021; Taylor et al., 2020), for this chapter we conducted further data analysis and present new results focusing on the implementation of ECM.

Designing Our Action Steps

We proposed a randomized controlled trial to evaluate both student learning and teachers' teaching practices in seventh grade (the final year of primary schooling) science, comparing everyday teaching against the effects of supporting teachers in the implementation of scientific, higher order thinking skills oriented teaching practices through the provision of ECM. Two experimental groups were defined: the control group, in which "business as usual" teaching practices and learning outcomes would be observed, and the intervention group, which we provided with ECM.

The study was set in the City of Buenos Aires, the federal capital of Argentina (CABA, for its Spanish acronym). CABA is the most densely populated city and also one of the districts with better socioeconomic and educational indicators at the national level. However, this is within a context of low levels of academic performance and high degrees of educational inequity to the detriment of students from disadvantaged contexts (Di Virgilio & Serrati, 2019; Meschengieser & Otero, 2016).

Mimicking Argentine education policies more generally, participating in this research program was mandatory yet not enforced, and there were no formal consequences if schools or teachers chose not to abide by what was requested (i.e., there were no financial or administrative incentives or penalties). Also, given that conducting research within school contexts is not a well-established practice in the country, this required involving and consulting senior officials of the CABA Ministry of Education, stakeholders from different ministerial areas, and school district superintendents and principals, as their support was essential to access schools and promote teachers' participation.

A sample of 47 state primary schools located across six (out of 21) school districts in the city participated in the study. These schools were representative of schools in CABA in terms of socioeconomic conditions, their total school and seventh grade enrollment, pass and repetition rates, overage rate, and previous district test scores (see Albornoz et al., 2019 for full sample balances).

As shown in Table 1, 24 schools were randomly assigned to the control group and another 23 to the intervention group. Within each group, all seventh grade science teachers participated in the study. After analyzing the differences in the means along with p-values from two-tailed t-tests of equality of means, no significant differences were observed between each group in student-level variables (e.g., gender, socioeconomic background, and academic performance), teacher-level variables (e.g., gender, seniority in seventh grade teaching, and seniority in the current school), nor school-level variables (e.g., location, total enrollment, amount of

Table 1 Sample description

	Control group	Intervention group
Number of schools	24	23
Total seventh grade classrooms	50	44
Total seventh grade students	1086	917
Total seventh grade science teachers	36	32

seventh grade classrooms, percentage of schools with double school day, and school dropout and overage student rates).

We chose to focus our study on one particular science seventh grade curricular topic: the human body, in order to limit the scope of the intervention and favor the comparison between both experimental groups. Besides being one of the prescribed contents in the national and district curricular guidelines (Argentine Ministry of Education, 2005; CABA Ministry of Education, 2012), the human body is regularly taught in primary level science lessons for it's a well-versed topic, often valued by teachers as relevant and interesting for students.

Teachers in the control group addressed the topic of "the human body" as they regularly did in science lessons, allowing us to capture common teaching practices and learning outcomes, while teachers in the intervention group were provided with science ECM to address the topic, focusing on the basics of the digestive, circulatory, and respiratory systems and how they contribute to cellular nutrition.

ECM were developed by the team of researchers based on curriculum guidelines and informed by research-based, best practice science principles (Davis et al., 2014; Davis & Krajcik, 2005; Harlen, 2013) and then discussed with nonparticipating primary science teachers and validated by the curricular team of experts at the CABA Ministry of Education. They ought to support teachers by organizing science content, with detailed indications on how to guide science activities with specific learning goals. Activities were designed to have students explore different topics and carry out diverse tasks following an inquiry-based approach (Harlen & Qualter, 2018). As Table 2 shows, as well as some more traditional activities such as reading from informative texts or diagrams and answering simple questions, the ECM included activities promoting higher order thinking skills, associated with scientific thinking (Anderson & Krathwohl, 2001; Li & Klahr, 2006). The ECM were designed for implementation in an estimated 38.5 hours of science lessons over a maximum of 12 weeks (the original Spanish version of the ECM can be found here (shorturl.at/gnsAR)).

The intervention began at the beginning of the school year, when all participating science teachers were summoned to one in-service 4-hour informative meeting aimed at presenting the research project, encouraging teachers' commitment to participate, and establishing what they were expected to do. All teachers were asked to teach the "human body" in their science lessons over the following 12 weeks, at the end of which students would complete an external assessment in the form of a written test.[1]

The written test was developed by our research team based on the content and skills covered in the local curriculum regarding the topic, so that students of teachers who did not use the ECM, but followed the national and district curriculum frameworks as expected, should be able to achieve good learning outcomes. It was reviewed by science teaching experts and then piloted in two seventh grade

[1] At the end of the intervention, teachers in the control group received a copy of the ECM.

Table 2 Characterization of the activities present in the provided ECM, according to their cognitive demand

	Activities associated with lower order science thinking skills	Activities associated with higher order science thinking skills
General description	Reading and writing activities based on content-based texts on different aspects related to the human body, specially focused on the learning of scientific vocabulary and facts, and activities involving observing and describing phenomena	Activities based on the analysis of history of science episodes and on designing and conducting experiments, as well as where students have to explain more complex scientific phenomena and reflect on their own understanding
Scientific thinking skills involved	Recall Read information from simple tables and diagrams Observe Describe	Analyze Explain Apply Interpret and draw conclusions from data Design a valid experiment Synthesize information Metacognition (reflect on one's own learning)
Examples of activities present in the provided science ECM	Read and answer simple questions about the importance of a balanced diet Write a sentence about the role of each organ in the digestive system Observe a heart dissection	Design and carry out an experiment on changes in human heart rate under different conditions Read about the experiments conducted by Robert Boyle on the role of oxygen in living cells, identifying his research questions and procedures and explaining how his experiments led to changes in understanding at the time Compare the role of the three systems (digestive, respiratory, and circulatory), and create a concept map on how they work together

classrooms from comparable, nonparticipating schools, based on which necessary adjustments on language, clarity, and content levels were made.

The result was an 11-item, 90-minute written science test, consisting of both multiple-choice and open-ended questions. This combination allowed evaluators to capture a wider range of student responses, including stronger evidence of critical thinking skills, than is typically associated with only multiple-choice tests (Stanger-Hall, 2012). In addition, items aimed to capture learning gains with different levels of complexity according to the demanded thinking skills (see Table 3).

Tests were conducted in class with the classroom teacher and an external researcher present following international examination standards (such as completing the examination individually on separate tables in silence and teachers only allowed to answer questions relating to clarifying the question). Then, student tests were anonymized and randomly allocated to members of the research team for the marking and grading process. The tests were graded on a 10-point scale according to a shared rubric, which had been piloted by researchers and shared with graders.

Table 3 Description and examples of the science test items, according to the cognitive demand involved in each question

Cognitive demand of evaluated thinking skills	Complexity level of the questions, according to the demanded thinking skills	Examples of test items for each level	Maximum score given to correct answers per question
Lower order thinking skills	Level 1: Recalling scientific concepts. Reading data from simple tables and diagrams (4 items)	John ate a plate of ravioli and is digesting. The following diagram shows the organs that form John's digestive system. Name each organ and briefly describe its function	1
Higher order thinking skills	Level 2: Analyzing and explaining a given situation involving scientific phenomena based on the concepts learned and establishing relationships between them (5 items)	7th grade students were in a physical education class. The teacher asked them to run for 5 minutes. When they finished, everyone noticed that their hearts were beating faster than at the beginning. Why does our heart beat faster when we exercise?	2
	Level 3: Interpreting science-related problematic situations, proposing feasible solutions based on evidence. Designing valid experiments to answer research questions (2 items)	7th graders wanted to investigate whether there are differences in the average heart rate between males and females when they are sitting and resting. Describe in as much detail as you can what experiment they could do to find it out	3

Graders participated in a training session and practiced marking a set of students' examinations, achieving inter-marker reliability of over 85%.

As shown in Table 3, the science test questions were weighted according to difficulty, with higher order questions scoring 2 or 3 points and lower order questions scoring 1 point. Answers were classified as either "correct" (given full marks), "partially correct" (given half the maximum marks), "incorrect" (no marks given), or "omitted" when no answer was provided by students (no marks given). Overall test results were calculated, and average levels of performance for each group were compared using ordinary least squares, including control variables for student, teacher, and school characteristics (Albornoz et al., 2019).

Reaching a Milestone: A Look at Students' Science Learning

So what did we find in terms of student learning? Students in the intervention group, in which teachers received ECM, learned more on the topic of the human body than their peers in the control group. When considering the total average test score for each group, as shown in Fig. 1, we found that the intervention group students significantly outperformed ($p < 0.01$) those in the control group. While students in the control group had an average score of 3.74 points over 10 (SD = 2.08), the average score in the intervention group was 4.79 points over 10 (SD = 2.33). Considering that the pass score in CABA is 4 (out of 10), this means that control group students did not reach the minimum level of performance considered acceptable and that the treatment produced a gain of 0.55 standard deviations in the average test score.

Our next step was to try to understand what happened in terms of students' science thinking skills, for which we analyzed the types of questions students in each group could solve correctly. We found that students in the intervention group achieved a significantly ($p < 0.01$) greater average percentage of correct answers for questions involving both lower and higher order science thinking skills (see Fig. 1). Coincidentally, the average percentage of omitted answers, frequently regarded as an indication of what students find too distant or complex to even risk a possible

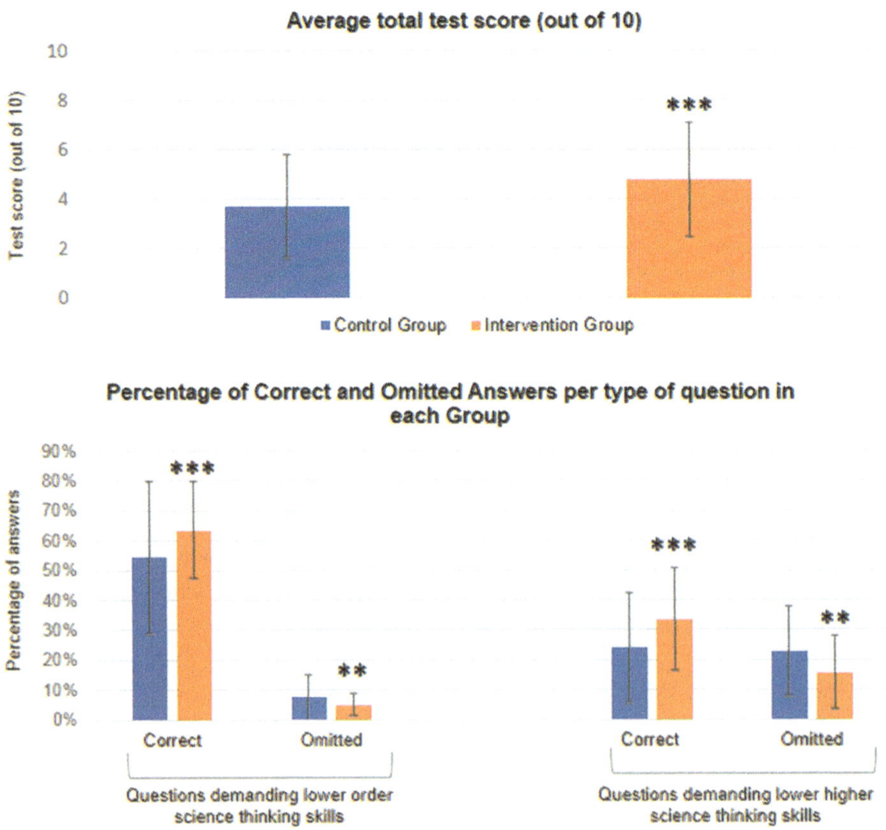

Fig. 1 Average total test score and percentage of correct and omitted answers per type of question for each group (*** = $p < 0.01$; ** = $p < 0.05$). Error bars show standard deviation

answer (Jakwerth et al., 1999; Köhler et al., 2015) was significantly lower in the intervention group ($p < 0.05$).

Encouragingly, our results show that when teachers are provided with science ECM to guide their teaching practices in ways that support higher order science thinking, student learning increases. Yet, overall results remain low and still far from what could be expected. Thus, here, we encountered a new major "think point": although the provision of ECM contributed to student learning, why were the effects limited?

Perhaps, as accounted for in the literature, we were in the presence of an "implementation gap", where the ways in which ECM are implemented in classroom contexts differ from their original design (Davis et al., 2016; Penuel et al., 2009). Analyzing ECM implementation requires addressing questions such as the following: Are the proposed activities implemented during instruction? Do teachers choose to implement certain activities in particular over others? How do they carry out these activities? Are the ways in which teachers use ECM consistent with the intended pedagogical goals and rationale?

These led us to want to "open the black box" of the ECM implementation. As we describe below, doing so implied characterizing and comparing what science teaching practices are usually like (as observed in the control group), against teaching practices when teachers are provided with ECM. Thus, a new milestone on the road emerged for us in our quest to understand the current state and opportunities to enhance science education.

Looking Deeper: Understanding Science Teachers' Teaching Practices

To understand teachers' teaching practices in each group, we decided to use student notebooks as the main data source. In general, student notebooks are endorsed as a legitimate data source for teaching practices for, despite capturing only written work, particularly at the primary level, they are extensively used during lessons to record classroom activities (Badanelli Rubio & Mahamud Angulo, 2007; del Pozo Andrés & Ramos Zamora, 2012; Gvirtz, 1997). Therefore, at the end of the intervention, all teachers were asked to choose a student notebook they considered to be most complete and thus representative of their science teaching. Student notebooks were photographed page by page and then analyzed by researchers. All information regarding the students' and schools' identities was blinded to contribute to performing an unbiased analysis.

Considering an "activity" (i.e., a distinct task specified by the teacher, with a specific learning objective – Cañal de León, 2000) as our basic observation unit, we determined (a) an estimation of the time dedicated to each activity and (b) what types of teaching activities were present in the notebooks. The estimation of the time dedicated to each activity was determined based on previous, similar interventions and in consultation with experienced science teachers. Three possible values of time were allocated to each activity: for example, 0.5 hour for closing activities aimed at reflecting on learning, 1 hour to complete a short set of questions based on a science text, or 2 hours for experimental activities. In this way, it was possible to calculate the total time destined to science teaching in each class during the intervention.

Also, following the same criteria used to classify ECM activities and test items, the activities found in student notebooks were identified as either promoting lower or higher order science thinking skills (Albornoz et al., 2019). Then, by calculating the percentage of time dedicated to activities demanding higher order science thinking skills versus those demanding lower order science thinking skills, we determined what we called the "cognitive blueprint" of science lessons. Making an analogy with fingerprints, this allowed us to characterize the "cognitive identity" of each science classroom, with a particular focus on what thinking skills students are learning, for both the control and the intervention groups.

Comparing the Cognitive Blueprints of Control and Intervention Group Lessons

When comparing the cognitive blueprint in the control and intervention groups, and that of the proposed science ECM, we found some clear differences, which led us to two important insights. For one, the cognitive blueprints of science lessons in the control and intervention groups were very different. As Fig. 2 shows, in the control group, on average only 20% of the time was dedicated to activities demanding higher order science thinking skills, and 80% of the time was focused on activities involving lower order thinking skills. In other words, "business as usual" science

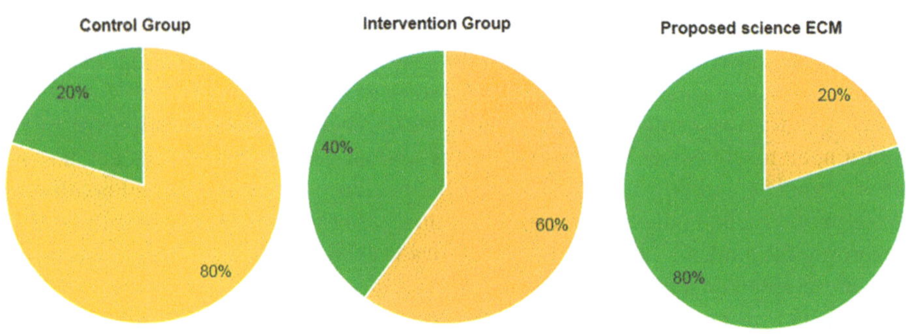

Fig. 2 Cognitive blueprint. Proportion of time dedicated to activities demanding higher and lower order science thinking skills during science lessons in the control and intervention groups and in the proposed science ECM

lessons are characterized by a prevalence of lower order science thinking, with a striking cognitive blueprint of 20/80.

On the other hand, in the intervention group, we found a cognitive blueprint of 40/60, meaning that 40% of the time was dedicated to activities demanding higher order thinking skills, exactly double that seen in the control group. However, this was still far from the original cognitive blueprint in the proposed science ECM (80/20).

So what did this mean in terms of how science lessons are being taught? In the control group, students spend most of their time conducting activities that involved lower order thinking skills: filling their notebooks with information, typically definitions of concepts, and even copying informative texts from their textbooks by hand. Activities based on reading and working with texts were also one of the most predominant across all notebooks, but in the vast majority of cases, the posed questions only demanded for students to reproduce explicit information with a particular focus in specific terminology. The latter was also seen in the great presence of word search puzzles, anagrams, crosswords, and "fill-in-the-blanks" exercises (see, e.g., Fig. 3).

On the other hand, in the intervention group notebooks, there was more frequent evidence of activities that are associated with the promotion of higher order science thinking skills. For example, even when they involved reading texts and answering questions, these usually included questions that implied analyzing and/or inferring information, explaining, and/or drawing conclusions. Experimental activities also had greater presence in the intervention group notebooks and were distinctive, in that in general they followed an inquiry-based approach.

Using the ECM: Patterns of Implementation by Teachers

These results led us once again to a new "think point," around how teachers used the ECM. Given that the cognitive blueprint in the intervention group was 40/60, still far from the original 80/20 of the ECM, we realized that teachers adapted the ECM by lowering its cognitive load.

Considering that previous research raises certain "red flags" around teachers' implementation of ECM, pointing out that lowering the cognitive load of proposed activities is a regular practice (see, e.g., Davis et al., 2016), we wondered: With what frequency and intensity did teachers use the ECM provided? Which activities did they select to implement with their students? How did they adapt them? Did they propose other activities besides those from the ECM?

A closer look at how teachers in the intervention group used the ECM allowed us to address these questions. For this, we analyzed which of the activities proposed in the ECM teachers chose to implement, identifying whether activities present in the intervention group student notebooks had come from the provided ECM or if they used activities from elsewhere.

Overall, we found great diversity in how much teachers used the ECM in their science lessons but, at the same time, certain common patterns as to the ways in

A. Illustrative examples of lower order activities that predominated in the Control Group student science notebooks

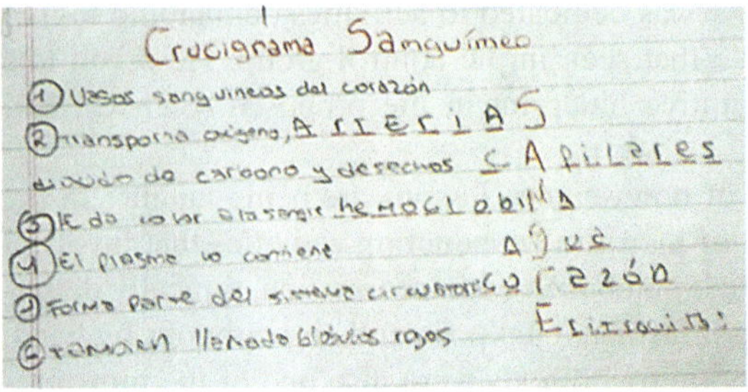

Example of lower order, terminology centered activity. In this case, a crossword puzzle on the topic of the circulatory system. Based on the six provided definitions (on the left), students had to complete the crossword on the right.
For example:
(1) Blood vessels in the heart - answer: "arteries"

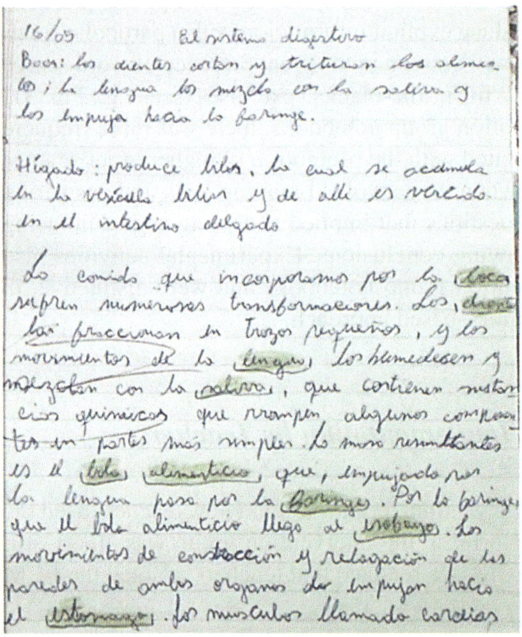

Example of an activity based on copying information (typically from textbooks, the board or dictated by the teacher). In this case, under the title "The digestive system", it defines the functions of the mouth and the liver. E.g.: "Mouth: teeth cut and grind food; the tongue mixes them with saliva and pushes them into the pharynx".

Then, the digestive process is described. Terms referring to the involved organs ("mouth", "tongue", "stomach") and other keywords ("saliva", "alimentary bolus") are highlighted.

Fig. 3 Illustrative examples of activities present in the control group and intervention group science student notebooks

which they were implemented. On average, teachers implemented 27.17% (SD = 17.98) of the activities in the ECM, ranging from teachers that implemented 0% ($n = 3$) to over 60% ($n = 2$) of the total activities. Most teachers predominantly used the ECM (as opposed to other activities of their own choice) in their science lessons. When calculating what percentage of the activities present in each notebook belonged to the provided ECM, the median was 73.2% (SD = 30.59). This means that, although teachers used a limited percentage of the activities proposed in the ECM, in general, they did not add other activities beyond those present in the

B. Illustrative examples of higher order activities found in the Intervention Group student science notebooks

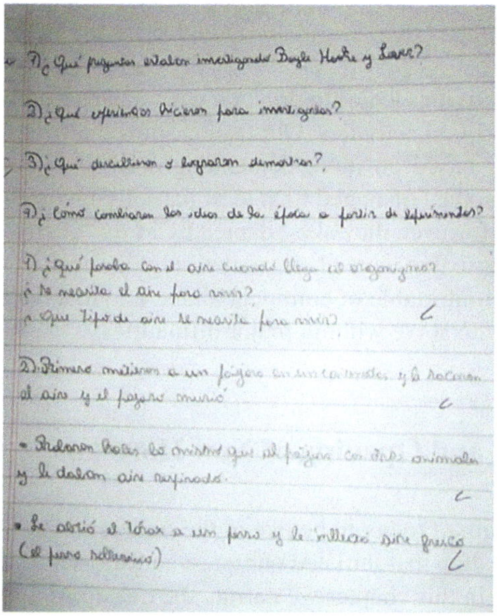

Questionnaire based on a history of science episode, in this case, the experiments performed by Boyle, Hooke and Lower on the respiratory system.
Questions included: (1) What questions were Boyle, Hooke and Lower investigating? (2) What experiments did they perform to investigate them? (3) What did they find out and managed to show? (4) How did the ideas from their time change based on their experiments?

Posters produced by the students, showing the results of an experimental activity. In groups, they had to define a research question (e.g. Is there a difference in the average height of female and male students? Are left-handed students' hands bigger than right-handed students?), acquire the necessary data (measuring students´ heights and length of their hands), analyze them, take averages, build graphs, discuss results in their teams, and draw conclusions.

Fig. 19.3 (continued)

ECM. They simply used fewer activities than those suggested in the ECM to teach the human body.

But which activities did they choose to implement? Interestingly, we found salient patterns in the types of activities that teachers selected (and those that they did not). In this way, we distinguished between what we called "popular" activities, that is, those that the vast majority of teachers implemented, and "unpopular" activities, which were rarely chosen. Popular activities were mainly those based on reading and answering questions on texts, while metacognitive (i.e., students reflecting on their own learning) and some experimental activities were identified as unpopular. Given that the popular activities involved lower order thinking skills, this explains why the cognitive blueprint of the intervention group showed a higher proportion of lower order thinking activities than originally intended by the ECM (40/60 versus 80/20).

Identifying popular and unpopular activities led us to think on what makes teachers choose to implement or skip certain ECM activities. Our results show that when teachers received ECM predominantly based on more demanding activities than those they regularly propose to students, they managed to at least partially incorporate them into their teaching practices. However, they also adapted ECM by "cherry-picking" activities, possibly favoring those that were closer to their regular practice, with which they might have felt most comfortable or confident about, and that they considered their students would be able to solve without major difficulties.

Conclusions: Lessons Learned and Opening of New Horizons

"What do we want the children and youth of today to be like in the future, and how do we achieve this?" This is the big question that initiated our quest and still sets the course of our work as a research team. In this chapter we narrate the questions and insights that emerged when conducting a research project on primary level student learning and teaching practices in science, in a context where there is a pressing need to enhance science education as an opportunity to foster higher order thinking skills. Along this way, we learned many valuable things, relevant in Argentina as in other non-WEIRD contexts.

First, we learned that opening the "black box" of science lessons in a rigorous manner contributes to understanding why student learning outcomes in the area are lower than expected. In Argentina, as in much of Latin America, research on the insides of classrooms and science teaching practices is still insufficient. Beyond those based on national or international standardized assessment programs, in general, few studies attend to student learning, and even less so that experimentally evaluate the effects of teaching and learning improvement interventions. In this sense, our study broadens the field of knowledge, potentially favoring the design, implementation, and evaluation of evidence-based interventions, so necessary in contexts such as ours.

Our findings when analyzing student notebooks and the cognitive blueprint of science lessons reveal that, in its regular form (as shown in the control group), science teaching in CABA is far from what is promoted by the literature and curricular policy, with a clear predominance of lower order activities focused on recalling factual information. They present an alarming picture of what and how our students are learning in science, showing that their opportunities to develop higher order science thinking skills are very limited, thus explaining the observed learning results. Having students spend 80% of their time in science lessons performing tasks that involve lower order science thinking skills is a matter that needs to be urgently addressed. This is of particular importance within non-WEIRD populations, considering that developing higher order science thinking skills is key to promoting critical thinking; citizens' informed and active participation in social, health, environmental, and economic issues; and the centrality of science and technology in economic development.

From our results it becomes evident that there is still a long way to go for science teaching practices to contribute to attaining the goal of fostering scientific thinking for all. Perhaps this is not entirely surprising, given that it has been reported that preservice teaching programs in the region are deficient in this regard. They predominantly focus on teaching contents also through traditional approaches rather than modeling and promoting the enactment and reflection on reform-based teaching practices oriented toward the learning of higher order science thinking skills (Cofré et al., 2015; Furman & Luzuriaga, 2017). Therefore, our study points toward the need to provide support to both preservice and in-service science teachers to strengthen their knowledge and practices.

Along this line, the second thing we learned is that research-based ECM are effective to provide this needed support to teachers to move toward higher order thinking skills oriented science teaching practices and that this has positive effects on student learning. What we identified as an alarming state of affairs of science teaching in the control group was improved in the intervention group where the cognitive blueprint of science lessons showed a higher proportion of activities demanding higher order science thinking skills. In turn, these changes in teaching practices are consistent with the improvement in the intervention group students' performance in the science test, both overall and specifically in the higher order questions. In other words, even when we are faced with a critical scenario, the condition of science education in CABA and, arguably, in similar contexts is not a lost cause. We found proof that it is possible to start up the path of improvement in the short term through the provision of ECM.

However, we also learned that ECM are not enough to profoundly transform science teaching practices and maximize student learning. Similarly to what was found in other contexts and educative levels (Beyer & Davis, 2012; Davis et al., 2016; Furman et al., 2017), in this study teachers in the intervention group selected and adapted the provided ECM in ways that resulted in lowering the cognitive load of the suggested activities. This scenario opened up new questions that are worth addressing in future research.

In particular, one of the main concerns that emerged is what is needed to go "the extra mile" toward enhancing the science education we aspire to offer our students? Is it a matter of time, in the sense that these types of interventions based on the provision of ECM need to be longer and sustained overtime? Are there other strategies that can be put in place to complement and deepen the effects of teacher professional development interventions? What other changes are called for?

As both previous research and our own study point out, the need to provide significant and sustained support for teachers is of clear importance. On the one hand, this has implications for in-service teacher professional development programs based on ECM. Particularly, to consider, design, and evaluate professional development strategies that may enhance the use of ECM, providing tools and learning opportunities to teachers that, while contributing to their professional autonomy, foster their abilities to select, curate, adapt, and design curriculum materials focused on the development of higher order science thinking skills. For instance, mentoring and instructional coaching programs, which provide teachers with individualized, relationship-based, context-specific, intensive, and sustained support (Desimone & Pak, 2017; Knight, 2007; Tschannen-Moran & Carter, 2016), may be promising to facilitate teachers to embody knowledge and discrete skills associated with research-based instructional practices (Joyce & Showers, 2002).

There is also evidence that suggests the need to take into account the institutional dimension of educational change. Considering the importance of evaluating the long-lasting effect of teacher professional development efforts, 1 year after the end of our intervention, we performed a follow-up study to reveal if participating teachers continued using the provided ECM. Surprisingly, we found that only 27% of the teachers in the intervention group continued to teach science in the seventh grade the following year. Even when all of those remaining teachers reported that they had used again the given ECM to teach the topic of the human body, which advocates the sustainability of our intervention, concerns arise regarding teaching turnover, which may dissipate part of the science teaching reform efforts. This could be indicating the importance of framing these types of interventions within the schools' institutional projects with the principals' support and endorsement to help sustain changes in teaching practices and thus student learning.

In non-WEIRD contexts such as ours, investing in in-service science teacher professional development is crucial as a large-scale, economically viable solution to ensure that students who are currently going through their schooling do not miss valuable learning opportunities while working together with teachers to strengthen their knowledge and skills (Bruns & Luque, 2015). But this is only part of the equation; deeper, systemic, long-term changes are also necessary, which demand attending to preservice teacher education and improving teachers' working conditions.

In all, what did we learn as researchers invested in science education in non-WEIRD contexts? The stakes are high; the need for change is urgent. But it is also possible, and that will continue to be our quest.

References

Albornoz, F., Anauati, M. V., Furman, M., Luzuriaga, M., Podestá, M. E., & Taylor, I. (2019). Training to teach science: Experimental evidence from Argentina. *The World Bank Economic Review, 34*(2), 393–417.

Albornoz, M., Marchesi, A., & Arana, L. (2009). *Cultura científica en Iberoamérica. Encuesta en grandes núcleos urbanos*. Madrid, España: FECYT, OEI, RICYT.

Anderson, L. W., & Krathwohl, D. R. (Eds.). (2001). *A taxonomy for learning, teaching and assessing: A revision of Bloom's taxonomy of educational objectives*. Longman.

Argentine Ministry of Education. (2005). *Núcleos de Aprendizaje Prioritarios Séptimo Año*. Buenos Aires, Argentina: Consejo Federal de Educación, Ministerio de Educación.

Ball, D. L., & Cohen, D. K. (1996). Reform by the book: What is—Or might be—The role of curriculum materials in teacher learning and instructional reform? *Educational Researcher, 25*(9), 6–14.

Badanelli Rubio, A. M., & Mahamud Angulo, K. (2007). Posibilidades y limitaciones del cuaderno escolar como material curricular: un estudio de caso. *Avances en Supervisión Educativa, 6*(junio), 1–16.

Beyer, C. J., & Davis, E. A. (2012). Developing preservice elementary teachers´ pedagogical design capacity for reform-based curriculum design. *Curriculum Inquiry, 42*(3), 386–413.

Bloom, B., Englehart, M., Furst, E., Hill, W., & Krathwohl, D. (1956). Taxonomy of educational objectives: The classification of educational goals. In *Handbook I: Cognitive domain*. Longmans.

Bruns, B., & Luque, J. (2015). *Profesores Excelentes*. Cómo mejorar el aprendizaje en América Latina y el Caribe.

CABA Ministry of Education. (2012). *Diseño curricular para la Escuela Primaria (segundo ciclo)*. CABA, Argentina: Ministerio de Educación del Gobierno de la Ciudad Autónoma de Buenos Aires.

Cañal de León, P. (2000) Las actividades de enseñanza. Un esquema de clasificación. *Investigación en la escuela, 40,* 5–21.

Cofré, H., González-Weil, C., Vergara, C., Santibáñez, D., Ahumada, G., Furman, M., ... Pérez, R. (2015). Science teacher education in South America: The case of Argentina, Colombia and Chile. *Journal of Science Teacher Education, 26*(1), 45–63.

Davis, E. A., & Krajcik, J. S. (2005). Designing educative curriculum materials to promote teacher learning. *Educational Researcher, 34*(3), 3–14.

Davis, E., Palincsar, A. S., Arias, A. M., Bismack, A. S., Marulis, L., & Iwashyna, S. (2014). Designing educative curriculum materials: A theoretically and empirically driven process. *Harvard Educational Review, 84*(1), 24–52.

Davis, E. A., Janssen, F. J. J. M., & Van Driel, J. H. (2016). Teachers and science curriculum materials: Where we are and where we need to go. *Studies in Science Education, 52*(2), 127–160. https://doi.org/10.1080/03057267.2016.1161701

del Pozo Andrés, M. D. M., & Ramos Zamora, S. (2012). Reconstruir la historia de la escuela a través de los cuadernos escolares. *Revista de ciencias de la educación, 231–232*(julio-diciembre), 415–434.

Desimone, L. M., & Pak, K. (2017). Instructional coaching as high-quality professional development. *Theory Into Practice, 56*(1), 3–12.

Di Virgilio, M., & Serrati, P. (2019). *Las desigualdades educativas en clave territorial*. CABA, Argentina: Unidad de Evaluación Integral de la Calidad y Equidad Educativa, Ministerio de Educación e Innovación, GCABA y Organización de Estados Iberoamericanos para la Eduación, la Ciencia y la Cultura.

Duschl, R. A., Schweingruber, H. A., & Shouse, A. W. (2007). *Takin science to school: Learning and teaching science in grades K-8*. National Academies Press.

Forehand, M. (2005). Bloom's taxonomy: Original and revised. In M. Orey (Ed.), *Emerging perspectives on learning, teaching and technology*. Global Text Project.

Furman, M. (2020). *Aprender ciencias en las escuelas primarias de América Latina*. Montevideo, Uruguay: Oficina Regional de Ciencias para América Latina y el Caribe, Organización de las Naciones Unidas para la Educación, la Ciencia y la Cultura.

Furman, M., & Luzuriaga, M. (2017). Science teacher education in Argentina. In Pedersen, J.,Isozaki, T., y Hirano, T. (Eds.) *Model Science Teacher Preparation Programs. An International Comparison of What Works*. Charlotte, NC, United States of America: Information Age Publishing.

Furman, M., Luzuriaga, M., Taylor, I., Podestá, M. E., & Jarvis, D. (2017). From inception to implementation: An argentine case study of teachers enacting early years inquiry-based science. *Early Years, 39*(4), 408–425.

Furman, M., Luzuriaga, M., Taylor, I., & Podestá, M. E. (2021). How does coaching influence teacher implementation of a science programme? Evidence from an experimental study. *International Journal of Mentoring and Coaching in Education, 10*(4), 449–465. https://doi.org/10.1108/IJMCE-09-2020-0059

Gvirtz, S. (1997). *Del curriculum prescripto al curriculum enseñado. Una mirada a los cuadernos de clase.* Buenos Aires: Aique.

Gopnik, A., Sobel, D., Schultz, L., & Gylmour, C. (2001). Causal learning mechanisms in very young children: Two-, three-, and four year-olds infer causal relations from patterns of variation and covariation. *Developmental Psychology, 37*(5), 620.

Gweon, H., & Shultz, L. (2011). 16-month-olds rationally infer causes of failed actions. *Science, 332*, 1524.

Gweon, H., Tenenbaum, J. B., & Schulz, L. E. (2010). Infants consider both the sample and the sampling process in inductive generalization. *Proceedings of the National Academy of Sciences, 107*(20), 9066–9071.

Harlen, W. (2013). Inquiry-based learning in science and mathematics. *Review of Science, Mathematics and ICT Education, 7*(2), 9–33.

Harlen, W., & Qualter, A. (2018). *The teaching of science in primary schools*. Routledge.

Jakwerth, P. M., Stancavage, F. B., & Reed, E. D. (1999). *An investigation of why students do not respond to questions, NAEP validity studies, working paper series*. American Institutes for Research.

Joyce, B., & Showers, B. (2002). *Designing training and peer coaching: Our needs for learning.* Alexandria, VA, United States of America: ASCD.

Kisilevsky, M., Zapata, C., Fridman, M., & Rinaldi, L. (Eds.). (2019). *Estudio Nacional 2016-2017. Prácticas de enseñanza en el contexto del aula en los niveles primario y secundario*. CABA, Argentina: Instituto Nacional de Formación Docente, Secretaría de Innovación y Calidad Educativa, Ministerio de Educación, Cultura, Ciencia y Tecnología de Argentina.

Klahr, D., Zimmerman, C., & Jiorut, J. (2011). Educational interventions to advance Children's scientific thinking. *Science, 333*, 971–975.

Knight, J. (2007). *Instructional coaching: A partnership approach to improving instruction.* Thousand Oaks, CA, United States of America: Corwin Press.

Köhler, C., Pohl, S., & Carstensen, C. H. (2015). Investigating mechanisms for missing responses in competence tests. *Psychological Test and Assessment Modeling, 57*(4), 499–522.

Kuhn, D. (2010). What is scientific thinking and how does it develop? In Goswami, U. (Ed.) *Handbook of childhood cognitive development.* West Sussex, United Kingdom: Blackwell Publishing Ltd.

Li, J., & Klahr, D. (2006). The psychology of scientific thinking: Implications for science teaching and learning. In J. Rhoton & P. Shane (Eds.) *Teaching science in the 21st century.* Arlington, VA, United States of America: NSTA Press.

Maiztegui, A., González, E., Tricárico, H., Salinas, J., de Carvalho, A. P., & Gil, D. (2000). La formación de los profesores de ciencias en Iberoamérica. *Revista Iberoamericana de Educación, 24*, 163–187.

Meschengieser, C., & Otero, M. P. (2016). *Informe de resultados TIMSS 2015. Ciudad Autónoma de Buenos Aires.* CABA, Argentina: Unidad de Evaluación Integral de la Calidad y Equidad Educativa.

Mourshed, M., Chijioke, C., & Barber, M. (2010). *How the world's most improved school systems keep getting better*. McKinsey & Company.

OECD. (2019). *PISA 2018 results. What students know and can do.* Paris, France: OECD.

Penuel, W., Fishman, B. J., Gallagher, L. P., Korbak, C., & Lopez-Prado, B. (2009). Is alignment enough? Investigating the effects of state policies and professional development on science curriculum implementation. *Science Education, 93*(4), 656–677. https://doi.org/10.1002/sce.20321

Scott, C. L. (2015). *El futuro del aprendizaje 2 ¿Qué tipo de aprendizaje se necesita en el siglo XXI? Investigación y prospectiva en educación*. Documentos de Trabajo.

Stanger-Hall, K. F. (2012). Multiple-choice exams: An obstacle for higher-level thinking in introductory science classes. *CBE-Life Sciences Education, 11*(3), 294–306.

Szenkman, P., & Lotitto, E. (2020). *Mujeres en STEM: cómo romper con el círculo vicioso. Documento de políticas públicas 224.* Buenos Aires, Argentina: CIPPEC.

Taylor, I., Sánchez, M., Luzuriaga, M., Podestá, M. E. & Furman, M. (2020). Marking classwork in primary science: written feedback practices in state schools in Buenos Aires, Argentina, *Assessment in Education: Principles, Policy & Practice*, https://doi.org/10.1080/0969594X.2020.172969

Tedesco, J. C. (2009). Prioridad a la enseñanza de las Ciencias: Una decisión política. In *Educación, Ciencia, Tecnología y Sociedad* (Vol. 03). Madrid, España: OEI, Centro de altos estudios universitarios.

Tschannen-Moran, M., & Carter, C. B. (2016). Cultivating the emotional intelligence of instructional coaches. *International Journal of Mentoring and Coaching in Education, 5*(4), 287–303.

UNESCO. (2015). *Informe de resultados TERCE. Logros de aprendizaje.* Santiago de Chile, Chile: OREALC/UNESCO.

Vegas, E., Ganimian, A., & Bos, M. S. (2014). *América Latina en PISA 2012: ¿Cuántos estudiantes tienen bajo desempeño?* Washington, DC: Banco Interamericano de Desarrollo (BID).

Part V
Future Perspectives

Latin American Perspectives in the Study of Childhood and Adolescence Poverty Through the Lenses of Neural Sciences

Sebastian Javier Lipina and María Soledad Segretin

Conceptualization of the Categories *Childhood* and *Development* in Latin America Poverty Studies

In Latin America, the scientific study of the impact of poverty on emotional, cognitive, and learning development and of the evaluation of interventions aimed at optimizing it has been approached mainly by economists, pediatric epidemiologists, developmental psychologists, and to a lesser extent neuroscientists (e.g., Aurino & Burchi, 2017; Bernal & Fernández, 2013; Canet-Juric et al., 2020; Lopez-Boo & Creamer, 2019; Rubio-Codina et al., 2015; Schady et al., 2015; Segretin et al., 2016). In general, these scientific approaches refer to categories of childhood, which assume that the fragility and dependency of children require specific, natural, relatively universal care practices and socialization processes, in accordance with the norms and habits that regulate kinship relationships considered hegemonic in Western societies (Nilsen, 2017). In tune with such academic proposals, different governmental, non-governmental, and multilateral organizations tend to sustain this perspective in their intervention and policy proposals (e.g., Keller, 2020; Lipina, 2021; Villalta & Tiscornia, 2014). In such cases, it is also possible to verify the use of notions of unique determinants of brain development that condition people throughout the life cycle (Lipina, 2021). Although in this perspective, the categories *childhood* and *development* usually involve the notion of integration of different aspects of human development, from the molecular to the cultural level, in general, they have a normative character based on milestones and the achievement of developmental aims supposedly consensual and universal (e.g., attachment, adult productivity) (Black et al., 2017; Gibson et al., 2015; Keller, 2020; Lipina, 2021; Nilsen, 2017). These notions,

S. J. Lipina (✉) · M. S. Segretin
Unidad de Neurobiología Aplicada (UNA, CEMIC-CONICET) and Instituto Universitario CEMIC (IUC), Buenos Aires, Argentina
e-mail: slipina@cemic.edu.ar; msegretin@cemic.edu.ar

M. V. Alves et al. (eds.), *Cognitive Sciences and Education in Non-WEIRD Populations*, https://doi.org/10.1007/978-3-031-06908-6_20

which are explicitly or implicitly proposed as prototypical, are usually based on (a) the type of predominant samples used in the academic studies, which are usually middle-class Western individuals; (b) an instrumental use of the category of *childhood*; and (c) the optimization of individual values (i.e., individualism) related to a certain independence of the individual in relation to his/her environment, in correspondence with contemporary Western capitalist culture, which globalization processes reproduce through the social and economic organization in different societies (Salazar Pérez et al., 2017; Yelland & Saltmarsh, 2013). Psychologically, such perspective sustains the notion of an individual autonomy in terms of independent mental states, with a focus on individual preferences for self-determination. The socialization format corresponding to this assumption usually refers to dyadic experiences in which a child participates in encounters with a single caregiver who attends his/her exclusively. Likewise, the emphasis on the face-to-face interaction format, complemented with verbalization and mentalization processes, usually also depends on primary caregivers with high degrees of formal education (Keller, 2020).

On the other hand, the Latin American anthropological, historical, and sociological explorations in the field of social studies of childhood verified that different social groups can propose distinct kinship systems, care practices, and socialization processes with high degrees of variability (e.g., Canosa & Graham, 2020; Keller, 2020; Villalta & Tiscornia, 2014). From these perspectives, there would not be a single category of childhood, but multiple childhoods, which are considered socially – not natural – constructions that combine several biological, social, and historical factors, including demographic categories, experiences, and a locus for human rights and political interventions. Such perspectives contribute to the deconstruction of the category *childhood* proposed by the Western middle-class culture. Complementary, several studies in the field of developmental psychology have evidenced that in caregivers with fewer years of formal education and higher socioeconomic deprivations, conceptions of childhood and child development may vary with respect to the emphasis on the individual (Canosa & Graham, 2020). Consequently, these perspectives give rise to other values and practices of social regulation in which children have greater autonomy in the construction of their subjectivities and are cared for by networks of polydiadic caregivers, in which development expectations may vary in terms of the individual's commitments to his/her social group (Keller, 2020; Lavelli et al., 2019; Lillard, 1998; Mejía-Arauz et al., 2007; Stagno, 2011). From these perspectives, the categories *childhood* and *development* are necessarily relational to the extent that any aspect of development is signified by its interdependence with social and cultural phenomena and processes of specific socio-historical moments (Lerner, 2018). In this context, the notion of integration is associated with equi- and multifinality phenomena, unlike the same notion in the more individualistic perspectives that generally limit it to accumulating information about different dimensions without operating theoretically and/or methodologically on the involved interdependent processes (e.g., Black et al., 2017).

Both types of perspectives (i.e., more individualistic and more relational) are far from representing opposite poles but rather positions on a continuum in which different types of conceptual integration operate on the categories *childhood* and *development*. Such continuum is the product of the historical and dynamic implementation of concepts and biases of different academic communities, with low to moderate degrees of dialogue and interdisciplinary work (Lipina, 2021). An aspect that is usually verified in some cases is the naturalization and reification of such categories (i.e., *childhood*, development), which does not guarantee their ontological condition, but rather a crystallization of reduced aspects of the relational processes that characterizes development. What is common to all of them is that although all children have emotional, cognitive, and learning adaptive skills, the ways of conceiving how their developmental contexts modulate their trajectories vary depending on how their caregivers and care systems conceive the categories *childhood* and *development*. In this sense, given the expectation of the multiplicity of trajectories and developmental needs, it is pertinent to ask what it does mean to approach the study of childhood poverty from a neuroscientific perspective in Latin America in particular, given that it is a culturally heterogeneous region with respect to care and socialization practices. In the following sections, we will address the synthesis of the available neuroscientific evidence of poverty studies at the global level, a critical analysis of which are the categories of childhood and development that it assumes, to finally evaluate possible research topics that contribute to the construction of a Latin American perspective.

Summary of the Neuroscientific Evidence in Poverty Studies During Childhood and Adolescence

Associations Between Poverty, Structural, and Functional Neural Networks

The neuroscientific study of poverty during childhood and adolescence is an area of recent development. The first works that implemented neuroimaging techniques began to be published in the 2000s (Farah, 2018). Approximately 80% of such evidence was generated in the United States; synchronous designs were applied in 75% of the published studies; 50% of them report structural information; and less than 5% address issues related to learning or interventions aimed at optimizing aspects of emotional, cognitive, or learning development in samples from poor homes. Although this evidence includes the consideration of factors that mediate and moderate the association between different indicators of poverty and structural and/or functional neural aspects, the character of such relationships continues to be essentially associative (Farah, 2018; Lipina, 2020; Lipina & Segretin, 2015). Although the specificity of the neuroscientific approach is still under construction, the incorporation of its technological advances contributes to deepening the knowledge of the effects of poverty at levels of analysis that had not previously been explored (i.e., neural). In such a context, it is worth to ask about the use of categories *childhood* and *development* in this literature.

A summary of the evidence from studies that have applied structural magnetic resonance imaging techniques indicates that socioeconomic deprivation evaluated in terms of low family income, lower parental education and occupation, and/or poor neighborhoods can be associated with changes in the volume, thickness, surface, connectivity, and functional segregation of cortical and subcortical networks in frontal, parietal, temporal, occipital, and basal ganglia networks, from the prenatal stage to at least the third decade of life (e.g., Assari, 2020; Assari & Boyce, 2021; Avants et al., 2015; Barch et al., 2021a, b; Betancourt et al., 2015; Brito et al., 2016; Ellwood-Lowe et al., 2018; Gonzalez et al., 2020; Hair et al., 2015; Leonard et al., 2019; Lu et al., 2021; Mackey et al., 2015; Merz et al., 2019a, b; Noble et al., 2015; Piccolo et al., 2016; Spann et al., 2020; Sripada et al., 2014; Takeuchi et al., 2021; Tomasi and Volkow, 2021; Ursache & Noble, 2016; Uy et al., 2019; Vanderauwera et al., 2019; Vargas et al., 2020; Weissman et al., 2018; Ziegler et al., 2020). In several of these studies, structural changes were also associated with performance on tasks with demands for cognitive control, language, and learning (e.g., Brito et al., 2016; Hair et al., 2015; Leonard et al., 2019; Mackey et al., 2015; Noble, et al., 2015; Takeuchi et al., 2021; Tooley et al., 2020; Tomasi & Volkow, 2021; Ursache & Noble, 2016). The evidence from one of the first published systematic review shows that early deprivations associated with experiences of poverty are associated with the thinning of frontoparietal, default, and visual networks (e.g., Colich et al., 2020). Recently, it has also been proposed that such associations would not only involve structural changes but also in the pace of development, so that socioeconomic status (SES) would be associated with an acceleration of cortical thinning and segregation of functional networks in children from homes with low SES (Tooley et al., 2021).

Regarding the evidence from studies in which functional magnetic resonance imaging techniques were applied, it has been verified that the same type of poverty indicators mentioned in structural studies can be associated with changes in the activation of (a) occipito-temporal networks during tasks with demand for rhyme discrimination and combination of sounds, in children from 4 to 8 years old; (b) prefrontal networks during the performance of tasks with associative learning demands, in children aged 4–8 years; (c) prefrontal and parietal networks during working memory and arithmetic processing tasks in different samples of children, adolescents, and young adults between the ages of 8 and 24; (d) amygdala networks during tasks in which threatening faces have to be processed during the first 5 years of life and in adults aged 23–25 years with a history of child poverty in their first decade of life; and (e) spontaneous activity of neural networks involved in language processing and emotional self-regulation between the ages of 6 and 14 (e.g., Biazoli et al., 2020; Finn et al., 2016; Gard et al., 2020; Javanbakht et al., 2015; Noble et al., 2006; Oshri et al., 2019; Raizada et al., 2008; Sheridan et al., 2012; Tomlinson et al., 2020; Turesky et al., 2019). Likewise, in some studies, it has been observed (a) the association of different frontal or occipital-parietal neural networks during the solution of arithmetic subtractions according to the SES of children aged 8–13 years (Demir-Lira et al., 2016) and (b) different connectivity patterns according to the SES of children of the same ages during the performance of standardized reading tasks (Gullick et al., 2016). This last group of evidence suggests that differences in SES would be associated with different types of recruitment of neural resources for solving mathematical and reading tasks, suggesting that the mechanisms involved in such modulations would not be necessary and only deficient but adaptive. On the other hand, from the evidence analyzed in the first meta-analysis studies, the conclusion emerges that SES is usually associated with the hypo- and hyperactivation of the neural networks involved in cognitive control and reward processes, respectively. Furthermore, levels of variability of such associations have been also verified according to different profiles of individual susceptibility (i.e., vulnerability versus resilience) (Yaple & Yu, 2020).

In relation to the studies in which electroencephalography (EEG) techniques have been applied, the evidence indicates that low family income, low parental education, and/or occupation levels have been associated with changes in (a) resting state electrical activity between 6 and 24 months; (b) evoked-related potentials (ERP) associated with attention and inhibitory control processing in children between 3 and 8 years old; (c) EEG activity associated with the processing of speech and environmental sounds in infants and pre-adolescents; (d) frontal ERP related to error detection and theta power, associated with cognitive control processes, in children aged 16 to 18 months and 4 years old; and (e) the prediction of cognitive performance at 15 months based on resting state EEG activity at 1 month of life (e.g., Brito et al., 2016; Cantiani et al., 2019; Conejero et al., 2018; D'Angiulli et al., 2012; Jensen et al., 2021; Skoe et al., 2013; St. John et al., 2019; Stevens et al., 2009; Tomalski et al., 2013). Likewise, from the implementation of other neural function exploration techniques (i.e., fNRIS, MEG), associations have been verified between low family income, low maternal education, and low subjective socioeconomic status with changes in the hemodynamic neural activity of frontal networks during performing tasks with working memory demands in children aged 4–48 months (Wijeakumar et al., 2019), face recognition of faces in children from 3 to 36 months (Perdue et al., 2019), and phonological processing in children aged 9–11 years (Anwyl-Irvine et al., 2021).

Mediation and Moderation Mechanisms in the Associations Between Poverty and Neural Structure and Function

The available evidence from neuroscientific studies of poverty during the first two decades of life shows (1) the moderation of the SES in the association between neural structures and functions and self-regulatory performance; (2) the moderation of neural structures and functions in the association between SES, emotional, cognitive, and learning performances; and (3) the mediation of different risk/protective factors in the associations between SES and neural structure and function (Farah, 2017; Stevens et al., 2020).

Some examples of such mediations and moderations are (a) epigenetic changes in the expression of genes that encode the reception of glucocorticoids in populations of preadolescents from poor homes exposed to abuse during their childhood, as a mediating factor between exposure to stress and gene methylation (Romens et al., 2015), and the mediation of cortisol in the association between parental education and the volume of the CA3 and dentate gyrus of the hippocampus between 5 and 9 years of age (Merz et al., 2019a, b); (b) volume, thickness, and cortical surface of different neural networks, as mediators of the association between income and academic performance in populations aged 4–22 years (Hair et al., 2015; Noble et al., 2015) and cognitive performance and high-risk behaviors in children aged 3–5 years (Barch et al., 2021a, 2021b); (c) connectivity between the hippocampus and amygdala, as mediators of the association between income and symptoms of depression in preschool-age children living in low-income homes (Barch et al., 2016), between maternal education, cognitive control, and academic achievement (Hair et al., 2015; Noble et al., 2013), or reasoning and vocabulary (Johnson et al., 2021) in school-aged children; (d) SES as a mediator of the relationship between gamma power and expressive language in 24-month-old children (Cantiani et al., 2019); (e) the moderation of neighborhood SES in the association between age and the functional structure (i.e., connectivity) of the limbic, somatosensory, and attentional neural networks between 8 and 22 years of age (Tooley et al., 2020); and (f) the moderation of social support (i.e., partner status) in the relationship between SES and frontal, temporal, and occipital neural networks in the neonatal stage (Spann et al., 2020).

In summary, the mediation and moderation mechanisms involve a diverse set of risk/protective at different levels of organization, which allow supporting hypotheses generated in other disciplines regarding the importance of early care environments (Ursache and Noble, 2016). Of greater neuroscientific specificity are the potential mediation and moderation of structural and epigenetic changes. However, neuroscientific research must still increase its efforts to incorporate different sources of variability involved in the processes of neural development in contexts of adversity due to poverty (e.g., sample compositions and size, longitudinal designs) (Lipina, 2021; Lipina and Segretin, 2015; Stevens et al., 2020).

Potential for Neural Change Due to Interventions

Neuroscientific approaches in the area of interventions for children living in poverty began in the past decade. In this section we will mention three examples that illustrate such a preliminary area of research. The first of these studies corresponds to an intervention aimed at optimizing selective attention processes for preschool children who lived in low-SES homes in Eugene-Oregon (United States), through the implementation of two weekly intervention modules, for 8 weeks, at school, after school hours (Neville et al., 2013). One of the modules consisted of attentional training activities for children through manual and group games. The other consisted of holding meetings with families, during which they discussed parenting and stress management issues and communication at home with experimenters. In addition, the families were given guides to carry out different activities at home, in order to stimulate self-regulatory behaviors in their children and to reduce the factors that induce stress in daily family communication in such a cultural context. The researchers compared performance before and after this intervention (PCMC) with that of children participating in two other interventions, one consisted only in attentional training and the other the usual as business Head-Start curriculum. The results showed that the PCMC intervention group improved their cognitive performance at a behavioral level but also electrophysiological for an ERP component of selective attention. Specifically, these children expressed a neurophysiological pattern in which the activation of different neural resources could be clearly differentiated for both relevant and irrelevant stimuli of the implemented attention paradigm. In a later study, the same researchers found that the children who benefited the most from the intervention were those who had a specific polymorphism in a gene that codes for serotonin transport (Isbell et al., 2017). This evidence addresses on the importance of consider different levels of organization in the analysis of the impact of the interventions and the specific neuroscientific contribution.

The second of these examples corresponds to a study that implemented an experimental design consisting of computerized tasks aimed at training attention, inhibitory control, cognitive flexibility, and planning processes in a sample of 5-year-old children from low-SES homes from the city of Buenos Aires (Argentina) (Pietto et al., 2018, 2021). After a 12-week period with biweekly training sessions, it was observed that the children in the intervention group expressed a greater differentiation in the amplitude of the ERP N2 component when comparing the trials with the highest and lowest inhibitory demands for a Go/NoGo paradigm.

The third example consists of a follow-up evaluation of the Abecederian project,[1] during which measurements of the cortical volume of different neural networks were made in a subsample of participants from the intervention and control groups, four decades after the intervention ended. In this case, a greater increase in this variable was verified in the male participants (Farah et al., 2021).

In summary, this evidence shows that it is possible to change neural resources in children from low-SES contexts through different intervention approaches. Furthermore, these changes can be possible to generate beyond the first 2 or 3 years of age and in different countries. Nonetheless, as was mentioned, it is still necessary to advance in the accumulation of more intervention evidence to improve the understanding of the mechanisms of plasticity in the context of a relational perspective of development (Lipina, 2020).

---

[1] Model RCT intervention aimed at optimizing emotional, cognitive, and learning development in the first 5 years of age, through multiple modules implemented in a responsive and positive child care environment (https://abc.fpg.unc.edu/abecederian-project)

Conceptions of Childhood and Development Based on the Neuroscientific Studies of Poverty

Although the researchers in the area do not make explicit which categories of childhood and development they support in their research and in the interpretation of their findings, it is possible to extract some aspects of them that could contribute to a discussion on the issue. On the one hand, in the theoretical frameworks that precede the research proposals, children are often referred to as active agents who, in interaction with their caregivers and with different factors present or absent in their developmental contexts, modulate self-regulation and learning trajectories. Although such assumptions imply variability of trajectories and socialization contexts, the implicit category of childhood tends to represent a universal version of what a child and self-regulatory competencies are in such a way that what would vary would be a typical general expectation (e.g., Johnson et al., 2016). Psychologically, persists the notion of an individual autonomy in terms of independent mental states, with a focus on individual preferences for self-determination. Complementary, the socialization values refer to care experiences in which are emphasized face-to-face interactions complemented with verbalization and mentalization processes.

In such discussions, at the beginning of the research in this field, the effects of poverty-related deprivations on neural structure and function were assumed as phenomena of deficit and immutability of long-term effects in adult emotional, cognitive, and learning functioning (D'Angiulli et al., 2012). In this respect, it is a fact that the structural and functional changes verified in different neural networks involved in emotional, cognitive, and learning processes have been observed in populations of children and adolescents from different countries and continents, which would imply a common effect. It has also been verified that such associations can be mediated and/or moderated by different individuals and contextual factors and that the longitudinal evidence available supports the hypothesis of an acceleration of cortical thinning and segregation of functional modules in association with deprivation of income and parental education (Tooley et al., 2021). Although the experiences of development and deprivation, and their individual experimentation, would vary between individuals, one aspect of the effects would occur on similar neural systems of different individuals involving distinct adaptive mechanisms, eventually regressive (i.e., deficit) and/or progressive (i.e., acceleration) (Colich et al., 2020; Kraaijenvanger et al., 2020). Thus, part of the evidence suggests similar neural impacts for different children and adolescents in distinct societies, which is consistent with the notion that the species *homo sapiens* shares at least some neural systems involved in adaptive and regulatory contextual demands. In addition, in recent years evidence has begun to accumulate suggesting that some of these effects could be the result of adaptive processes that can be the result of the recruitment of different neural resources in distinct individuals (e.g., Demir Lira et al., 2016; Gullick et al., 2016) and that different individuals would also preserve plausible levels of plasticity if affected by different types of interventions (e.g., Neville et al., 2013; Pietto et al., 2018, 2021).

This evidence suggests that beyond sharing adaptive and regulatory neural mechanisms, there are some qualitative aspects of them that can be modulated by individual and contextual differences. It could be conjectured that the field of neuroscientific studies on child poverty began from modern conceptions of the categories of childhood and development and that in the last decade such conceptions have begun to change in relational senses based on the sequential and simultaneous interdependent phenomena illustrated in the recent literature. However, such constructions do not necessarily follow an interdisciplinary effort aimed at discussing the categories *childhood* and *development*. In this sense, Latin America could constitute a propitious setting to initiate such efforts explicitly, in order to contribute to the construction of notions that honor the complexity and variety of childhood experiences in contexts of poverty. On the one hand, its cultural diversity imposes the need to consider specific neural and cognitive development experiences in terms of the effects of the combination of exposure and co-occurrence of deprivations and threats, and the modulation of different child care practices on different physiological systems involved in regulatory processes. In addition, the progressive implementation of portable neuroimaging technologies in combination with regional home and school visit programs will allow the generation of neural information in contexts in which participants cannot access research facilities. Finally, to the extent that the evidence is communicated in interdisciplinary forums on child development, the opportunities for processes of deconstruction of reductionist categories and the interdisciplinary construction of others of a relational nature will expand. In such a sense, future directions aimed at deepening knowledge constructions that contribute to more relational conceptions of development should consider: (1) Expanding the analysis of deprivations considering the ways in which children and adolescents experience them in different contexts, which implies complementing the classic indicators of income and SES with specific information of developmental experiences and contexts, since different indicators can be with distinct impacts, and their use in the Latin American can be controversial (Lipina, 2017). (2) The identification and differentiation of common and specific aspects of the effects of deprivation on different neural networks in mechanistic, structural, and functional terms and the conceptual interpretation of such evidence in terms of developmental theories. And (3) the identification of diverse socialization practices that contribute to experimentally exploring different intervention platforms aimed at capitalizing on neural plasticity, attending to not only individual but also contextual particularities.

References

Anwyl-Irvine, A., Dalmaijer, E. S., Quinn, A. J., Johnson, A., & Astle, D. E. (2021). Subjective SES is associated with children's neurophysiological response to auditory oddballs. *Cerebral cortex Communications, 2*(1), tgaa092. https://doi.org/10.1093/texcom/tgaa092

Assari, S. (2020). Race, ethnicity, family socioeconomic status, and Children's hippocampus volume. *Research in Health Sciences, 5*(4), 25–45. https://doi.org/10.22158/rhs.v5n4p25

Assari, S., & Boyce, S. (2021). Race, socioeconomic status, and cerebellum cortex fractional anisotropy in pre-adolescents. *Adolescents, 1*, 70–94. https://doi.org/10.3390/adolescents1020007

Aurino, E., & Burchi, F. (2017). Children's multidimensional health and medium-term cognitive skills in low- and middle-income countries. *The European Journal of Development Research, 29*(2), 289–311. https://doi.org/10.1057/ejdr.2016.7

Avants, B. B., Hackman, D. A., Betancourt, L. M., Lawson, G. M., Hurt, H., & Farah, M. J. (2015). Relation of childhood home environment to cortical thickness in late adolescence: Specificity of experience and timing. *PLoS One, 10*(10), e0138217. https://doi.org/10.1371/journal.pone.0138217

Barch, D., Pagliaccio, D., Belden, A., Harms, M. P., Gaffrey, M., Sylvester, C. M., … Luby, J. L. (2016). Effect of hippocampal and amygdala connectivity on the relationship between preschool poverty and school-age depression. *The American Journal of Psychiatry, 173*(6), 625–634. https://doi.org/10.1176/appi.ajp.2015.15081014

Barch, D. M., Donohue, M. R., Elsayed, N. M., Gilbert, K., Harms, M. P., Hennefield, L., … Luby, J. L. (2021a). Early childhood socioeconomic status and cognitive and adaptive outcomes at the transition to adulthood: The mediating role of gray matter development across 5 scan waves, biological psychiatry. *Cognitive Neuroscience and Neuroimaging*. https://doi.org/10.1016/j.bpsc.2021.07.002

Barch, D. M., Donohue, M. R., Elsayed, N. M., Gilbert, K., Harms, M. P., Hennefield, L., … Luby, J. L. (2021b). Early childhood socioeconomic status and cognitive and adaptive outcomes at the transition to adulthood: The mediating role of gray matter development across 5 scan waves. *Biological Psychiatry: Cognitive Neuroscience and Neuroimaging, 14*, S2451-9022(21)00196–8. https://doi.org/10.1016/j.bpsc.2021.07.002

Bernal, R., & Fernández, C. (2013). Subsidized childcare and child development in Colombia: Effects of Hogares Comunitarios de Bienestar as a function of timing and length of exposure. *Social Science & Medicine, 97*(C), 241–249. https://doi.org/10.1016/j.socscimed.2012.10.029

Betancourt, L. M., Avants, B., Farah, M. J., Brodsky, N. L., Wu, J., Ashtari, M.,… Hurt, H. (2015). Effect of socioeconomic status (SES) disparity on neural development in female African-American infants at age 1 month. *Developmental Science, 19*(6), 947–956. https://doi.org/10.1111/desc.12344

Biazoli, C. E., Salum, G. A., Gadelha, A., Rebello, K., Monteiro Moura, L., Pan, P. M., … Sato, J. R. (2020). Socioeconomic status in children is associated with spontaneous activity in right superior temporal gyrus. *Brain Imaging and Behavior, 14*, 961–970. https://doi.org/10.1007/s11682-019-00073-z

Black, M.M., Walker, S.P., Fernald, L.C.H., Andersen, C.T., DiGirolamo, A.M., Lu, C., . . . & Grantham-McGregor, S. (2017). Early childhood development coming of age: Science through the life course. The Lancet, 389 (10064), 77–90. https://doi.org/10.1016/S0140-6736(16)31389-7.

Brito, N. H., Fifer, W. P., Myers, M. M., Elliott, A. J., & Noble, K. G. (2016). Associations among family socioeconomic status, EEG power at birth, and cognitive skills during infancy. *Developmental Cognitive Neuroscience, 19*(6), 144–151. https://doi.org/10.1016/j.dcn.2016.03.004

Canet-Juric, L., García-Coni, A., Andrés, M. L., Vernucci, S., Aydmune, Y., Stelzer, F., & Richard's, M.M. (2020). Intervención sobre Autorregulación Cognitiva, Conductual y Emocional en Niños: Una Revisión de Enfoques Basados en Procesos y en el Currículo Escolar, en Argentina. *Revista Argentina de Ciencias del Comportamiento, 12*(1), 1–25.

Canosa, A., & Graham, A. (2020). Tracing the contribution of childhood studies: Maintaining momentum while navigating tensions. *Childhood, 27*(1), 25–47. https://doi.org/10.1177/0907568219886619

Cantiani, C., Piazza, C., Mornati, G., Molteni, M., & Riva, V. (2019). Oscillatory gamma activity mediates the pathway from socioeconomic status to language acquisition in infancy. *Infant Behavior and Development, 57*, a101384. https://doi.org/10.1016/j.infbeh.2019.101384

Colich, N. L., Rosen, M. L., Williams, E. S., & McLaughlin, K. A. (2020). Biological aging in childhood and adolescence following experiences of threat and deprivation: A systematic review and meta-analysis. *Psychological Bulletin, 146*(9), 721–764. https://doi.org/10.1037/bul0000270

Conejero, A., Guerra, S., Abundis-Gutiérrez, A., & Rueda, M. R. (2018). Frontal theta activation associated with error detection in toddlers: influence of familial socioeconomic status. *Developmental Science, 21*(1), e12494. https://doi.org/10.1111/desc.12494

D'Angiulli, A., Lipina, S. J., & Olesinsak, A. (2012). Explicit and implicit issues in the developmental cognitive neuroscience of social inequality. *Frontiers in Human Neuroscience, 6*, Article 254. https://doi.org/10.3389/fnhum.2012.00254

Demir-Lira, O. E., Prado, J., & Booth, J. R. (2016). Neural correlates of math gains vary depending on parental socioeconomic status (SES). *Frontiers in Psychology, 7*, 892. https://doi.org/10.3389/fpsyg.2016.00892

Ellwood-Lowe, M. E., Humphreys, K. L., Ordaz, S. J., Camacho, M. C., Sacchet, M. D., & Gotlib, I. H. (2018). "Time-varying Effects of income on hippocampal volumen trajectories in adolescent girls". *Developmental Cognitive Neuroscience, 30*(4), 41–50. https://doi.org/10.1016/j.dcn.2017.12.005

Farah, M. J. (2017). The neuroscience of socioeconomic status: Correlates, causes, and consequences. *Neuron, 96*(1), 56–71. https://doi.org/10.1016/j.neuron.2017.08.034

Farah, M. J. (2018). Socioeconomic status and the brain: Prospects for neuroscience-informed policy. *Nature Reviews Neuroscience, 19*(7), 428–438. https://doi.org/10.1038/s41583-018-0023-2

Farah, M. J., Sternberg, S., Nichols, T., Duda, J. T., Lohrenz, T., Luo, Y., ... Ramey, C. T. (2021). Randomized manipulation of early cognitive experience impacts adult brain structure. *Journal of Cognitive Neuroscience, 33*(6), 1197–1209. https://doi.org/10.1162/jocn_a_01709

Finn, A. S., Minas, J. E., Leonard, J. A., Mackey, A. P., Salvatore, J., Goetz, C., ... Gabrieli, J. D. E. (2016). Functional Brain organization of working memory in adolescents varies in relation to family Income and academic achievement. *Developmental Science, 20*(5), e12450. https://doi.org/10.1111/desc.12450

Gard, A. M., Maxwell, A. M., Shaw, D. S., Mitchell, C., Brooks-Gunn, J., McLanahan, S. S., ... Hyde, L. W. (2020). Beyond family-level adversities: Exploring the developmental timing of neighborhood disadvantage effects on the brain. *Developmental Science, 24*, e12985. https://doi.org/10.1111/DESC.12985

Gibson, M., McArdle, F., & Hatcher, C. (2015). Governing child care in neoliberal times: Discursive constructions of children as economic units and early childhood educators as investment brokers. *Global Studies of Childhood, 5*(3), 322–332. https://doi.org/10.1177/2043610615597149

Gonzalez, M. R., Palmer, C. E., Uban, K. A., Jernigan, T. L., Thompson, W. K., & Sowell, E. R. (2020). Positive economic, psychosocial, and physiological ecologies predict brain structure and cognitive performance in 9–10-year-old children. *Frontiers in Human Neuroscience, 14*, 578822. https://doi.org/10.3389/fnhum.2020.578822

Gullick, M. M., Demir-Lira, O. E., & Booth, J. R. (2016). Reading skill-fractional anisotropy relationships in visuospatial tracts diverge depending on socioeconomic status. *Developmental Science, 19*(4), 673–685. https://doi.org/10.1111/desc.12428

Hair, N. L., Hanson, J. L., Wolfe, B. L., & Pollak, S. D. (2015). Association of Child Poverty, brain development, and academic achievement. *JAMA Pediatrics, 169*(9), 822–829. https://doi.org/10.1001/jamapediatrics.2015.1475

Isbell, E., Stevens, C., Pakulak, E., Hampton Wray, A., Bell, T. A., & Neville, H. J. (2017). Neuroplasticity of selective attention: Research foundations and preliminary evidence for a gene by intervention interaction. *Proceedings of the National Academy of Sciences (PNAS), 114*(35), 9247–9254. https://doi.org/10.1073/pnas.1707241114

Javanbakht, A., King, A. P., Evans, G. W., Swain, J. E., Angstadt, M., Phan, K. L., & Liberson, I. (2015). Childhood poverty predicts adult amygdala and frontal activity and connectivity in response to Emotional faces. *Frontiers in Behavioral Neuroscience, 9*, 154. https://doi.org/10.3389/fnbeh.2015.00154

Jensen, S. K. G., Xie, W., Kumar, S., Haque, R., Petri, W. A., & Nelson, C. A. (2021). Associations of socioeconomic and other environmental factors with early brain development in Bangladeshi infants and children. *Developmental Cognitive Neuroscience, 50*(8), e100981. https://doi.org/10.1016/j.dcn.2021.100981

Johnson, A., Bathelt, J., Akarca, D., Crickmore, G., & Astle, C. E. (2021). Far and wide: Associations between childhood socio-economic status and brain Connectomics. *Developmental Cognitive Neuroscience, 48*, 100888. https://doi.org/10.1016/j.dcn.2020.100888

Johnson, S. B., Riis, J. L., & Noble, K. G. (2016). State of the art review: Poverty and the developing brain. *Pediatrics, 137*(4), e20153075. https://doi.org/10.1542/peds.2015-3075

Keller, H. (2020). Children's socioemotional development across cultures. *Annual Review of Developmental Psychology, 2*, 27–46. https://doi.org/10.1146/annurev-devpsych-033020-031552

Kraaijenvanger, E. J., Pollok, T. M., Monninger, M., Kaiser, D., Banaschewski, T., & Holz, N. E. (2020). Impact of early life adversities on human brain functioning: A coordinate based meta-analysis. *Neuroscience and Biobehavioral Reviews, 113*, 62–76. https://doi.org/10.1016/j.neubiorev.2020.03.008

Lavelli, M., Carra, C., Rossi, G., & Keller, H. (2019). Culture-specific development of early mother-infant emotional co-regulation: Italian, Cameroonian and west African immigrant dyads. *Developmental Psychology, 55*(9), 1850–1867. https://doi.org/10.1037/dev0000696

Lerner, R. M. (2018). Concepts and theories of human Development, 4th Edition. New York: Routledge.

Leonard, J. A., Romeo, R. R., Park, A. T., Takada, M. E., Robinson, S. T., Grotzinger, H., … Mackey, A. P. (2019). Associations between cortical thickness and reasoning differ by socioeconomic status in development. *Developmental Cognitive Neuroscience, 36*(4), a100641. https://doi.org/10.1016/j.dcn.2019.100641

Lillard, A. (1998). Ethnopsychologies: Cultural variations in theories of mind. *Psychological Bulletin, 123*(1), 3–32. https://doi.org/10.1037/0033-2909.123.1.3

Lipina, S. J. (2017). Critical considerations about the use of poverty measures in the study of cognitive development. *International Journal of Psychology, 52*(3), 241–250. https://doi.org/10.1002/ijop.12282

Lipina, S. J. (2020). *Pobre cerebro. Lo que la neurociencia nos propone pensar y hacer acerca de los efectos de la pobreza sobre el desarrollo cognitivo y emocional* (2nd ed.). Siglo Veintiuno Editores.

Lipina, S. J. (2021). Conceptions of childhood development in Latin America: Between the modern and relational perspectives. In M. S. Ferrari, H. E. Offerdal, & M. A. Kania (Eds.), *Why Latin America matters* (pp. 222–241). Edinburgh: Centre for Contemporary Latin American Studies, University of Edinburgh.

Lipina, S. J., & Segretin, M. S. (2015). Strengths and weakness of neuroscientific investigations of childhood poverty: Future directions. *Frontiers in Human Neuroscience, 9*, Article 53. https://doi.org/10.3389/fnhum.2015.00053

Lopez-Boo, F., & Creamer, J. (2019). *Cash, conditions, and child development: Experimental evidence from a cash transfer program in Honduras* (IDB WORKING PAPER SERIES N° IDB-WP-952). Social Protection and Health Division Inter-American Development Bank. https://doi.org/10.18235/0001397

Lu, Y. C., Kapse, K., Andersen, N., Quistorff, J., Lopez, C., Fry, A., … Limperopoulos, C. (2021). Association between socioeconomic status and in utero fetal brain development. *JAMA Network Open, 4*(3), e213526. https://doi.org/10.1001/jamanetworkopen.2021.3526

Mackey, A. P., Finn, A. S., Leonard, J. A., Jacoby-Senghor, D. S., West, M. R., Gabrieli, J. D. E. (2015). Neuroanatomical correlates of the income-achievement gap. *Psychological Science, 26*(4), 925–933. https://doi.org/10.1177/0956797615572233

Mejía-Arauz, R., Rogoff, B., Dexter, A., & Najafi, B. (2007). Cultural variation in children's social organization. *Child Developoment, 78*(3), 1001–1014. https://doi.org/10.1111/j.1467-8624.2007.01046.x

Merz, E. C., Desai, P. M., Maskus, E. A., Melvin, S. A., Rehman, R., Torres, S. D., … Noble, K. G. (2019a). Socioeconomic disparities in chronic physiologic stress are associated with brain structure in children. *Biological Psychiatry, 86*(12), 921–929. https://doi.org/10.1016/j.biopsych.2019.05.024

Merz, E. C., Desai, P. M., Maskus, E. A., Melvin, S. A., Rehman, S. D., Torres, S. D., . . . Noble, K. G. (2019b). Socioeconomic disparities in chronic physiologic stress are associated with brain structure in children. *Biological Psychiatry, 86*(12), 921–929. https://doi.org/10.1016/j.biopsych.2019.05.024.

Neville, H. J., Stevens, C. Pakulak, E., Bell, T. A. Fanning, J. Klein, S.,… Isbell, E. (2013). Family-based training program improves brain function, cognition, and behavior in lower socioeconomic status preschoolers. *Proceedings of the National Academy of Sciences USA, 110*(29), 12138–12143. https://doi.org/10.1073/pnas.1304437110.

Nilsen, A. C. E. (2017). The expansion of early childhood development services and the need to reconceptualize evidence. *Contemporary Issues in Early Childhood, 18*(3), 269–280. https://doi.org/10.1177/1463949117731021

Noble, K. G., Korgaonkar, M. S., Grieve, S. M., & Brickman, A. M. (2013). Higher education is an age-independent predictor of white matter integrity and cognitive control in late adolescence. *Developmental Science, 16*(5), 653–664. https://doi.org/10.1111/desc.12077

Noble, K. G., Houston, S. M., Brito, N. H., Bartsch, H., Kan, E., J. M., Kuperman, J. M., … Sowell, E. R. (2015). Family income, parental education and brain structure in children and adolescents. *Nature Neuroscience, 18*(5), 773–778. https://doi.org/10.1038/nn.3983

Noble, K. G., Wolmetz, M. E., Ochs, L. G., Farah, M. J., & McCandliss, B. D. (2006). Brain-behavior relationships in reading acquisition are modulated by socioeconomic factors. *Developmental Science, 9*(6), 642–654. https://doi.org/10.1111/j.1467-7687.2006.00542.x

Oshri, A., Hallowell, E., Liu, S., MacKillp, J., Galván, A., Kogan, S. M., & Sweet, L. H. (2019). Socioeconomic hardship and delayed reward discounting: Associations with working memory and emotional reactivity. *Developmental Cognitive Neuroscience, 37*(6), e100642. https://doi.org/10.1016/j.dcn.2019.100642

Perdue, K. L., Jensen, S. K. G., Kumar, S., Richards, J. E., Kakon, S. H., Haque, S., … Nelson, C. A. (2019). Using functional near-infrared spectroscopy to assess social information processing in poor urban Bangladeshi infants and toddlers. *Developmental Science, 22*(5), e12839. https://doi.org/10.1111/desc.12839

Piccolo, L. R., Merz, E. C., He, X., Sowell, E. R., & Noble, K. G. (2016). Pediatric Imaging, Neurocognition, Genetics Study. Age-related differences in cortical thickness vary by socioeconomic status. *PLoS One, 11*(9): e0162511. https://doi.org/10.1371/journal.pone.0162511

Pietto, M. L., Giovannetti, F., Segretin, M. S., Kamienkowski, J. E., & Lipina, S. J. (2021). Conflict-related brain activity after individualized cognitive training in preschoolers from poor homes. *Journal of Cognitive Enhancement, 6*, 74. https://doi.org/10.1007/s41465-021-00223-6

Pietto, M. L., Giovannetti, F., Segretin, M. S., Belloli, L. M. L., Lopez-Rosenfeld, M., Goldin, A. P., … Lipina, S. J. (2018). Enhancement of inhibitory control in a sample of preschoolers from poor homes after cognitive training in a kindergarten setting: Cognitive and ERP evidence. *Trends in Neuroscience and Education, 13*, 34–42. https://doi.org/10.1016/j.tine.2018.11.004

Raizada, R. D., Richards, T. L., Meltzoff, A., & Kuhl, P. K. (2008). Socioeconomic status predicts hemispheric specialisation of the left inferior frontal gyrus in Young children. *Neuroimage, 40*(3), 1392–1401. https://doi.org/10.1016/j.neuroimage.2008.01.021

Romens, S. E., McDonald, J., Svaren, J., & Pollak, S. D. (2015). Associations between early life stress and gene methylation in children. *Child Development, 86*(1), 303–309. https://doi.org/10.1111/cdev.12270

Rubio-Codina, M., Attanasio, O., Meghir, C., Varela, N., & Grantham-McGregor, S. (2015). The socioeconomic gradient of child development: Cross-sectional evidence from children 6–42 months in Bogota. *The Journal of Human Resources, 50*(2), 464–483. https://doi.org/10.3368/jhr.50.2.464

Salazar Pérez, M., Saavedra, C. M., & Habashi, J. (2017). Rethinking global north onto-epistemologies in childhood studies. *Global Studies of Childhood, 7*(2), 79–83. https://doi.org/10.1177/2043610617708875

Schady, N., Behrman, J., Araujo, M., Azuero, R., Bernal, R., Bravo, D., … Vakis, R. (2015). Wealth gradients in early childhood cognitive development in five Latin American countries. *The Journal of Human Resources, 50*(2), 446–463. https://doi.org/10.3368/jhr.50.2.446

Segretin, M. S., Hermida, M. J., Prats, L. M., Fracchia, C. S., Ruetti, E., & Lipina, S. J. (2016). Childhood poverty and cognitive development in Latin America in the 21st century. *New Directions for Child and Adolescent Development, 2016*(152), 9–29. https://doi.org/10.1002/cad.20162

Sheridan, M. A., Sarsour, K., Jutte, D., D'Esposito, M., & Boyce, W. T. (2012). The impact of social disparity on prefrontal function in childhood. *PLoS ONE, 7*(4), e35744. https://doi.org/10.1371/journal.pone.0035744

Skoe, E., Krizman, J., & Kraus, N. (2013). The impoverished brain: disparities in maternal education affect the neural response to sound. *Journal of Neuroscience, 33*(44), 17221–17231. https://doi.org/10.1523/JNEUROSCI.2102-13.2013

Spann, M. N., Bansal, R., Hao, X., Rosen, T. S., & Peterson, B. S. (2020). Prenatal socioeconomic status and social support are associated with neonatal brain morphology, toddler language and psychiatric symptoms. *Child Neuropsychology, 26*(2), 179–188. https://doi.org/10.1080/09297049.2019.1648641

Sripada, R. K., Swain, J. E., Evans, G. W., Welsh, R. C., & Liberzon, I. (2014). Childhood poverty and stress reactivity are associated with aberrant functional connectivity in default mode network. *Neuropsychopharmacology, 39*(9), 2244–2251. https://doi.org/10.1038/npp.2014.75

St. John, A. M., Finch, K., & Tarullo, A. R. (2019). Socioeconomic status and neural processing of a go/no-go task in preschoolers: An assessment of the P3b. *Developmental Cognitive Neuroscience, 38*, a100677. https://doi.org/10.1016/j.dcn.2019.100677

Stagno, L. (2011). El descubrimiento de la infancia, un proceso que aún continúa. In S. Finocchio & N. Romero (Eds.), *Saberes y prácticas escolares* (pp. 1–15). Homo Sapiens Editores – FLACSO.

Stevens, C., Pakulak, E., Segretin, M. S., & Lipina, S. J. (Eds.). (2020). *Neuroscientific perspectives on poverty (1° edition)*. International School of Mind, Brain and Education – Ettore Majorana Foundation.

Stevens, C., Lauinger, B., & Neville, H. (2009). Differences in the neural mechanisms of selective attention in children from different socioeconomic backgrounds: an eventrelated brain potential study. *Developmental Science, 12*(4), 634–646. https://doi.org/10.1111/j.1467-7687.2009.00807.x

Takeuchi, H., Taki, Y., Asano, K., Asano, M., Sassa, Y., Yokota, S., … Kawashima, R. (2021). Childhood socioeconomic status is associated with psychometric intelligence and microstructural brain development. *Communications Biololgy, 4*, e470. https://doi.org/10.1038/s42003-021-01974-w

Tomasi, D., & Volkow, N. D. (2021). Associations of family income with cognition and brain structure in USA children: Prevention implications. *Molecular Psychiatry*. https://doi.org/10.1038/s41380-021-01130-0

Tomalski, P., Moore, D. G., Ribeiro, H., Axelsson, E. L., Murphy, E., Karmiloff-Smith, A.… Kushnerenko, E. (2013). Socioeconomic status and functional brain Development: associations in early infancy. *Developmental Science, 16*(5), 676–687. https://doi.org/10.1111/desc.12079

Tomlinson, R. C., Burt, S. A., Waller, R., Jonides, S., Miller, A. L., Gearhardt Scott, A. N., … Hyde, L. W. (2020). Neighborhood Poverty predicts altered neural and behavioral response inhibition. *Neuroimage, 209*(4), 116536. https://doi.org/10.1016/j.neuroimage.2020.116536

Tooley, U. A., Bassett, D. S., & Mackey, A. P. (2021). Environmental influences on the pace of brain development. *Nature Reviews Neuroscience, 22*, 372–384. https://doi.org/10.1038/s41583-021-00457-5

Tooley, U. A., Mackey, A. P., Ciric, R., Ruparel, K., Moore, T. M., Gur, R. C., ... Bassett, D. S. (2020). Associations between neighborhood SES and functional brain network development. *Cerebral Cortex, 30*(1), 1–19. https://doi.org/10.1093/cercor/bhz066

Turesky, T. K., Jensen, S. K. G., Yu, X., Kumar, S., Wang, Y., ... Gaab, N. (2019). The relationship between biological and psychosocial risk factors and resting-state functional connectivity in 2-month-old Bangladeshi infants: A feasibility and pilot study. *Developmental Science, 22*(5), e12841. https://doi.org/10.1111/desc.12841

Ursache, A., & Noble, K. G. (2016). Neurocognitive development in socioeconomic context: Multiple mechanisms and implications for measuring socioeconomic status. *Psychophysiology, 53*(1), 71–82. https://doi.org/10.1111/psyp.12547

Uy, J. P., Goldenberg, D., Tashjian, S. M., Do, K. T., & Galván, A. (2019). Physical home environment is associated with prefrontal cortical thickness in adolescents. *Developmental Science, 22*(6), e12834. https://doi.org/10.1111/desc.12834

Vanderauwera, J., van Setten, E. R. H., Maurits, N. M., & Maassen, B. A. M. (2019). The interplay of socioeconomic status represented by paternal educational level, white matter structure and reading. *PLoS One, 14*(5), e0215560. https://doi.org/10.1371/journal.pone.0215560

Vargas, T., Damme, K. S. F., & Mittal, V. A. (2020). Neighborhood deprivation, prefrontal morphology and neurocognition in late childhood to early adolescence. *NeuroImage, 220*, e117086. https://doi.org/10.1016/j.neuroimage.2020.117086

Villalta, C., & Tiscornia, S. (2014). Un vasto campo de estudios: la familia y el parentesco desde la perspectiva antropológica. *Revista Derecho de Familia, 63*, 75–84.

Weissman, D. G., Conger, R. D., Robins, R. W., Hastings, P. D., & Guyer, A. E. (2018). Income change alters default mode network connectivity for adolescents in poverty. *Developmental Cognitive Neuroscience, 30*(4), 93–99. https://doi.org/10.1016/j.dcn.2018.01.008

Wijeakumar, S., Kumar, A., Delgado Reyes, L. M., Tiwari, M., & Spencer, J. P. (2019). Early adversity in rural India impacts the brain networks underlying visual working memory. *Developmental Science, 22*(5), e12822. https://doi.org/10.1111/desc.12822

Yaple, Z. A., & Yu, R. (2020). Functional and structural brain correlates of socioeconomic status. *Cerebral Cortex, 30*(1), 181–196. https://doi.org/10.1093/cercor/bhz080

Yelland, N., & Saltmarsh, S. (2013). Ethnography, multiplicity and the global childhoods project: Reflections on establishing an interdisciplinary, transnational, multi-sited research collaboration. *Global Studies of Childhood, 3*(1), 2–11. https://doi.org/10.2304/gsch.2013.3.1.2

Ziegler, G., Moutoussis, M., Hauser, T. U., Fearon, P., Bullmore, E. T., Goodyer, I. M., ... Dolan, R. J. (2020). Childhood socio-economic disadvantage predicts reduced myelin growth across adolescence and young adulthood. *Human Brain Mapping, 41*(12), 3392–3402. https://doi.org/10.1002/hbm.25024

Index

M. V. Alves et al. (eds.), *Cognitive Sciences and Education in Non-WEIRD Populations*, https://doi.org/10.1007/978-3-031-06908-6

The manufacturer's authorised representative in the EU is Springer Nature Customer Service Centre GmbH, Europaplatz 3, 69115 Heidelberg, Germany. If you have any concerns regarding our products, please contact ProductSafety@springernature.com

Printed and bound by CPI Group (UK) Ltd, Croydon, CR0 4YY
30/09/2026
02225239-0014